Technological Advances in Organ Transplantation

Satish N. Nadig • Jason A. Wertheim
Editors

Technological Advances in Organ Transplantation

Springer

Editors
Satish N. Nadig
Medical University of South Carolina
Charleston, SC, USA

Jason A. Wertheim
Northwestern University
Chicago, IL, USA

ISBN 978-3-030-09673-1 ISBN 978-3-319-62142-5 (eBook)
DOI 10.1007/978-3-319-62142-5

Printed on acid-free paper

This Springer imprint is published by the registered company Springer International Publishing AG part of Springer Nature.
The registered company address is: Gewerbestrasse 11, 6330 Cham, Switzerland

To my parents Pushpa and N.H. Nadig and the two lights of my life – my wife Nandita and my daughter Janavi. Thank you for your patience and support,
I know it is not always easy.

–Satish N. Nadig

To my wife, Christine, and son, Sam, and my parents for all of their support along the way.

–Jason A. Wertheim

Foreword

The ability to replace the organs of one individual with those of another is one of the most dramatic and effective therapies of modern medicine. Thousands have been rescued with organ transplants and thousands more are waiting for them. Over 60 years after the first successful genetically identical kidney transplant performed by Dr. Joseph Murray in 1954, transplantation might be considered a mature therapy that has achieved its goals and reached its limits. Unfortunately, numerous problems still limit its widespread application. This volume describes some of the current scientific studies to ameliorate these difficulties.

Lack of high-quality donor organs for transplantation is one of the major problems facing clinical transplantation. Half of this book highlights areas of experimentation directed to increasing donor organ availability. Pretransplant machine perfusion of donor organs, a technique known since the 1960s, is undergoing repeat examination to ascertain whether marginal organs can be resuscitated. Recent clinical evidence suggests that normothermic machine perfusion can salvage and improve marginal renal transplants and predict satisfactory posttransplant function, thus making organs transplantable that would have otherwise been discarded. Liver, heart, and lung machine perfusion has been introduced clinically and evidence for clinical efficacy is accumulating. Development of the ex vivo lung perfusion (EVLP) system, in which donor lungs are preserved under normothermic perfusion conditions, has enabled clinicians to accurately evaluate lung function prior to transplant and has increased donor lung availability and utilization worldwide. The concept of combining enhanced EVLP treatments with targeted tissue repairs and molecular therapies, i.e., cell- based and gene therapy, to further resuscitate injured lungs and also possibly enhance graft survival and long-term outcomes is currently being examined. All organ transplants sustain varying degrees of tissue injury due to surgical trauma, transient ischemia, and reperfusion injury (the so-called ischemia-reperfusion syndrome) that contributes to long-term malfunction and chronic rejection. Innovative preservation techniques involving the use of nanotechnology are being studied to address this problem. Creation of organs in the laboratory (organogenesis) and the use of animal organs (xenotransplantation) could dramatically increase organ transplants. Experimental organogenesis strategies currently utilize implantation of living cells onto a synthetic structural matrix or employ the process of decellularization of an existing organ followed by addition of living cells. The use of 3D printing to transfer living cells onto an extracellular matrix in a tridimensional fashion is a significant bioengineering advance in this area. There is significant, renewed interest in xenotransplantation. Pigs are the most likely candidate source given their anatomic and physiologic similarities with humans and favorable breeding characteristics, their ability to undergo genetic modification (which could modulate immune reactions), and the fact that transfer of serious disease from this species to humans has not been observed.

The other half of this volume discusses innovative activity in three diverse areas: transplant surgery, the possible role of newer communication technologies in transplantation and vascularized composite allotransplantation, and the induction of immunological tolerance. The successful arteriovenous fistula (AVF) rate used for hemodialysis has increased to 65% through advances in vascular and endovascular surgery, thereby reducing the morbidity from the use of intravenous dialysis catheters. Although laparoscopic surgery has been adopted for donor nephrectomy, it has not been embraced for complicated transplant procedures. Recently, the minimally invasive da Vinci Robotic Surgical System has been used to perform complex transplantations with two apparent advantages: reduced surgical site infections and improved ability to operate on patients with extreme BMIs. The application of telemedicine and/or mobile health (mHealth) – defined as medical and public health practice supported by the use of mobile phones and related technology – to enhance efficiency of transplant care delivery is reviewed. mHealth as a medium for telemedicine is attractive in its accessibility for both patients and practitioners. Another chapter provides a primer on the basic concepts of big data methodology in which the authors discuss the applicability of big data techniques in clinical transplantation. The continued immunosuppression required to prevent rejection is associated with severe toxicities and morbid complications. Establishment of operational tolerance – survival of the organ transplant with normal function without rejection and no IS requirement – intuitively should improve long-term patient and graft survival. Current clinical tolerance strategies involve infusion of various donor bone marrow cell populations to induce transient or permanent chimerism or the adoptive transfer of in vitro expanded T regulatory cells, both followed by additional immunosuppressive therapy. Early results are very encouraging. Vascularized composite allotransplantation (VCA) is a valid restorative option for patients with extensive tissue defects and whole limb loss. These procedures are not life-saving and the risks of required lifelong immunosuppression are difficult to evaluate. The ability to induce tolerance and eliminate IS would catalyze clinical VCA application.

Solid organ and cellular transplantation is the standard of care for most diseases with end-stage organ failure, but it certainly has not achieved all its goals or reached its limits. Success in any one of several innovative areas described herein will have a quantum positive impact on its ability to improve human health and quality of life. This book presents a vision of the future of clinical transplantation. It is a bright one.

Anthony Peter Monaco
The Peter Medawar Professor of Surgery, Harvard Medical School
Beth Israel Deaconess Medical Center, Boston, MA, USA

Preface

Today, transplant surgery represents a mainstream therapeutic option for those suffering with end-stage organ failure. The field has progressed dramatically since its inception; to get to this point, however, was not easy. We enjoy the spoils of decades of battles fought by many of the generals of our discipline. Surgical pioneers such as Joseph Murray, Thomas Starzl, Sir Peter Morris, Sir Roy Calne, and Anthony Monaco (to name a few) forged forward during times wherein their actions were viewed as barbaric and often unethical. Keeping their laser focus on the ultimate prize of surgical organ replacement remained the foremost concern, and the rest of the world was forced to catch up. Thankfully, it did. After the first successful transplant between the Herrick twins at the Peter Bent Brigham Hospital by Dr. Murray and colleagues in 1954, transplantation struggled to achieve large-scale success due mostly to the lack of compatibility between individual immune systems. Thus, the *era of immunosuppression* began. Throughout the 1960s, 1970s, and early 1980s, surgical scientists were working feverishly on methods to dampen the immune response so as to allow for allograft acceptance. It was during this time that lymphocyte depletion and cellular antiproliferative strategies were developed along with the advent of calcineurin inhibition, which ultimately allowed for longer-term success in allograft survival and growth of multiorgan transplantation.The discipline had officially gained some acceptance, and it was necessary to allow for its growth by providing an environment wherein organs could travel distances and remain viable. This brought us into the *era of preservation*. Seminal works by Belzer and colleagues allowed for solutions to be developed to help "preserve" these organs in cold storage during their transport. This era of preservation allowed for transplant programs to sprout all over the country and begin working together as a community. Much of this work was done in the 1980s, and by the 1990s many programs were established and training programs began to mature. Surgical success in organ transplant had been proven and refinement of the procedure itself had begun. Minimally invasive approaches to donor nephrectomies, for example, started to become the norm by the late 1990s. These improvements, along with progress in allocation policies, highlight the *era of technique*.

It is on the shoulders of the giants mentioned here, and also those whom we have not made mention of, that we are now at the precipice of the next era. The tools and technologies are in place, the infrastructure is built, and the expertise exists for us, as a community, to usher in the *era of technology*. This textbook entitled *Technological Advances in Organ Transplantation* represents the first of its kind and seeks to expose the various advances that are not only novel but, in many cases, have been implemented with success in centers across the globe. As transplant surgeons, we have historically been at the center of scientific and technological progress in transplantation. This textbook is intended to showcase the remarkable work being done by our community to move our field into the next generation and build upon the foundation laid by our collective mentors. The book begins with a historical homage in Chap. 1 and goes on to discuss the advancements that we have made and those still in progress with regard to machine perfusion, mobile health technology, dialysis access, and minimally invasive and robotic surgery (Chaps. 2, 3, 4, and 5). We then underscore the experimental and innovative therapies undergoing animal and clinical trials regarding cellular and nanotherapy in transplantation (Chaps. 6 and 7). The integration of bioengineering and transplantation is the subject of discussion in Chaps. 8, 9, 10, and 11 wherein we learn about tissue engineering, composite tissue grafting, ex vivo organ repair, and 3D bioprinting. Exciting advancements in xenotransplantation are revealed in Chap. 12. Chapter 13 brings the text to a close with a discussion on the utilization of artificial intelligence in the processing of big data as it pertains to transplant.

The readers will surely see the caliber of the authors who have contributed to this text, many are pioneers in their own right and for their contributions and collaboration we thank them.

Our field represents the best of mankind. The willingness to give one's organs, either in life or death, to save another is the ultimate sacrifice and is what wakes each and every one of us every morning. Our patients are our heartbeat and our conscience. We continually strive to do better for them. For this gift that they give us – our livelihood and our passion – we as a community thank them. This book is a humble representation of what we as transplant surgeons obsess upon in order to make the lives of our patients better.

Charleston, SC, USA — Satish N. Nadig
Chicago, IL, USA — Jason A. Wertheim

Acknowledgements

We would like to thank our mentors who have helped us along our personal journey through transplantation. Drs. Prabhakar Baliga and Michael Abecassis have helped build our careers, supported us clinically, and fostered our passion to seek new therapies in transplantation via translational science. They, along with our institutions (The Medical University of South Carolina and Northwestern University), have supported our endeavors and have created an environment for our success.

Contents

Chapter 1
Historical Review of Solid Organ Transplantation

Bernard J. DuBray and Ronald W. Busuttil

Abstract One of the most remarkable achievements in modern medicine has been the ability to replace the failing organs of one individual with those from another. Over the last 60 years, solid organ transplantation has evolved from an experimental concept to a standard of care with increasing clinical efficacy. The collaborative effort across multiple disciplines brought transplantation from the laboratory into hospitals where it has transformed the lives of thousands. Today over 100,000 individuals await a lifesaving organ transplant with an ever-widening gap between the supply of suitable organs and demand for them. While the field works to address these challenges, it is important to recognize the pioneers who have led the journey.

1.1 Ancient Concepts of Tissue Transfer

Documentation of autotransplantation and allotransplantation of nonvisceral tissues can be found as early as 3000 BC [1]. Archeological records from Hindu text provide detailed accounts of skin autografting in order to reconstruct noses mutilated by punishment [1]. Other accounts of nonvisceral tissue transfer include transplantation of teeth and skull fragments in ancient Egypt, Greece, and Rome. In 1668, Dutchman Job van Meeneren is credited with repairing a human skull defect with the skull fragment of a dog [1]. While these accounts bare no evolutionary connection to the modern science of solid organ transplantation, they represent the conceptual genesis of tissue transfer among individuals.

B.J. DuBray (✉) • R.W. Busuttil
Division of Liver and Pancreas Transplantation, Department of Surgery,
David Geffen School of Medicine at the University of California Los Angeles,
757 Westwood Plaza, Los Angeles, CA 90095, USA
e-mail: bernard.j.dubray@vanderbilt.edu; rbusuttil@mednet.ucla.edu

S.N. Nadig, J.A. Wertheim (eds.), *Technological Advances in Organ Transplantation*, DOI 10.1007/978-3-319-62142-5_1

Fig. 1.1 Charles Claude Gutherie (From: Friedman [2]). The collaborative effort of Charles Gutherie and Alexis Carrel in the early twentieth century led to multiple technical advancements that established the field of vascular surgery and provided the technical basis for solid organ transplantation. Permission to use image granted by the author under license number 4257661395407

1.2 The Technical Basis for Modern Transplantation

At the turn of the twentieth century, surgery focused on hemostasis with limited ability to repair or replace injured vessels. A new technique to join vessels was the combined effort of two surgeons whose collaboration established the field of vascular surgery.

Born in St. Charles County, MO, Charles Claude Guthrie earned his medical degree from the University of Missouri in 1901 (Fig. 1.1). He studied physiology as a medical student and developed techniques to re-anastomose severed cadaveric canine arteries [3]. In 1905, he departed for Chicago to establish his own laboratory. This is where he met Alexis Carrel, a French surgeon who also had an interest in blood vessel surgery with multiple publications on vascular techniques. Carrel sealed the ends of vessels through triangulation at three equidistant points [3], although it was only after Guthrie recommended full-thickness bites with intimal inclusion that his experiments succeed [2, 4] (Fig. 1.2). While their collaboration only lasted 6 months, the 29 publications that followed helped establish the field of vascular surgery. Alexis Carrel went on to make several other important observations regarding tissue ischemia and cold storage that helped establish the basis of organ preservation [5, 6]. For his contributions, Carrel was awarded the Nobel Prize in American Medicine or Physiology in 1912.

1.3 The First Attempts at Clinical Transplantation

With many of the requirements for solid organ transplantation established, the first attempts at solid organ transplantation were made. Using the techniques developed by Carrel, Jaboulay of France was the first individual to attempt human kidney

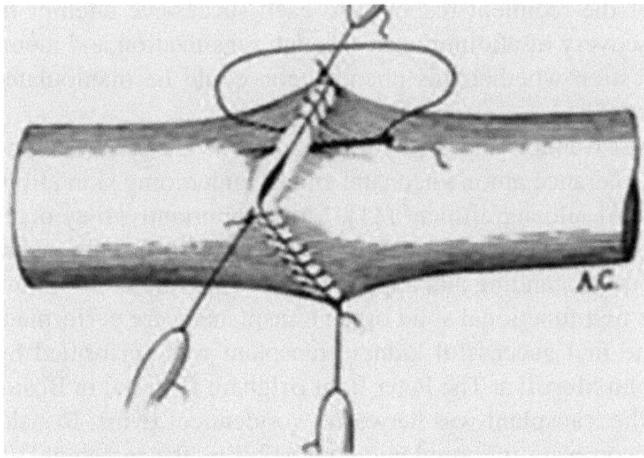

Fig. 1.2 The triangulation technique of vascular anastomosis (From: Carrel [4]). By placing stay sutures in each corner, Alexis Carrel was able to divide the suture line in thirds and run fine suture between three equidistant points

transplantation [7]. In 1906, he joined the renal vessels of sheep and pig kidneys into the brachial vessels of two uremic patients. While these kidneys never produced urine, they represented a technical achievement to build upon.

Further experimentation with kidney transplantation occurred in 1936 when Russian Surgeon, Voronoy, attempted the first human to human cadaveric renal allotransplantation. In 1951, Kuss and Dubos were the first to place a transplanted kidney extraperitoneally in the pelvic fossa [8]. While none of these kidneys were functional, they each took incremental steps toward achieving a technical foundation for transplantation that would await a better understanding of the physiologic barriers impeding function.

1.4 Early Insights Foreshadow Immunologic Challenges

While experimenting with kidney reimplantation in a canine model, Alexis Carrel made keen observations that foreshadowed the impending physiologic barrier to successful solid organ allotransplantation.

> *Should an organ, extirpated from an animal and replanted into its owner by a certain technique, continue to functionate normally, and should it cease to functionate normally when transplanted into another animal by the same technique, the physiologic disturbance could not be considered as brought about by the organ but would be due to the influence of the host, that is, the biological factors* [1]

These observations were later proven through a series of experiments performed by British surgeons, Peter Medawar and Thomas Gibson in the early 1940s [9, 10]. They discovered that rejection was an immunologic phenomenon evidenced by acceleration of the recipient response to each successive attempt from the same donor. The discovery of alloimmunity through sensitization and memory led investigators to question whether this phenomenon could be manipulated to facilitate alloengraftment.

In 1952, Jack Cannon and William Longmire of UCLA were the first to report any degree of tolerance among neonatal chicks undergoing skin allotransplantation with a 6% rate of alloengraftment [11]. More importantly, they demonstrated that the rate of permanent alloengraftment rose to 20% with the administration of cortisone. With an understanding and appreciation for the role of alloimmunity in transplantation, the first functional solid organ transplants were performed.

In 1954, the first successful kidney transplant was performed by Drs. Joseph Murray and John Merrill at The Peter Bent Brigham Hospital in Boston. Essentially an autograft, the transplant was between two identical twins, Ronald and Richard Herrick. With complete immunologic compatibility, the recipient survived 9 years with intact renal function before succumbing to cardiovascular disease. This proof of concept that prior failures were primarily immunologic provided the impetus to search for ways to overcome genetic incompatibility.

1.5 Early Immunosuppression

Based on the success of bone marrow transplantation in the 1950s, the first attempts at immunomodulation were aimed at total body irradiation with complete cytoreduction. While two recipients of kidneys from a nonidentical twin did result in some function, the consequences of total immunosuppression proved to be extremely morbid with infectious sequelae [12]. Of ten patients treated with sublethal total body irradiation in Boston, nine died within a month of transplant due to the effects of the radiation [1]. Alternatives to total body irradiation included myelosuppression with cyclophosphamide and methotrexate. Although long-term survival was not achieved, crossing the genetic compatibility barrier was a historic achievement that set the stage for less cytotoxic and more specific immunosuppression.

1.6 Pharmacological Immunosuppression

With the concept of cytoreduction, the cornerstone of early immunosuppression efforts was soon underway to selectively inhibit individual arms of the immune system. Serendipitously, concurrent development of chemotherapeutic agents became the pharmacologic answer to selective cytoreduction. Gertrude B. Elion and George H. Hitchings developed azathioprine in 1962 [1]. A purine analogue derived from 6-mercaptopurine, azathioprine incorporates into replicating DNA and halts purine synthesis and lymphocyte production. Murray utilized this new drug to break the immunocompatibility barrier and demonstrated prolonged kidney allograft survival among unrelated individuals [13, 14]. For their efforts and contribution to transplantation, Elion and Hitchings shared the 1988 Nobel Prize. Success, however, was limited using the single drug regimen. Improvement in clinical efficacy would be dependent on the efforts to match genetically similar individuals and the utilization of multidrug immunosuppressive cocktails.

1.7 Treatment of Rejection

In a series conducted in 1960 at the University of California Los Angeles, Willard Goodwin was able to demonstrate transient reversal of clinical rejection with the use of prednisone [15]. This finding was confirmed by Dr. Thomas E. Starzl who reported that high dose prednisone in patients treated with azathioprine could reverse episodes of rejection [16]. Prior to this, rejection was universally associated with graft loss. Dissemination of this new discovery led to the formation of nearly 50 transplant centers in the United States. Additionally, improved clinical efficacy led to transplantation of other solid organs, including liver, heart, and pancreas.

With increasing levels of immunosuppression, however, more complications were seen. In a landmark paper by Thomas E. Starzl in 1967, he characterized the outcomes of the first 60 recipients of solid organ transplants at the University of Colorado [17]. In patients treated for rejection, increasing immunosuppression led to a predictable course highlighted by overwhelming depression of host defense mechanisms. Many patients succumbed to opportunistic pathogens, such as Cytomegalovirus and Pneumocystis carinii. While some in the medical community questioned the ethical basis of transplantation, the "azathioprine era" led to steady improvement in 1-year patient survival. By the end of the 1970s, 1-year patient survival following kidney transplantation had climbed to more than 90% [7]. Further advances in this era included the development of antithymocyte globulins synthesized from the serum of horses. These preparations were utilized to treat steroid-resistant rejection and later were part of an induction regimen. Finally, the discovery of calcineurin inhibitors in late 1970s and early 1980s drastically reduced the rates of rejection across all domains of transplantation and improved graft survival. With the ability to selectively modulate the alloimmune response pharmacologically, this opened the door for experimentation with other more immunogenic solid organs, such as lung and intestine.

1.8 Immunocompatibility

Unbeknownst to the early practitioners of solid organ transplantation, members of the same species had variable immunocompatibility that was critically important to their clinical outcome. While early studies demonstrated hyperacute humoral-mediated rejection in the setting of ABO mismatched kidneys, the determinants of acute cell-mediated processes were less clear [18]. Discovery of the major histocompatibility complex (MHC) and human leukocyte antigen (HLA) by Jean Dausset in the 1950s was a landmark in immunology research that garnered him a share of the 1980 Nobel Prize in medicine [1]. With the knowledge of leukocyte antigens, Paul Terasaki from UCLA developed the first microassay to detect preformed, cytotoxic antibodies in recipient serum [19]. The so-called crossmatch would become a standard practice prior to performing kidney transplantation in order to prevent hyperacute antibody-mediated rejection.

1.9 Liver Transplantation

By the late 1950s, experimental models for all intra-abdominal organs had been developed (Fig. 1.3) [20]. The most vital and lifesaving of these combinations involved the liver. Through preclinical experimentation with large animal models, a great deal was learned regarding the metabolic and endocrine functions of the liver and how they are supported by the other intra-abdominal viscera. Several approaches to liver replacement were studied in the early 1950s (Fig. 1.4).

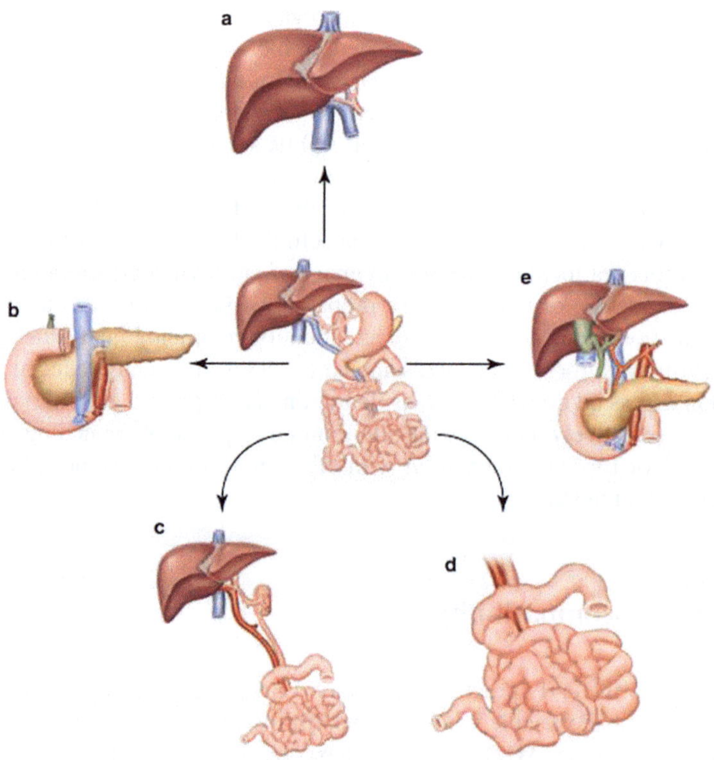

Fig. 1.3 Models of intra-abdominal solid organ transplantation (From Starzl [20]). By the late 1950s, preclinical models for multiple solid organ transplants had been developed. These included (**a**) liver, (**b**) pancreas, (**c**) liver and intestine, (**d**) isolated intestine, (**e**) liver and pancreas

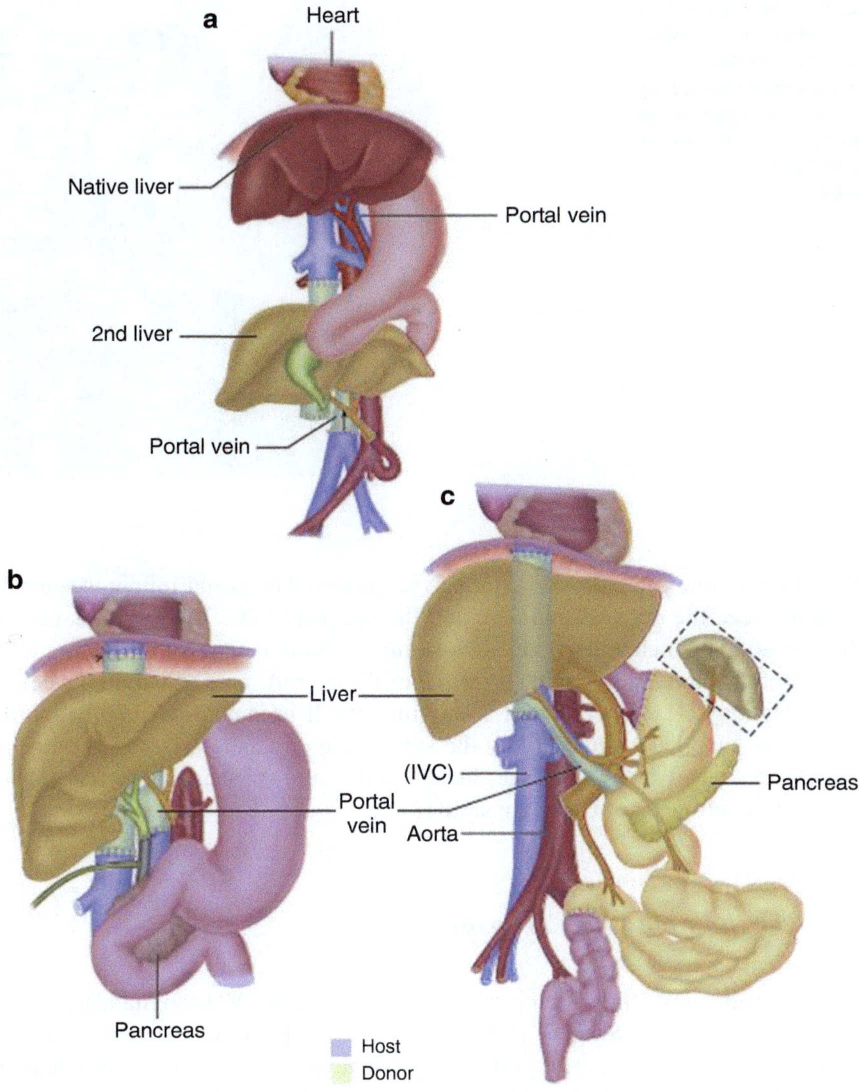

Fig. 1.4 Models of liver transplantation (From Starzl [20]). Multiple models of liver transplantation were studied in preclinical models. These included (**a**) auxiliary liver transplantation, (**b**) orthotopic liver transplantation, and (**c**) multivisceral transplantation

1.9.1 Auxiliary Liver Transplantation

In 1955, C. Stuart Welch developed a model of auxiliary liver transplantation [20]. Without removing the diseased liver, Welch placed an allograft liver into the right paravertebral space. The suprahepatic vena cava was anastomosed to the recipient IVC, whereas the infrahepatic cava was oversewn. The hepatic artery was sewn to

Fig. 1.5 Professor Vittorio Staudacher (From Busuttil et al. [22]). Professor Staudacher of Milan, Italy was the first to perform an orthotopic liver transplant in 1952 utilizing a canine model

either directly onto the aorta or external iliac artery. The portal inflow was established between the donor portal vein and the recipient IVC. Biliary drainage was achieved utilizing a choledochoduodenostomy. Without heterotrophic factors from the splanchnic circulation, rapid shrinkage of these grafts ensued within a week of transplantation [21]. Unbeknownst at the time, the liver is nourished by heterotrophic factors supplied exclusively by the splanchnic circulation. Decades later this would be described as the "pancreas factor" and would spawn a whole new area of investigation into hepatotrophic physiology.

1.9.2 Orthotopic Liver Transplantation

Liver replacement was first attempted in 1952 by Professor Vittorio Staudacher [22] (Fig. 1.5). Professor of Surgery at the University of Milan, Staudacher provided a detailed five-step procedure for performing orthotopic liver transplantation in dogs (Fig. 1.6). His description is similar to the classic bicaval technique with a few notable exceptions. Staudacher performed the caval anastomoses over a plastic stent and arterialized the portal vein by the native hepatic artery. Biliary reconstruction was deferred in favor of donor cholecystostomy. Soon afterward other groups attempted hepatic replacement including Jack Cannon in 1956 at UCLA and Francis D. Moore in 1959 at the Brigham hospital.

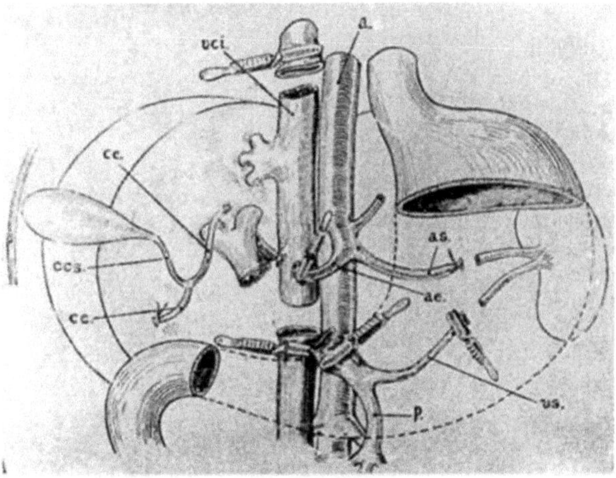

Fig. 1.6 "Five-step" procedure for performing orthotopic liver transplantation (From Busuttil et al. [22]). The technique described by Staudacher in 1952 closely resembles the classic bicaval technique. Transplantation was accomplished by anastomosing the suprahepatic IVC, followed by the infrahepatic IVC and the portal vein, which was arterialized to the recipient hepatic artery via the recipient splenic vein

1.9.3 The Problem of Venous Reconstruction

While it has been known for centuries that the liver had a dual blood supply, the physiologic significance of the portal blood flow has only recently been appreciated [23]. With its nutrient-rich supply, portal blood flow is essential for liver function and had many implications in the development of liver transplantation. In 1959, Thomas Starzl began to address the problems with venous reconstruction using canine models of orthotopic liver transplantation [24]. First, during the hepatectomy, Starzl described decompressing the splanchnic circulation through a combination of a portacaval shunt with passive femoral to internal jugular external bypass (Fig. 1.7). This technique was an important development in order to promote hemodynamic stability during the hepatectomy and engraftment. Additionally, Starzl demonstrated three potential ways for venous reconstruction (Fig. 1.8). The first method employed a "reversed Eck fistula," which diverted both portal and systemic flow through the liver. This invariably leads to portal hypertension, venous congestion, and early graft compromise. Later experiments resulted in normal venous pathways with and without ligation of the portacaval shunt. In these initial canine experiments, Starzl noted the value of external bypass and anatomic venous reconstruction, both of which are still standard approaches to orthotopic liver transplantation today.

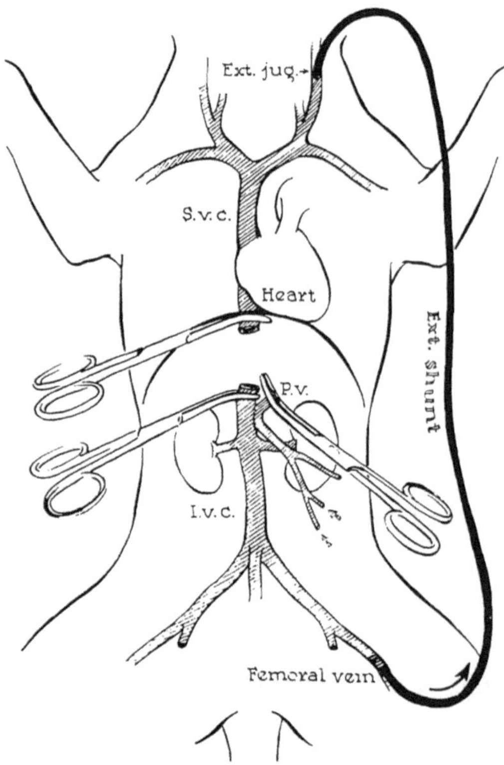

Fig. 1.7 Passive venovenous bypass (From Starlz et al. [24]). In order to decompress the splanchnic circulation while clamping the portal vein, Dr. Starzl described a combination of portacaval shunt with passive diversion of blood flow from the femoral vein to the internal jugular vein

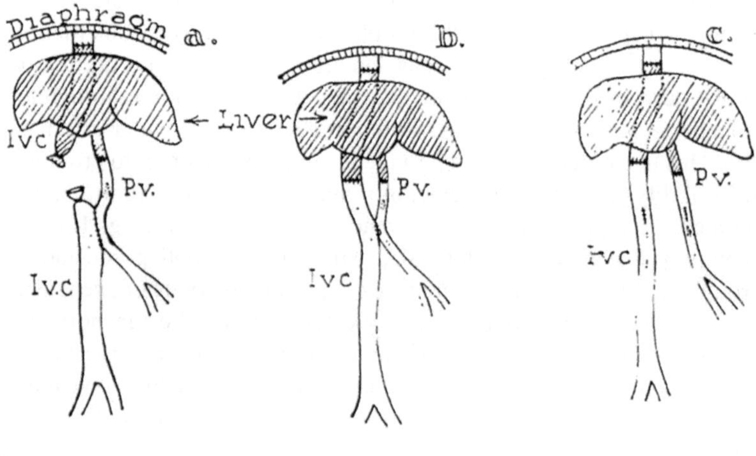

Fig. 1.8 Venous reconstruction in orthotopic liver transplantation (From Starlz et al. [24]). Thomas Starzl proposed three potential venous reconstructions in liver transplantation: (**a**) The reversed Eck fistula which diverts both portal and systemic flow through the liver; (**b**) normal venous pathways without (**b**) and with (**c**) portacaval ligation

1.9.4 The First Clinical Attempts at Liver Transplantation

Despite hundreds of preclinical canine experiments and the perceived technical feasibility of liver transplantation, nothing could prepare surgical and anesthesia teams for the clinical manifestations of portal hypertension and coagulopathy associated with chronic liver disease. The first cases of attempted liver transplantation were undertaken at the University of Colorado by Dr. Thomas Starzl in 1963 [25]. The first patient was a 3-year old boy with a history of biliary atresia and multiple prior abdominal operations. Bleeding from massive venous collaterals in the setting of extreme coagulopathy and thrombocytopenia resulted in exsanguinating hemorrhage. Two more cases were attempted in the next 4 months in Colorado with recipient survival of 22 and 7 days, respectively. Attempts to treat coagulopathy with blood products and e-aminocaproic acid led to clotting of the bypass circuits with delayed pulmonary emboli and lung abscesses [20]. Around this time, there were other ill-fated attempts at liver transplantation in Boston and Paris. The practical utility of liver transplantation was questioned and pessimism surrounding the practice grew. A self-imposed worldwide moratorium on further attempts at clinical liver transplantation was instituted in 1964 due to these outcomes [26]. Further application of liver transplantation would need to await advancements in operative technique, organ preservation, and immunosuppression in order to expand its clinical utility (Fig. 1.9) [26].

1.10 Advances in Organ Procurement and Preservation

Interruption of an organ's blood supply during recovery creates an obligatory ischemic injury that can manifest in varying degrees of postoperative organ dysfunction. Mitigating the effects of ischemia upon reperfusion has been the result of several advancements.

1.10.1 Hypothermia

The knowledge that temperature influences the degree of metabolic activity and cellular demand for oxygen led to the implementation of hypothermic protocols in clinical kidney transplantation [27]. While originally described for cooling of individual organs following procurement, the in situ technique involved continuous hypothermic perfusion of cadaveric donors before acceptance of brain death [20] (Fig. 1.10). In situ methods were later simplified to allow for the cooling of all thoracic and abdominal organs without continuous perfusion [28] (Fig. 1.11). By 1987, standardization of multiorgan procurement across the country and throughout the world allowed for the growth of transplantation.

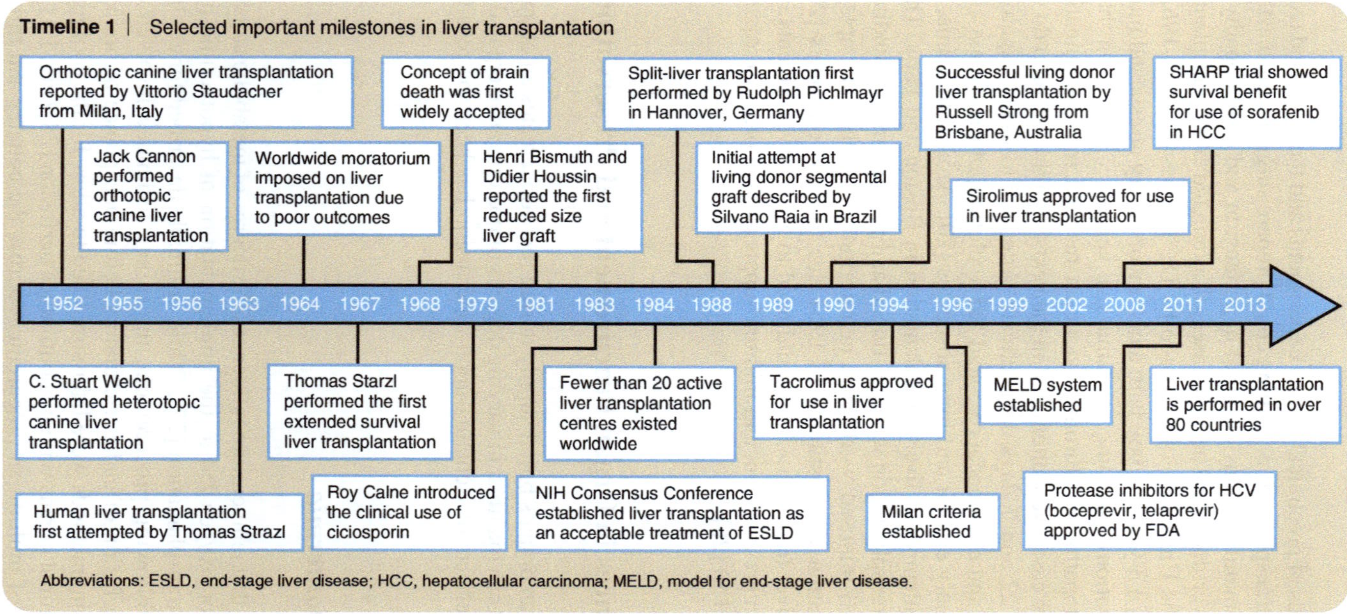

Fig. 1.9 Milestones in liver transplantation (From Zarrinpar et al. [26]). Several key advances and notable events have been milestones in the field of liver transplantation

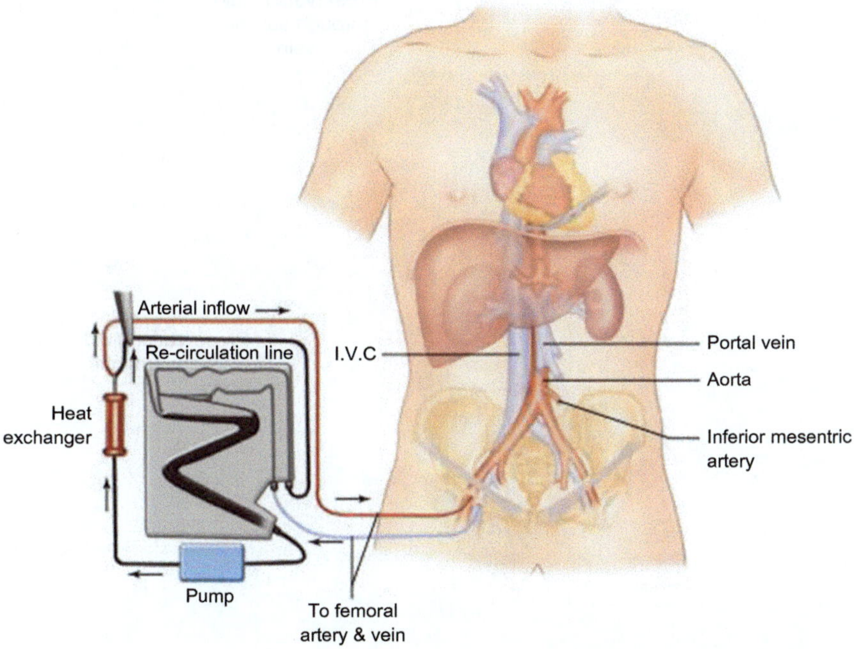

Fig. 1.10 In situ continuous hypothermic perfusion (From Starzl [20]). Continuous hypothermic cooling was an active process that involved cannulation of the aorta and IVC via the femoral vessels soon after cardiopulmonary death in order to pump blood through a heat exchanger

1.10.2 Organ Preservation

In order to expand the physical radius between donors and recipients, organs for transplantation needed additional measures of preservation. While donor hypothermia mitigated in situ ischemia, organ immersion or "slush" technique post procurement allowed for cold storage during transport [1]. A number of solutions have been formulated to mimic the intracellular compartment to prevent cell swelling and lysis during storage. The development of the Euro-Collins solution and University of Wisconsin solution was major milestones in transplantation that significantly lengthened preservation time [1]. Kidneys were extended to 48–72 h, while livers were viable up to 24 h and hearts up to 12 h. This significantly expanded the donor pool for individual recipients and facilitated the clinical growth of solid organ transplantation (Table 1.1) [20].

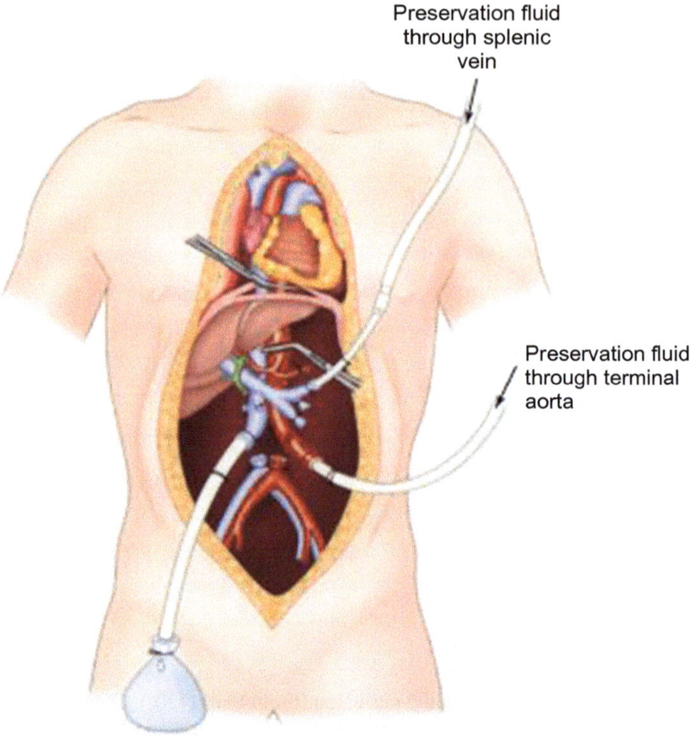

Fig. 1.11 Passive in situ cooling for multiorgan procurement (From Starzl [20]). With direct cannulation of the aorta and splenic vein, in situ cooling takes place passively and allows for simultaneous cold preservation of all abdominal organs

1.11 Organ Donation and Allocation

1.11.1 Defining Death

With increasing clinical efficacy and expanding indications for solid organ transplantation, the field needed ethical clarity for how to identify and consent potential cadaveric donors. Up until 1968, death was defined by cardiopulmonary arrest. However, as critical care and life-support technology improved throughout the mid-twentieth century, questions surrounding the definition of death grew. The line between sustaining life and prolonging death was blurred. The principle governing organ donation at the time was the "dead donor rule," which simply stated that organ retrieval itself cannot cause death [29]. An alternative concept of death by neurological criteria was examined in 1968 by the Ad Hoc Committee of the Harvard Medical School. They established criteria for declaring death in cases of irreversible devastating neurologic injury. These included patients who, despite mechanical

Table 1.1 Milestones in liver transplantation [20, 26]

Year	Event	Contributor
1952	First attempt at orthotopic liver transplantation in canine model	Vittorio Staudacher
1955	First auxiliary liver transplant	C. Stuart Welch
1963	First attempt at human liver transplant	Thomas E. Starzl
1963	Discovery of azathioprine, prednisone "cocktails"	Thomas E. Starzl
1979	Cyclosporine introduced	Sir Roy Calne
1981	First reduced size allograft for pediatric patient	Henry Bismuth
1984	Standardization of multiple organ procurement	Thomas E. Starzl
1988	First split liver transplants	Rudolf Pichlmayr
1989	FK-506-steroid immunosuppression introduced	Thomas E. Starzl
1989	First attempt at living liver transplant	Silvano Raia

ventilation with cardiopulmonary function, did not respond, have brainstem reflexes, or spontaneous respirations [30]. This definition of death by neurologic rather than cardiopulmonary criteria gave clarity to the transplant community in regards to the adherence to the "dead donor rule." This allowed recovery of organs from brain dead, heart beating donors. With advancements in surgical technique, organ preservation, and procurement, the first extended survival following liver transplantation was realized by Dr. Starzl in Colorado and Sir Roy Calne of the UK. When cyclosporine was added to the immunosuppressive armamentarium, the 1-year recipient survival rose to 70% for the first time [26].

1.11.2 Fair and Just Organ Allocation

As the practice of transplantation grew, federal legislation and regulation intervened to help ensure equitable and just allocating and sharing of donor organs. In 1984, the National Organ Transplant Act (NOTA) was passed which established a network of local organ procurement organizations to organize the recovery and equitable sharing of organs. The United Network for Organ Sharing (UNOS) won the federal contract in 1986 to oversee the OPTN and ensure fair and equitable system for organ allocation.

1.12 Success Creates Organ Shortage

The success of solid organ transplantation led to expansion of indications to include almost all causes of end-stage liver disease. Unfortunately, the number of donors has remained relatively constant creating an ever-widening gap between supply and demand. Efforts to combat the critical organ shortage have been multifaceted and include several surgical innovations that have improved utilization of the existing donor pool (Fig. 1.12).

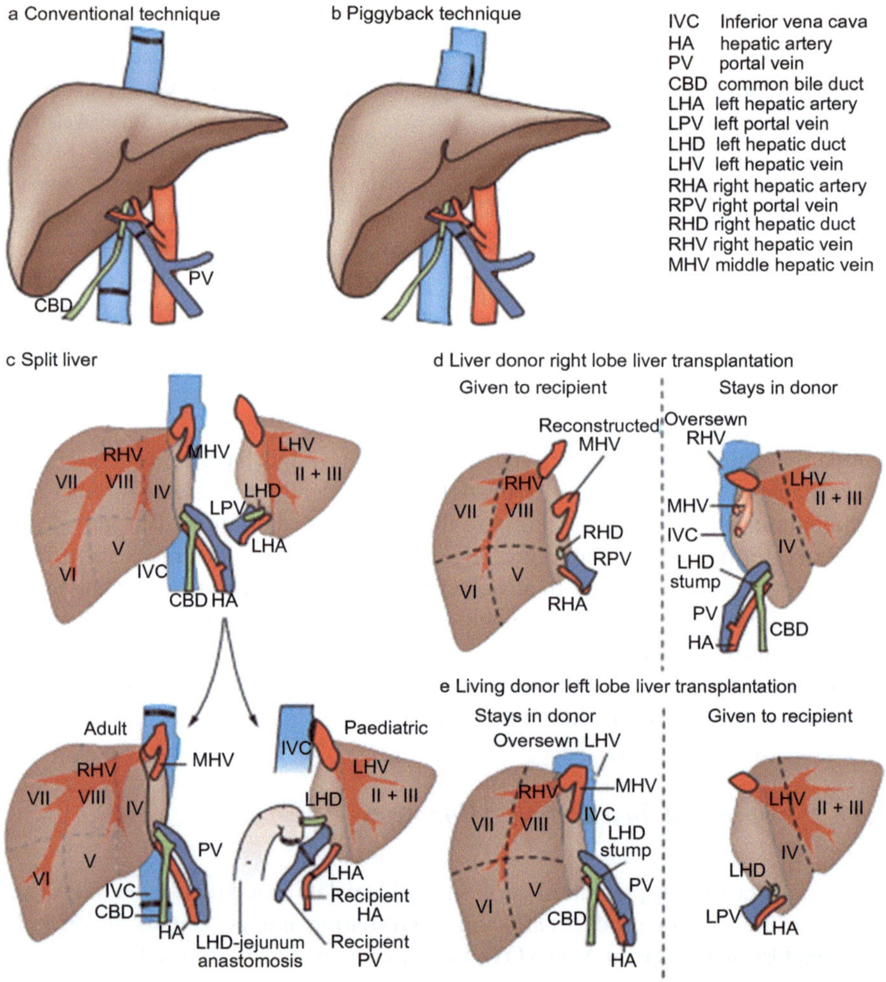

Fig. 1.12 Techniques of liver transplantation (From Zarrinpar et al. [26]). In addition to the classic bicaval technique (**a**), the piggyback technique (**b**) preserves the recipient IVC. Additionally, efforts to optimize utilization have resulted in a number of reduced size grafts including (**c**) in situ split liver allografts, (**d**) living donor right lobe, and (**e**) living donor left lobe

1.12.1 Reduced Size and Split Allografts in Liver Transplantation

Children often have some of the longest wait times due to size constraints and relative rarity of pediatric donors. With this in mind, Henry Bismuth performed the first reduced size allograft transplant in 1981 [31]. He utilized the left lobe of an adult liver for a child with a caval sparing approach. The success of this operation led to in situ split liver transplantation. First performed by Rudolf Pichlmayr in Hannover, Germany, in 1988, split liver transplantation yielded two grafts from a single donor. Segments II and III went to a child, while I, IV, and V–VIII went to an adult. A caval sparing approach was utilized in each recipient. These technical achievements helped optimize the utilization of the existing deceased donor pool for liver transplantation.

1.12.2 Living Donation in Liver Transplantation

While living donation has been a standard approach in kidney transplantation from inception, its utilization in liver transplantation has been measured due to concerns for donor safety. The first attempt at living liver donation was in 1989 by Silvano Raia in Brazil [32]. This was followed the successful transplantation of the left lobe from a living adult to her child by Strong and colleagues in Brisbane, Australia [33]. The first adult-to-adult living liver transplant was performed in 1993 by Professor Makuuchi from Japan using a left lobe graft [34]. In 1997, C.M. Lo and S.T. Fan performed the first right lobe living donor transplant in Hong Kong [35]. Today, living donor transplants are steadily growing in carefully selected donor/recipient pairs, especially in countries where cadaveric donation is limited.

1.13 The Future of Transplantation

In a relative short time span, solid organ transplantation has made tremendous clinical progress transforming the lives of thousands of patients. The journey has been the result of both measured, scientific inquiry and serendipity. The challenges ahead include better and more specific immunomodulation with the ultimate goal of immune tolerance. With increasing clinical efficacy, efforts to alleviate the organ shortage include improved utilization of existing donors, expansion of living donation, xenotransplantation, and tissue engineering. Through a continued multidisciplinary collaborative effort, the field will continue to grow and meet the challenges ahead.

References

1. Linden, P. K. (2009). History of solid organ transplantation and organ donation. *Critical Care Clinics, 25*(1), 165–184.
2. Friedman, S. G. Credit where due. *Journal of Vascular Surgery.* [Internet]. (2016). Available from: http://dx.doi.org/10.1016/j.jvs.2016.04.033.
3. Dangoor, J. Y., Hakim, D. N., Singh, R. P., & Hakim, N. S. (2015). Transplantation: A brief history. *Experimental and Clinical Transplantation, 13*(1), 1–5.
4. Carrel, A. Anastomose bout a bout de la jugulaire et de la carotide primitive. Lyon Med. 1902;99:114.

5. Carrel, A., & Guthrie, C. C. (1906). Results of a replantation of the thigh. *Science, 23*(584), 393–394.
6. Carrel, A. (1907). Heterotransplantation of blood vessels preserved in cold storage. *The Journal of Experimental Medicine, 9*(2), 226–228.
7. Morris, P. J. (2004). Transplantation — A medical miracle of the 20th century. *The New England Journal of Medicine, 351*(26), 2678–2680.
8. Starzl, T. E. (1998). Surgery: Art or science? Birth of organ transplantation. *Chirurgie, 123*(3), 225–228.
9. Gibson, T., & Medawar, P. B. (1943). The fate of skin homografts in man. *Journal of Anatomy, 77*(Pt 4), 299–310.4.
10. Medawar, P. B. (1944). The behaviour and fate of skin autografts and skin homografts in rabbits: A report to the War Wounds Committee of the Medical Research Council. *Journal of Anatomy, 78*(Pt 5), 176–199.
11. Cannon, J. A., & Longmire, W. P., Jr. (1952). Studies of successful skin homografts in the chicken; description of a method of grafting and its application as a technic of investigation. *Annals of Surgery, 135*(1), 60–68.
12. Watson, C. J. E., & Dark, J. H. (2012). Organ transplantation: Historical perspective and current practice. *British Journal of Anaesthesia, 108*(Suppl 1), i29–i42.
13. Murray, J. E., Merrill, J. P., Harrison, J. H., Wilson, R. E., & Dammin, G. J. (1963). Prolonged survival of human-kidney homografts by immunosuppressive drug therapy. *The New England Journal of Medicine, 268*, 1315–1323.
14. Starzl, T. E. (1994). The early days of transplantation. *JAMA, 272*(21), 1705–1705.
15. Goodwin, W. E., Kaufman, J. J., Mims, M. M., et al. (1963). Human renal transplantation. I. Clinical experiences with six cases of renal homotransplantation. *The Journal of Urology, 89*, 13–24.
16. Starzl, T. E., Marchioro, T. L., & Waddell, W. R. (1963). The reversal of rejection in human renal homografts with subsequent development of homograft tolerance. *Surgery, Gynecology & Obstetrics, 117*, 385–395.
17. Hill, R. B., Dahrling, B. E., Starzl, T. E., & David, R. (1967). Death after transplantation. *The American Journal of Medicine, 42*(3), 327–334.
18. Starzl, T. E., Marchioro, T. L., Hermann, G., Brittain, R. S., & Waddell, W. R. (1963). Renal homografts in patients with major donor-recipient blood group incompatibilities. *Surgical Forum, 14*, 214–216.
19. Terasaki, P. I., & McClelland, J. D. (1964). Microdroplet assay of human serum cytotoxins. *Nature, 204*(4962), 998–1000.
20. Starzl, T. E. (2015). History of liver and other splanchnic organ transplantation. In Doria, Cataldo (ed.), *Contemporary liver transplantation* (pp. 1–28). Springer publishers.
21. Starzl, T. E., Noriko, M., Amadeo, M., & Fung, J. J. (2005). History of liver and multivisceral transplantation. In *Transplantation of the liver* (pp. 3–22). Elsevior publishers.
22. Busuttil, R. W., De Carlis, L. G., Mihaylov, P. V., Gridelli, B., Fassati, L. R., & Starzl, T. E. (2012). The first report of orthotopic liver transplantation in the western world. *American Journal of Transplantation, 12*(6), 1385–1387.
23. Starzl, T. E., Porter, K. A., Francavilla, J. A., Benichou, J., & Putnam, C. W. (1977). A hundred years of the hepatotrophic controversy. *Ciba Foundation Symposium, 55*, 111–129.
24. Starlz, T. E., Kaupp, H. A., Jr., Brock, D. R., Lazarus, R. E., & Johnson, R. V. (1960). Reconstructive problems in canine liver homotransplantation with special reference to the postoperative role of hepatic venous flow. *Surgery, Gynecology & Obstetrics, 111*, 733–743.
25. Starzl, T. E., Marchioro, T. L., Vonkaulla, K. N., Hermann, G., Brittain, R. S., & Waddell, W. R. (1963). Homotransplantation of the liver in humans. *Surgery, Gynecology & Obstetrics, 117*, 659–676.
26. Zarrinpar, A., Ali, Z., & Busuttil, R. W. (2013). Liver transplantation: Past, present and future. *Nature Reviews. Gastroenterology & Hepatology, 10*(7), 434–440.
27. Starzl, T. E. (1963). Renal transplantation in identical twins. *Archives of Surgery, 86*(4), 600.

28. Starzl, T. E., Hakala, T. R., Shaw, B. W., Jr., et al. (1984). A flexible procedure for multiple cadaveric organ procurement. *Surgery, Gynecology & Obstetrics, 158*(3), 223–230.
29. Robertson, J. A. (1999). The dead donor rule. *The Hastings Center Report, 29*(6), 6–14.
30. A definition of irreversible coma. (1968). Report of the Ad Hoc Committee of the Harvard Medical School to examine the definition of brain death. *JAMA, 205*(6), 337–340.
31. Bismuth, H., & Houssin, D. (1984). Reduced-sized orthotopic liver graft in hepatic transplantation in children. *Surgery, 95*(3), 367–370.
32. Raia, S., Nery, J. R., & Mies, S. (1989). Liver transplantation from live donors. *Lancet, 2*(8661), 497.
33. Strong, R. W., Lynch, S. V., Ong, T. H., Matsunami, H., Koido, Y., & Balderson, G. A. (1990). Successful liver transplantation from a living donor to her son. *The New England Journal of Medicine, 322*(21), 1505–1507.
34. Hashikura, Y., Yasuhiko, H., Masatoshi, M., et al. (1994). Successful living-related partial liver transplantation to an adult patient. *Lancet, 343*(8907), 1233–1234.
35. Lo, C. M., Fan, S. T., Liu, C. L., et al. (1997). Adult-to-adult living donor liver transplantation using extended right lobe grafts. *Annals of surgery, 226*(3), 261–9; discussion 269–70.

Chapter 2
Machine Perfusion of Organs

Matthew F. Blum, Qiang Liu, Basem Soliman, Toshihiro Okamoto, Bahar Bassiri-Gharb, Teresa Diago Uso, Laura D. Buccini, and Cristiano Quintini

Abstract Transplant organ supplies are insufficient to meet the demands, resulting in prolonged patient wait times and waitlist mortality. Offering improved organ preservation, assessment, and resuscitation, machine perfusion of organs can enable the use of marginal organs, thereby expanding the donor pool. The technology is based on the common principle of continuous provision of oxygen and nutrients. Perfusion settings and components vary widely in terms of pump pulsatility, temperature control, oxygen provision, oncotic agents, and pharmacologic supplementation. While this technology was first pioneered in the 1960s, it has seen a recent resurgence. Hypothermic renal perfusion is the most clinically advanced area of perfusion and a source of continued innovation. Liver, heart, and lung perfusion techniques have been introduced into the clinical realm as safe alternatives, and evidence demonstrating clinical efficacy and superiority is accumulating. Machine perfusion of the pancreas and small intestine is being explored predominantly in preclinical models. Machine perfusion of limbs offers improved opportunities for limb transplantation and autologous replantation. Machine perfusion is a promising option to salvage function in marginal organ grafts and may enable prediction of organ function or dysfunction after transplantation.

2.1 Introduction

Since the 1960s, organ preservation by machine perfusion (MP) has progressed alongside the birth and maturation of transplant science. Though it comes in many forms, at its core, perfusion technology involves the mechanical pumping of fluid through an organ's vasculature in a controlled environment to sustain a measure of physiologic function and provide superior organ preservation for transplantation. As organ transplantation has become a reliable treatment for end-stage organ

M.F. Blum • Q. Liu • B. Soliman • T. Okamoto • B. Bassiri-Gharb • T.D. Uso • L.D. Buccini
C. Quintini (✉)
Cleveland Clinic Transplant Center, Cleveland, OH, USA
e-mail: quintic@ccf.org

S.N. Nadig, J.A. Wertheim (eds.), *Technological Advances in Organ Transplantation*, DOI 10.1007/978-3-319-62142-5_2

disease, increasingly more patients await these operations, causing the demand for transplantation to outstrip the supply and transplant waiting times to increase. Accepting lesser quality organs for transplantation could increase organ supply, but this requires improved preservation modalities to adequately preserve these marginal organs. To demonstrate how MP can increase and improve transplantable organs, this chapter will explore the principles of MP in general and in relation to specific organs and composite tissues including the kidney, liver, pancreas, intestines, heart, lungs, and limbs.

2.2 Organ Shortage and the Limits of Cold Storage

The transplant community is facing an organ shortage crisis. The number of patients who require organ transplantation is growing faster than the supply of organs available [1], resulting in longer waiting times and higher waitlist mortality. From 2002 to 2009, the kidney transplant waiting list in the United States increased from 50,000 to 96,000 [142]. Liver waitlist mortality increased from 11.1 per 100 waitlist years in 2009 to 12.3 in 2013 [119]. In 2013 more adult patients than ever before were added to the lung transplant waitlist [226], and the heart transplant waitlist increased by 34% from 2003 to 2013 [44].

While high-quality organ grafts, such as living donor organs and organs donated after brain death (DBD), have traditionally been available for transplantation, efforts to make up this organ shortfall have led to a greater reliance on marginal quality grafts, including organs donated after cardiac death (DCD) and previously designated expanded criteria donors (ECD). Compared to DBD grafts, marginal organs are discarded more frequently and generally result in worse outcomes [120]. For example, ECD kidneys have a higher rate of delayed graft function (DGF) and worse graft and patient survival [182], and DCD livers experience biliary complications resulting in lower graft survival [153]. The organ shortage is further compounded by a reduction in rates of living donation and a trend toward the use of older, unrelated living donors [120].

Static cold storage (SCS) is the dominant graft preservation modality used today. Though invented after hypothermic machine perfusion (HMP), SCS is used in the preservation of most organs for its efficacy and simplicity. It is sufficient for the preservation of high-quality organs from standard deceased donors [122], but its utility in expanding the donor pool with marginal organ grafts is limited. As outlined by Vogel et al., SCS has four limitations: (1) prior organ injury is not reversed, (2) cold ischemic injury continues to occur during storage, (3) organ quality is not easily assessed, and (4) storage is time-limited [229].

The heart and lung are typically only viable up to 6 and 8 h of SCS, respectively, while the small bowel and liver can withstand up to 12 and 16 h of SCS [240]. The pancreas and kidney can tolerate up to 24 and 36 h, respectively [240]. Time limitation is due to the fact that the SCS graft exists in an ischemic environment. At low temperatures, metabolism and oxygen consumption are reduced [74], thus slowing but not eliminating the same destructive processes that occur in a warm ischemic environment. The sequelae of ischemia-reperfusion injury (IRI) occurs after implantation and includes acidosis, cellular swelling, enzymatic dysfunction, calcium accumulation, and reactive oxygen species production [167]. SCS is particularly detrimental when combined with warm ischemic injury as it results in extensive cellular death and creates a unique challenge for the DCD organ [122]. During SCS, grafts are in a nonfunctional state, making it difficult to measure organ quality prior to transplantation. For example, in liver preservation, where only SCS is performed clinically, 10% of grafts initially show poor function, and 5% have PNF, which results in death unless immediate re-transplantation can be performed [90]. Since marginal grafts have an even greater risk of failure, they are typically avoided in this SCS-only preservation paradigm. Therefore, an organ preservation modality is required that can salvage marginal organs and predict their likelihood of function after transplantation.

2.3 Benefits and Principles of Organ Perfusion

In light of these challenges, machine perfusion research is at the forefront of efforts to expand the organ pool. The advantages of machine perfusion fall within three main categories: preservation, assessment, and intervention. In many cases, evidence suggests that MP offers better and longer preservation than SCS by removing toxic metabolites, providing nutrients and oxygen, and in some cases avoiding cold temperatures [28, 69, 215]. By placing the organ in a dynamic environment where at least partial function is maintained, there are more opportunities to assess organ function during the preservation period, offering the ability to predict which organs will be viable for transplantation. MP, especially at higher temperatures, provides an opportunity to intervene on the graft prior to transplant in effort to improve quality. This could include pharmacologic, immunologic, or genetic interventions that, beyond slowing organ decline, can promote graft regeneration.

2.3.1 Pumps and Flow

Organ perfusion systems share functional principles with cardiopulmonary bypass, and many components overlap. As a general rule, the perfusate is propelled through non-thrombogenic tubing in a temperature-controlled and occasionally oxygenated system [74]. The pump, perfusate, oxygen, and temperature are important variables that determine the behavior of perfusion systems. Flow may be pulsatile, typically via roller or piston pumps, or continuous through centrifugal pumps [227]. The organ, temperature, and perfusate dictate the optimal propulsion pattern. Pulsatile pumps were initially employed in early hypothermic renal perfusion systems as they simulate the natural circulation of the cardiac cycle. This allows for a reduced mean capillary pressure with systolic bursts of pressure high enough to perfuse capillary beds [74, 126]. The necessity of pulsatile perfusion has been debated over the years

[130, 184], with some recent evidence suggesting that pulsatile flow mitigates preservation injury. On the other hand, recent HMP liver systems have employed continuous flow through the portal vein and hepatic artery [88]. In normothermic perfusion, where perfusates often contain red blood cells (RBCs), centrifugal pumps are typically preferred to avoid RBC trauma and hemolysis [229]. This has been the case in clinical trials for kidney [163], lung [49], and liver perfusion [188]. However, the claim that centrifugal pumps produce less hemolysis than roller pumps is controversial. Shear stress, negative pressure, and turbulent flow of centrifugal pumps contribute to hemolysis, and reducing the occlusivity of roller pumps can minimize hemolysis to levels equal to or less than centrifugal pumps [228].

2.3.2 Temperature

Temperature is categorized as hypothermic, subnormothermic, and normothermic. In HMP, temperature is reduced to just above freezing, typically 4 °C but often anywhere from 0 to 10 °C. At this point, organ metabolic rates are approximately 13-fold below normothermia [112, 140]. Since oxygen consumption is minimal, HMP often lacks supplemental oxygenation, though this paradigm is being challenged [62, 79]. Resistance is elevated at these temperatures, and flow and pressure are minimized to avoid edema [87]. Warm perfusion, including normothermic machine perfusion (NMP) and subnormothermic machine perfusion (SMP), is a more recent topic of research. NMP occurs at 35 to 38 °C. SMP temperatures are poorly defined with examples ranging from 20 to 33 °C. In an effort to standardize MP nomenclature, Karangwa and colleagues have proposed dividing this temperature range into midthermic (13–24 °C) and subnormothermic (25–34 °C) [115]. For clarity, however, this chapter will continue to the use of the traditional subnormothermic temperature range when discussing previous publications. While some SMP systems occur at temperatures low enough that dissolved oxygen is sufficient to support metabolic activity [35, 100], oxygen carriers are an essential component of warmer SMP and NMP systems. Many perfusion systems employ RBCs for this purpose [49, 163, 188]. Artificial oxygen carriers have also been explored, including perfluorocarbons [26] and pyridoxylated bovine hemoglobin [30]. Warm temperatures avoid hypothermic injury, enable observation of function, and support efficient graft interventions, but they also present a number of challenges including the potential need for substrate renewal as substrates are rapidly consumed and metabolites are deposited in the perfusate. Bacterial growth is a greater risk at warmer temperatures.

2.3.3 Colloids and Impermeants

While individual components vary between different perfusates, colloids are essential in most MP systems to mitigate edema. In the setting of IRI, endothelial disruption enables interstitial fluid accumulation, which acts in concert with endothelial swelling, vasoconstriction, and thrombosis to produce the no-reflow phenomenon, wherein circulation is impeded [200]. Colloid agents exert an oncotic force to draw extravascular volume into the intravascular space, opening vascular flow. A variety of oncotic agents have been explored. Plasma, which contains albumin and other osmotic protein agents, was used in the earliest successful HMP models [14]. Albumin alone was also explored [42], and this was replaced with the artificial colloid hydroxyethyl starch to avoid the risk of blood-borne disease [95]. Other colloids include polyethylene glycol (PEG) [20, 217], succinylated gelatin [188], and dextran both alone [38] and combined with albumin [237].

Ischemia depletes ATP, causing Na-K pumps to fail and Na and water to accumulate intracellularly as intracellular edema [126]. MP solutions often address this by featuring impermeants, high-molecular-weight molecules that are impermeable to most cellular membranes. They exit the vasculature and accumulate in the interstitial space, creating an intracellular-to-extracellular osmotic gradient that reverses the flow of water from edematous cells. Examples of impermeants include gluconate, which is a part of Belzer machine perfusion solution (B-MPS), and lactobionic acid, raffinose, sulfate, and phosphate, all of which are a part of University of Wisconsin (UW) solution [181]. Other common examples include histidine and mannitol of histidine-tryptophan-ketoglutarate and glucose of Euro-Collins solution.

2.3.4 *Perfusate Electrolytes, Supplements, and Pharmacologic Interventions*

A variety of other factors contribute to the function of MP systems. The electrolyte composition is an important component, particularly in synthetic perfusates. Collins solution (later Euro-Collins) and UW solution are two of the earliest SCS solutions and mimic the intracellular environment to minimize energy demands of the Na-K pump [43, 204]. In MP systems, electrolyte concentrations are often closer to physiologic concentrations. B-MPS contains a more physiologic Na/K ratio, though still hyponatremic (100 mEq/L) and hyperkalemic (25 mEq/L) [126, 167]. Warm perfusates also tend to have physiologic electrolyte concentrations [30, 163].

During IRI, reactive oxygen species form and produce oxidative damage. To prevent this, perfusates are supplemented with antioxidants. For example, B-MPS includes allopurinol and glutathione [204]. Newer antioxidants are being explored in the SCS setting, such as the free-iron chelator deferoxamine [5, 102].

The perfusate is a vehicle for medications that act on the graft during the MP preservation period. Drugs may be included to either guarantee the stability of the MP system or modify the graft prior to transplantation. Vasodilators can be included to minimize resistance and maximize flow during MP [98, 99, 161, 187]. Other medications may alter the nature of the graft prior to transplantation, such as the use of cobalt protoporphyrin in renal SMP to promote the endogenous expression of the protective molecule heme oxygenase-1 [24]. Supplementation of trophic factors like fibroblast growth factor promotes cellular recovery [27]. Gene therapy, another technology at the forefront of MP innovation, can induce localized immunosuppression.

This offers the theoretic possibility of reducing the systemic immunosuppression required for graft recipients. Cypel et al. demonstrated ex vivo transfection of IL-10 to discarded human lungs could shift cytokine expression from a pro-inflammatory to anti-inflammatory profile as discussed in Chapter 10 of this textbook [46]. Another strategy for localized immunosuppression is known as immunocloaking. Brasile et al. engineered an innovative nano-barrier, which, when applied to a renal graft during MP, hides antigenic surfaces from recipient immune mediators [25].

2.3.5 Viability Assessment

Preserving grafts in a dynamic state creates the opportunity for functional assessment. Some markers are common across organs and composite tissues including hemodynamic parameters (i.e., flow, resistance, resistive index), metabolic markers such as oxygen consumption and lactate levels, and edema measured with post-perfusion weight increase or wet-to-dry tissue weight ratios. Perfusate biomarkers, including both nonspecific and tissue-specific makers, are commonly measured to gather evidence of cellular function or death. In the warm perfusion environment especially, assessment of normal organ functions can inform graft quality. Examples include urine and bile production, insulin excretion, respiratory PaO_2/FiO_2 ratio, and muscle contraction. However, as the discussion of kidney HMP will highlight below, most clinically tested biomarkers have predicted transplant outcomes poorly thus far [21], and the need for multivariate scores that synthesize many factors may be necessary. Ideally, the predictive value of these biomarkers will be derived not from increasing the discard of possibly dysfunctional grafts, but rather, by directing ex vivo resuscitative efforts to restore such grafts.

As this discussion indicates, a multitude of factors contribute to the behavior of each perfusion system. The remainder of this chapter will explore the clinical evidence for MP in individual organs. It is important to consider that as researchers investigate and innovate MP systems, it is likely that the optimal preservation system will vary depending on the circumstances unique to each organ (i.e., organ, manner of death, anticipated storage time). For example, a renal HMP study demonstrated that kidneys without warm ischemia (WI) performed best with B-MPS, while after 75 min of WI, perfusion with UW was significantly better [129]. Rather than a one-size-fits-all approach, the search must be for the best preservation modality for a given situation.

2.4 Machine Perfusion of the Kidney Grafts

The kidney has proven to be an ideal organ for machine perfusion for a number of reasons. It was the object of initial forays into machine preservation, and therefore, the history of kidney preservation describes much of the history of organ preservation in general. The kidney is the only organ commonly perfused clinically, and it continues to be the target of many recent and innovative research efforts.

2.4.1 History of Kidney Perfusion

Kidney preservation became clinically necessary with the expansion of transplantation in the 1960s. Early kidney transplants used living-related donors or cadaveric donors following cardiac death [12]. Advancements in HLA tissue typing necessitated the organization of large pools of patients who were matched to donor kidneys, but for kidneys to be transported across these networks, improved organ preservation was necessary [91]. Initial research explored HMP. Humphries et al. carried out some of the first renal perfusions, first with diluted whole blood followed by plasma [103, 104].

A major breakthrough came in 1968 when Belzer discovered that his initial attempts to perfuse kidneys with plasma were complicated by the embolization of lipid microemboli that caused rising perfusion pressures [15]. By removing these precipitants through cryoprecipitation and filtration, Belzer and colleagues were able to preserve functional canine kidneys for 72 h of perfusion [13] and performed the first perfusion and transplantation of a human cadaver kidney [14]. This HMP system was further modified by a number of researchers. Silica gel filtration was employed to filter lipoproteins from plasma, resulting in longer perfusate shelf life and improved stability during perfusions lasting at least 48 h [219]. Human plasma was replaced with a physiologic electrolyte solution supplemented with human albumin, which supported intravascular volume and reduced tissue edema [42]. To eliminate any risk of spreading blood-borne pathogens, Belzer developed an entirely synthetic perfusate, using hydroxyethyl starch (HES) as an osmotic agent [17, 95].

At the same time, SCS emerged as an alternative to hypothermic perfusion. Collins developed a solution characterized by intracellular electrolyte profile along with a high concentration of glucose [43]. By flushing kidneys with this cold solution and storing them in a cold environment, kidneys could be preserved with less effort and cost while increasing increased graft survival [178]. This development was all the more important with the establishment of brain death [9], allowing for organ donation after brain death (DBD), avoiding the warm ischemic damage of donation after cardiac death. With higher quality kidneys coming from DBD donors and the relative ease of preservation and transportation with cold storage, there was less need for HMP, and SCS became the predominant means of kidney preservation [16, 239]. Belzer and his team developed the UW solution, which has since become the gold standard SCS solution for kidney preservation [185].

2.4.2 Clinical Evidence for HMP Kidney Preservation

The success of kidney transplantation brought increased demand for organs, necessitating a broader pool of kidneys. The transplantation community returned to DCD and ECD kidneys, but these marginal organs were associated with higher rates of DGF [41]. In the 1990s and early 2000s, there was an increasing belief that HMP minimized DGF compared to SCS, but numerous studies exploring this comparison yielded conflicting results. A meta-analysis comparing the risk of DGF between these competing preservation techniques found that HMP was associated with a 20% reduction in DGF compared to SCS and reduced healthcare costs [238, 239]. Analysis was limited by the high heterogeneity, low power, and minimal long-term follow-up of the included studies, but it drew attention to the need for long-term, randomized clinical trials.

Over the course of the decade, evidence accumulated demonstrating the superiority of HMP among standard and high-risk donors [123, 154, 189, 198], and a landmark multicenter RCT was performed [150]. Deceased donor kidney pairs of all types were randomized to SCS or HMP and transplanted within the Eurotransplant organ-exchange network. Kidneys preserved by HMP had both a reduced rate and shorter duration of DGF than SCS kidneys. In subgroup analyses, this relationship was consistent among SCD, ECD, DBD, and DCD donors. While 1-year graft survival rate was higher in the HMP group and the hazard of graft failure was reduced, there was no difference in PNF rates and 1-year patient survival (Table 2.1). Interestingly, when DGF was included as a covariate in their graft failure model, HMP was no longer a significant factor in predicting graft failure, suggesting that HMP promotes graft survival by reducing DGF. A follow-up report showed that 3-year graft survival was superior among HMP-preserved kidneys, especially for those that experienced DGF [149]. Despite the increased initial cost of HMP compared to SCS, a cost-effectiveness analysis of this trial demonstrated that HMP ultimately lowers healthcare costs [85].

Since this trial, additional analyses have confirmed the DGF reduction benefit among cadaveric kidney donors. A retrospective analysis of 52,000 patients in the United Network for Organ Sharing database also found a similar reduction in a propensity-matched cohort and paired kidney analysis [39]. Two meta-analyses comparing all types of deceased kidney donors also found reduced DGF with no difference in PNF, 1-year graft, and patient survival [124, 168]. Thus, evidence suggests that for cadaveric kidneys—not simply marginal ones—HMP provides a benefit by reducing the risk of DGF. Whether this ultimately translates to improved graft and patient survival still remains to be determined.

Other analyses have measured the benefits of HMP among marginal kidneys in particular. A separate ECD subgroup analysis of the Eurotransplant MP trial found that in addition to improved DGF and 1-year graft survival rates, in the broader analysis HMP reduced PNF incidence in ECD kidneys [220]. The graft survival advantage persisted after 3 years, especially among those who experienced DGF [77]. A separate analysis exploring the role of HMP in donors 65 years or older also showed a PNF reduction among HMP, and among those who developed DGF, 1-year graft survival improved [78]. More recently, a meta-analysis of seven trials involving over 10,000 ECD donor grafts showed a reduction in DGF and improvement in 1-year graft survival, but notably no change in PNF and patient survival [108].

The benefits of HMP in the DCD kidney are less certain. In an expanded DCD cohort of the Eurotransplant MP trial, HMP reduced the frequency and duration of DGF, but no difference in PNF, 1-year graft survival, or 1-year patient survival was found [111]. In the United Kingdom, Watson et al. performed a multicenter RCT

Table 2.1 Results of recent large-scale renal HMP trials and subgroup analyses

Author	Study	Donors (n)[a]	DGF[b]	PNF[b]	1-year graft survival[b]	1-year patient survival[b]
Moers et al. [150]	MP Trial – overall	336	20.8% vs. 26.5% ($P = 0.05$)	2.1% vs. 4.8% ($P = 0.08$	94% vs. 90% ($P = 0.04$)	97% vs. 97% (no P listed)
Treckmann et al. [220]	MP Trial – ECD	91	22% vs. 29.7% ($P = 0.27$)	3% vs. 12% ($P = 0.04$)	94% vs. 85% ($P = 0.164$)	93.4% vs. 96.7% ($P = 0.30$)
Gallinat et al. [78]	MP Trial – ≥ 65 years	85	29.4% vs. 34.1% ($P = 0.58$)	3.5% vs. 12.9% ($P = 0.02$)	89% vs. 81% ($P = 0.139$	94.2% vs. 95.0% ($P = 0.79$)
Jochmans et al. [111]	MP Trial – DCD	82	53.7% vs. 69.5% ($P = 0.007$)	2.4% vs. 2.4% ($P = 1.00$)	93.9% vs. 95.1% (no *P* listed)	96.3% vs. 97.6% (no *P* listed)
Watson et al. [234]	UK Trial – DCD	45	57.8% vs. 55.6% ($P = 0.99$)	2% vs. 0% (no *P* listed	93.3% vs. 98% ($P = 0.3$)	93% vs. 100% ($P = 0.08$)

MP trial refers to Eurotransplant machine perfusion trial
[a]Two recipients per donor
[b]Listed as HMP vs. SCS

comparing HMP and SCS among DCD donors and found no difference in DGF incidence, renal function, graft, or patient survival [234]. There are a number of potential reasons for why, unlike the Eurotransplant DCD extension study, this trial failed to show superiority of HMP. Notably, kidneys in the Eurotransplant study experienced slightly longer warm and cold ischemia on average. In addition, a portion of kidneys randomized to HMP in the UK trial underwent cold storage during transportation to a base transplant hospital, whereas in the Eurotransplant trial, all HMP kidneys were perfused immediately after procurement [109]. With these conflicting results, the utility of HMP for DCD kidney donors remains unclear.

2.4.3 *Kidney HMP Innovations*

As researchers test the efficacy of HMP in preserving organs, efforts are also focused on optimizing this technology (Fig. 2.1). In standard HMP study protocols, the perfusate is non-oxygenated since the low metabolic rate at 4 °C requires little to no oxygen, and oxygen can theoretically exacerbate cold-induced reactive oxygen species damage. Recently, however, a number of researchers have questioned this paradigm. Studies comparing oxygenated and non-oxygenated HMP in porcine models

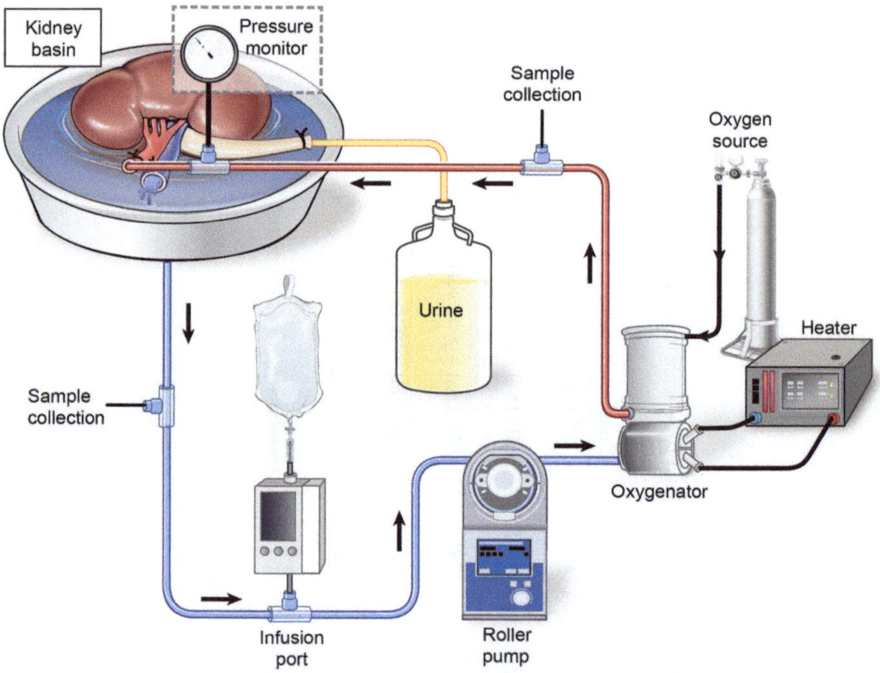

Fig. 2.1 Schematic of a normothermic perfusion system for the kidney. (Reprinted with permission, Cleveland Clinic Center for Medical Art & Photography © 2014-2016. All Rights Reserved)

have identified modest short- [101] and long-term [217] benefits of oxygenation among DCD but not DBD kidneys [79]. Clinical benefits remain to be seen.

As the UK HMP trial highlights, the timing of kidney perfusion is an important variable. The increased logistic complexity of HMP has led to the concept of hypothermic reconditioning, in which SCS is followed by a brief period of HMP. This practice is supported by two porcine studies. In one study, continuous and brief HMP similarly demonstrated faster recovery of renal function and mitigated expression of pro-inflammatory molecules compared to SCS [80]. A follow-up study demonstrated that hypothermic reconditioning is just as valuable if HMP limited to only 1 h at the end of SCS [76].

HMP is a platform for organ assessment prior to transplantation. Intrarenal vascular resistance is a common and accessible biomarker, and though used as a reason for rejecting allografts, it is imprecise. Initially, discard decisions were made based on empirically established threshold resistance levels, but the reliability of these limits has been questioned [155]. In a prospective analysis from the Eurotransplant trial, renal resistance at the end of perfusion was weakly predictive of DGF and 1-year graft failure [110]. Retrospective analyses are difficult in this field, since perfusion parameters are often used as criteria for discarding organs, introducing bias. However, a retrospective study DCD kidney HMP from Maastricht University, where perfusion parameters were not used as criteria to discard organs, found that renal resistance at the start of HMP was moderately predictive of PNF [53]. A retrospective analysis looking at changes in flow through the duration of perfusion found that in cases where flow increased after 2 h of perfusion, the discard rate was lower but DGF rate higher [183]. The information gleaned from renal resistance is minimally informative, and it appears unwise to base discard decisions solely off of this poorly predictive criteria. An alternative or supplement to this hemodynamic indicator is required.

Perfusate biomarker analysis provides another window into kidney status during MP. Various markers have been studied going back to the 1970s, and novel markers continue to be proposed. However, no reliable predictor of renal dysfunction has been found, and many recent studies have reached the same conclusion: while some markers are associated with DGF, and less frequently PNF, none predict function with sufficient sensitivity and specificity. The Eurotransplant MP trial provided a valuable opportunity for biomarker analysis because unlike many retrospective reviews, no kidneys were discarded on the basis of biomarker levels. Glutathione S-transferase (GST), N-acetyl-β-D-glucosaminidase, and heart-type fatty acid-binding protein (H-FABP) were found to be independent risk factors for DGF, but no markers predicted PNF or 1-year graft survival, and no markers correlated with renal resistance [151]. A comprehensive review of 12 studies explored urine perfusate biomarkers going back to 1972. Lactate dehydrogenase (LDH), GST, and aspartate aminotransferase were significant associated with DGF in the majority of studies, while GST was the only marker significantly associated with PNF in most studies [21]. Since this review, additional large-scale biomarker analyses have been performed. A DCD biomarker analysis found that LDH, IL-18, GST, H-FABP, redox iron, and neutrophil gelatinase-associated lipocalin were either poorly or fairly predictive of PNF, and the authors concluded that in DCD kidneys, elevated biomarker levels should not be grounds for discard [97]. Parikh et al. found that compared to transplanted kidneys, discarded kidneys had worse hemodynamic parameters and similar biomarker values, but those that were transplanted despite poor hemodynamic function or biomarker levels had acceptable 6-month kidney function nonetheless. Overall, no one marker can identify a dysfunctional kidney, but perhaps as increasingly more patient outcomes are available for analysis, allowing for a higher number of PNF occurrences, researchers can develop a comprehensive risk score featuring a variety of markers to identify kidneys with a high likelihood of failure.

2.4.4 Warm Perfusion of the Kidney

Over the last two decades, researchers have turned to warm kidney perfusion. As opposed to hypothermic suppression of metabolism, warm perfusion promotes active metabolism, thus necessitating supplemental oxygen, often alongside an oxygen carrier. Kootstra et al. carried out some of the earliest attempts at warm kidney perfusion. After initially finding success interrupting prolonged HMP with 3 h of normothermic whole-blood perfusion [190], they developed a cell culture media-like acellular perfusate to sustain organs at 32 °C. This model, known as exsanguinous metabolic support, has been remarkable in its ability to recover severely damaged kidneys and sustain organs for a prolonged period of time. For example, Brasile et al. restored function in canine kidneys exposed to 120 min of warm ischemia [30] and sustained DBD human kidneys for 48 h of perfusion [31]. It has also served as a springboard for interesting innovations including the kidney restoration using fibroblast growth factor [27], viral transfection [29], and immunocloaking [25]. SMP has also been employed in experimental models at cooler temperatures approaching room temperature [75, 100]. This is notable for it enables a higher level of organ function than hypothermic perfusion while avoiding the need for an oxygen carrier.

Kidney NMP has been employed in two models recently: short-term conditioning and prolonged NMP with STEEN solution. The conditioning model is the only circumstance in which warm kidney perfusion has been conducted clinically. To restore cellular energy stores, Nicholson et al. have used RBCs in Ringer's solution along with infused nutrients to perfuse kidneys for 1 h prior to transplantation. This led to a reduced rate of DGF among ECD kidneys compared to standard SCS [163]. Kaths et al. recently published a protocol in which diluted STEEN solution supplemented with RBCs supports NMP for 10 h [117].

2.5 Machine Perfusion of Liver Grafts

Livers from ideal donors tolerate up to 12 h of SCS, which is the current standard preservation method [60]. Livers from ECD and DCD donors are more susceptible to cold ischemia and reperfusion injury and should be stored by SCS as short a

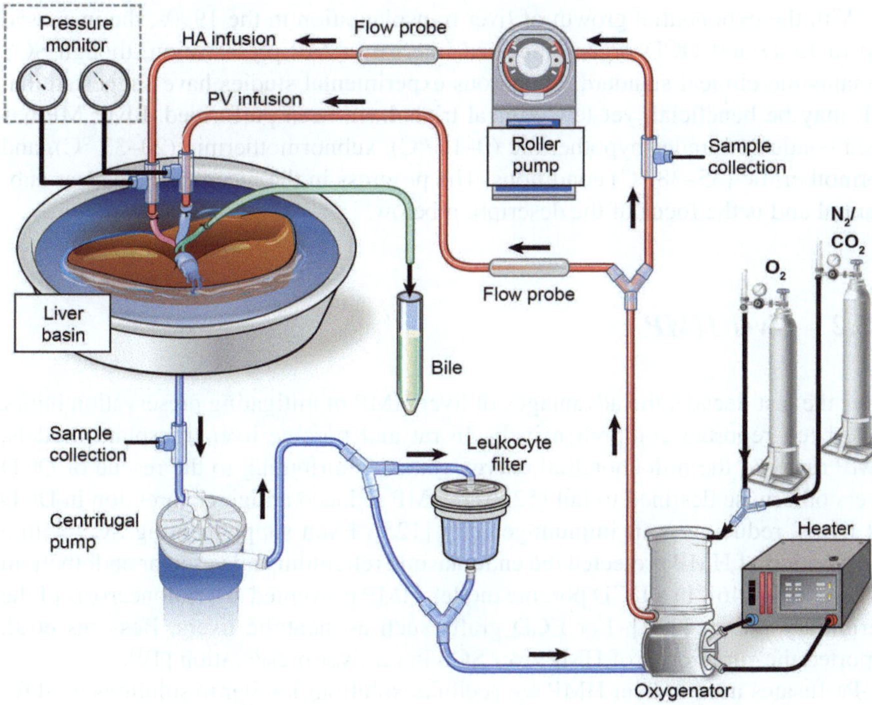

Fig. 2.2 Schematic of a normothermic perfusion system for the liver. (Reprinted with permission, Cleveland Clinic Center for Medical Art & Photography © 2014-2016. All Rights Reserved)

duration as possible [60]. With the increasing use of ECD and DCD livers comes the need to preserve and resuscitate livers and assess graft viability prior to transplant. Liver MP can fulfill these goals (Fig. 2.2).

2.5.1 History of Liver MP

The history describing development of liver MP can be divided into three eras: initial development, SCS dominance, and MP rival [152]. It began in the 1960s and was applied clinically by Starzl et al. on 11 human livers for 4–7 h with cold sanguineous perfusate [206]. MP preservation did not gain widespread acceptance at that time due to logistic challenges and the introduction of SCS. With the invention of Collins solution in the 1970s, along with UW solution and histidine-tryptophan-ketoglutarate solution (HTK) in the 1980s, the clinical practice of liver transplantation changed profoundly. UW solution has been the clinical standard preservation solution since the 1980s due to the simplicity and efficiency of UW-based SCS to safely preserve high-quality livers within the clinical time frame of transplantation. HTK solution was found to be as effective as UW solution for liver SCS preservation [67, 141].

With the exponential growth of liver transplantation in the 1990s, the increased use of ECD and DCD organs renewed interest in MP preservation, though SCS remains the clinical standard. Numerous experimental studies have suggested that MP may be beneficial, yet few clinical trials have been performed. Liver MP has been conducted under hypothermic (0–10 °C), subnormothermic (20–33 °C), and normothermic (35–38 °C) conditions. The progress in the recent decades was substantial and is the focus of the description below.

2.5.2 *Liver HMP*

Over the last decade, the advantages of liver HMP in mitigating preservation injury have been reported comprehensively. In rat and porcine liver transplant models, HMP restored the mitochondrial energy status, contributing to the rescue of DCD livers otherwise destined to fail [52, 61]. HMP reduced antigen expression in DCD rat livers, reducing graft immunogenicity [125]. Even supplementing SCS with a short period of HMP protected the endoplasmic reticulum and vascular endothelium of rat livers [146]. In a DCD porcine model, HMP prevented arteriolonecrosis of the peribiliary plexus [177]. For ECD grafts such as steatotic livers, Bessems et al. reported the superiority of HMP over SCS in rat liver preservation [19].

Perfusates used in liver HMP are acellular solutions similar to solutions used for SCS [133]. One unique challenge of liver MP is the dual vascular supply, from both the portal vein and hepatic artery. Dual infusion via both the portal vein and hepatic artery has been used in most preclinical studies on large animals. In clinical trials, the perfusion settings have varied widely.

Guarrera et al. reported the first clinical trial using HMP [89]. After a period of SCS, they used non-oxygenated Vasolol solution (a modified B-MPS) to perfuse 20 standard criteria livers and transplanted these into recipients. They demonstrated the feasibility and reliability of HMP and reported lower peak hepatocellular enzyme release, total bilirubin, and serum creatinine, and fewer biliary complications compared to SCS. They observed reduced pro-inflammatory cytokine expression in HMP, signifying less IRI compared to SCS [93]. Similar benefits were observed in 31 ECD livers [88].

A different HMP setting was developed in Zurich, Switzerland, and applied clinically both at a single center [63] and internationally [62]. Their system, named Hypothermic Oxygenated Perfusion or HOPE, has been used to resuscitate DCD livers. The HOPE system perfuses B-MPS through the portal vein only with active oxygenation. Using only 1–2 h of perfusion following a standard period of SCS, both the single-center study (8 patients) and the international study (25 patients) showed the benefits of HOPE compared to matched cases with only SCS preservation.

A number of studies have explored markers of liver viability during HMP. In canine liver HMP preservation, Uematsu et al. found oxygen consumption to be a viability marker of DCD livers [221]. Lockett et al. and Changani et al. scanned whole-liver parenchyma with phosphorus-31 magnetic resonance spectroscopy and found this noninvasive tool effective for analyzing ATP status and energy metabolism [40, 138]. Proton magnetic resonance spectroscopy can be used to study components of HMP perfusate and helped identify alanine and histidine as potential functional markers [137]. A functional index combining multiple biomarkers during liver HMP was developed with a porcine DCD model and demonstrated greater sensitivity and specificity than a single marker [136]. In a clinical trial, Guarrera et al. observed that the concentration of hepatocellular enzymes in the perfusate correlated with levels in the transplant recipient after transplantation [88]. A widely accepted liver graft viability score that can be utilized during HMP remains to be developed.

2.5.3 Liver SMP

SMP was rarely performed before the twenty-first century, but in the last 15 years, multiple research centers have established SMP models in small animals [68, 132, 174, 218, 223, 224], large animals [71, 84, 121, 162, 171], and discarded human livers [35]. Perfusion settings vary widely between systems, especially with regard to perfusate contents (Table 2.2). An oxygen carrier is present in most porcine models, but not rat models. Despite these variations, the benefits of SMP are consistent; it protects hepatobiliary function, restores energy storage, and reduces oxidative stress.

Knaak et al. established a promising SMP model in porcine DCD livers [121]. They used STEEN solution with supplemental RBCs at 33 °C, with dual perfusion through the hepatic artery and portal vein. With a brief 3-h SMP between two SCS phases, they observed less hepatobiliary injury and better-preserved histology in SMP livers compared to SCS controls after transplantation.

Table 2.2 Settings for liver SMP

Publication	Model	Perfusate
Vairetti M, et al. *Liver Transpl* (2008) [224]	Normal rat	Krebs-Henseleit solution
Vairetti M, et al. *Liver Transpl* (2009) [223]	Steatotic rat	Krebs-Henseleit solution
Olschewski P, et al. *Cryobiology* (2010) [174]	DCD rat	Lifor solution
Ferrigno A, et al. *Cryobiology* (2011) [68]	DCD rat	Belzer MPS
Tolboom H, et al. *J Surg Res* (2012) [218]	DCD rat	Cell culture medium + RBC
Liu Q, et al. *Transpl Proc* (2013) [132]	Steatotic rat	Cell culture medium
Gringeri E, et al. *Transpl Proc* (2012) [84]	DCD pig	Celsior solution
Bruinsma BG, et al. *Am J Transpl* (2014) [35]	Human (discarded)	Cell culture medium
Knaak JM, et al. *Liver Transpl* (2014) [121]	DCD pig	STEEN solution + RBC
Okada N, et al. *Transpl Proc* (2015) [171]	DCD pig	Cell culture medium + blood
Fontes P, et al. *Am J Transpl* (2015) [71]	DCD pig	Belzer MPS + cell-free O2 carrier
Nassar A, et al. *Int J Arti Org* (2016) [162]	DCD pig	Whole blood

A porcine SMP model with cell-free hemoglobin-based oxygen carrier (Hemopure®) was reported in Pittsburgh [71]. The oxygen carrier was mixed with B-MPS, and perfusion was performed at 21 °C. DBD porcine livers demonstrated superior posttransplantation graft function and recipient survival in SMP compared to SCS.

Currently, liver SMP has not been applied clinically, and only one discarded human liver study has been reported. In the absence of an oxygen carrier, Bruinsma et al. maintained the venous oxygen partial pressure above 200 mm Hg and observed continuous bile production, lactate clearance, and ATP regeneration in liver tissue and stable hepatocellular enzyme levels [35]. Since SMP was limited to only 3 h, the efficacy of this model in a prolonged perfusion setting remains to be seen.

2.5.4 Liver NMP

Only two porcine transplant models employing NMP have been performed in the twenty-first century, one in Berlin and another in Oxford [34, 199]. While both used RBCs as oxygen carrier, the Oxford group perfused whole blood at a near-physiologic flow rate, while the Berlin group mixed blood with a balanced electrolyte solution and perfused at lower flow rate. The Oxford group performed a 72-h NMP in a porcine model [37] and successfully resuscitated porcine livers exposed to 40 min of warm ischemia [34]. The Berlin group resuscitated porcine livers exposed to 60 min of warm ischemia.

Liver NMP has recently entered the clinical realm. Watson et al. documented a case of human liver NMP from a DCD donor. After a period of SCS, the liver was perfused with RBCs and a colloid solution (succinylated gelatin) and transplanted successfully [233]. The Oxford group published the first-in-human clinical trial [188]. Using a similar perfusate, they reported the safety and feasibility of NMP for liver transplantation in 20 patients. With the success of this trial, the expansion of this technology to multiple centers is anticipated.

The Oxford group's perfusion settings have been reproduced by a number of groups over the last 5 years [11, 176, 242]. In addition to protecting hepatocellular integrity, restoring energy content, and reducing inflammation posttransplant inflammation, NMP has improved the preservation and regeneration of biliary epithelial cells and peribiliary glands [135]. This is important since the risk of long-term biliary complications is one of the main concerns limiting the use of marginal donor livers, and this suggested for the first time that cellular regeneration occurs during liver NMP.

In contrast to the sanguineous perfusate of most NMP studies, it was reported that STEEN solution without an oxygen carrier was still better than SCS in a study on porcine DCD livers [23]. However, other groups have not yet successfully reproduced this perfusion setting. A comparison of NMP with various perfusates in porcine DCD livers found that (1) oxygen carrier is necessary to maximize DCD liver resuscitation, and (2) whole blood performed similarly to colloid-based solution mixed with packed red blood cells [134].

Since a large portion of livers are discarded due to steatosis, efforts to expand the liver donor pool have explored the use of NMP to reduce fat content in these livers prior to transplantation. The Oxford group employed NMP on a porcine model of hepatic steatosis and observed a reduction of steatosis after 48 h NMP [107]. Supplementing NMP with a pharmacologic intervention, known as a "defatting cocktail," has also been explored, with the goal of stimulating lipid metabolism and decreasing steatosis [159]. This significantly reduced steatosis after only 3 h of NMP. With the near-physiologic metabolic rates of NMP, this is a model for efficient pharmacologic interventions that can improve graft function during the preservation period, and for the liver in particular, these studies suggest a future in which fewer livers will be discarded on the basis of steatosis.

Biomarkers measured during NMP may serve as useful predictors of posttransplant liver function, but identifying reliable markers remains a challenge. In a porcine DCD liver transplant study, bile production, base excess, transaminase levels, and hemodynamic markers were used to predict the posttransplant viability [34]. In a discarded human liver study, liver quality was distinguished with a quantitative bile production rate threshold: livers that produced at least 30 g of bile through 6 h of NMP had generally better performance than those that produced less than 20 g [213]. In a non-transplant model, these markers are preliminary, and the validity and reliability of these markers must be tested in a clinical transplant setting.

The ideal temperature or setting for liver MP has not been defined. The utility of HMP, SMP, and NMP have been compared and debated [162, 196], but a gold standard has not been identified. Nevertheless, even with the diverse settings and protocols, evidence supporting MP preservation of liver grafts compared to SCS is promising across preclinical animal models and clinical trials.

2.6 MP of Small Intestine Grafts

The small intestine is the least commonly transplanted organ, and the challenge of adequate preservation limits intestinal transplant expansion. Despite advances in immunosuppression and surgical techniques, high rates of graft rejection occur, and recipient survival is suboptimal. The small intestines are highly sensitive to ischemic injury, and as a result, the tissue can withstand only minimal IRI [2]. Improved intestinal graft preservation is needed to protect the vulnerable mucosa and decrease IRI-associated injury. As a result, enhancing preservation through MP and optimizing the quality of intestinal tissue is a priority [157]. Intestinal MP is unique in that it requires both vascular and luminal perfusions (Fig. 2.3). Most previous intestinal MP studies were not performed for organ preservation, but rather, to study physiologic mechanisms, drug behavior, and intestinal diseases. Nonetheless, these studies are useful resources to guide the expansion of MP in this organ. Small intestine MP is in its infancy, but wide efforts are underway to establish a stable MP preservation model.

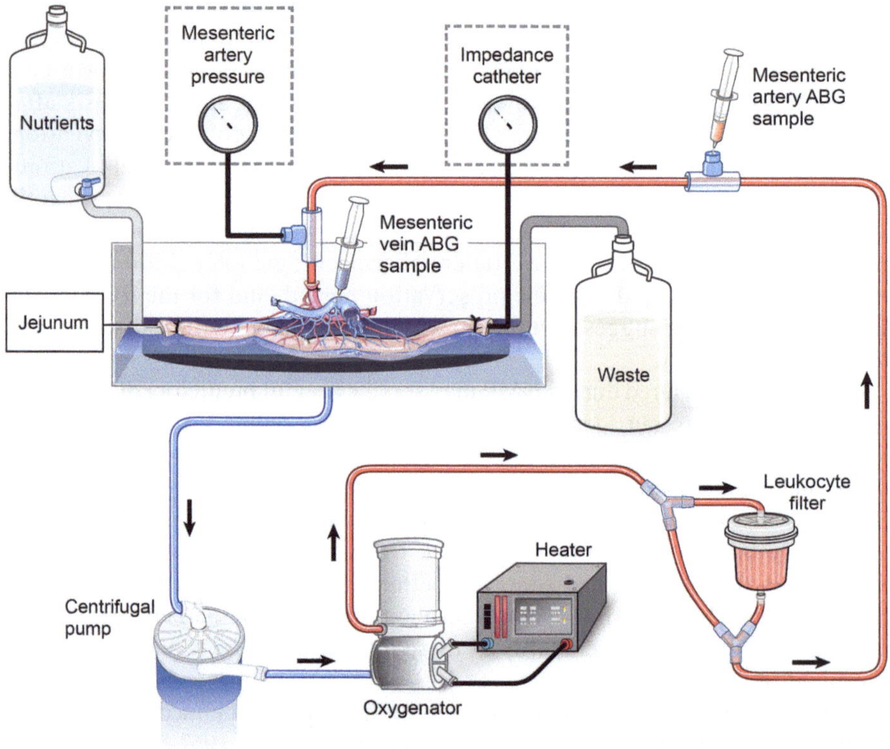

Fig. 2.3 Schematic of a normothermic perfusion system for the intestine. (Reprinted with permission, Cleveland Clinic Center for Medical Art & Photography © 2014-2016. All Rights Reserved)

2.6.1 History and Development of Small Intestinal Perfusion

The earliest reports of intestinal perfusion come from the 1940s, when the intestinal lumen was perfused with high fluid volumes to treat chronic uremia, an approach known as intestinal dialysis [191]. Subsequent attempts employed whole blood in the vascular perfusion of isolated intestinal loops in guinea pigs [81]. In 1969, Hohenleitner et al. established a normothermic ex vivo intestine perfusion model for metabolic analysis [96]. The vasculature of jejunoileal loops was perfused for 5 h with modified Ringer's solution and oxygenated RBCs, while a monosaccharide-based solution was used as a luminal perfusate. Viability was assessed with gross peristaltic activity, oxygen and glucose consumption, and carbon dioxide and lactate production. Edema and intraluminal fluid loss were noted when perfusion extended beyond 5 h. In 1971, Ruiz et al. developed a model in which intestines were perfused following SCS of varying duration [193]. With a perfusate of heparinized whole blood diluted with lactated Ringer's and dextran solution, intestinal function and histologic integrity were preserved following up to 12 h of SCS.

Thereafter, progress was slow and most perfusions were brief [166]. The majority of studies were performed on small animal models and explored specific technical issues such as the optimal pump settings [113] and the need for luminal perfusion [147]. Prolonged MP returned in the 1990s with the work of Braun et al., who developed porcine NMP model using heparinized whole blood at pressures up to 140 mmHg for up to 9 h [32]. However, perfusion at such high pressure resulted in congested capillaries, hemorrhage of the mucosa and mucosa-associated lymphatic tissue, and ulceration.

With minimal progress in intestinal MP, SCS has been the clinical choice for intestinal preservation in recent decades. However, in light of the renewed interest in intestinal MP, the following section will review considerations for model development.

2.6.2 *Features of Small Intestinal MP Models*

2.6.2.1 NMP vs. HMP

NMP was used in early experimental models as previously described. HMP has been tested less frequently. The first HMP model tested luminal only perfusion with oxygenated UW solution. One-hour perfusion before SCS resulted in better outcomes compared to continuous perfusion or SCS alone [245]. In a recent HMP model with discarded human intestines, Abraham et al. perfused both the vasculature and intestinal lumen HTK and demonstrated superior histologic integrity compared to SCS [157].

2.6.2.2 Pulsatile Versus Continuous Pump

Kachelhoffer et al. investigated optimal pump type by measuring the hemodynamic and metabolic behavior during alternating periods of pulsatile and nonpulsatile pumping [113]. While both patterns produced the same vascular resistance, pulsatile flow increased oxygen consumption. Other investigators have reached similar conclusions, finding that pulsatile flow supports myogenic autoregulation and preservation of intestinal microcirculation [202].

2.6.2.3 Vascular and Luminal Perfusates

In attempts to establish an MP model, a number of different perfusates have been tested. An acellular isotonic electrolyte solution was stable for 1–2 h [166] but produced high vascular resistance beyond 2 or 3 h [96, 186]. Whole-blood perfusion supported stable hemodynamics through 9 h, though histologic evidence of ischemic injury was apparent after 5 h [32]. Diluted whole blood with normal saline or

Krebs-Henseleit solution produced edema and intraluminal perfusate accumulation after only 2 h of perfusion [205]. RBCs with modified Krebs-Henseleit solution and albumin preserved histologic integrity through 90 min of perfusion [57]. Luminal perfusion serves dual purposes. Luminal perfusion with UW solution preserved mucosal integrity [245], and with the addition of nutrient substrates such as D-xylose, luminal perfusion enables measurement of absorptive function [32].

2.6.2.4 Viability Assessment Tools

A variety of tools are available to assess intestinal viability. Subjectively, intestinal color and presence of peristalsis are used to visually assess organ function. Traditionally informative parameters include perfusate pH, oxygen consumption, lactate, and histological assessment of the intestinal tissues before and after perfusion using the Park/Chiu classification for grading intestinal ischemia-reperfusion injury [175]. Other markers of intestinal integrity and ischemic injury include perfusate levels citrulline [56], neurotensin [201], and stool calprotectin [144, 212]. The reliability of these markers has not been validated.

Intestinal preservation with MP is preliminary. Its efficacy must still be tested against SCS in discarded human intestine models. No clinical trials have been conducted to test the feasibility and safety of MP before intestinal transplantation.

2.7 Machine Perfusion of Pancreas Grafts

Pancreas and beta-islet cell transplantation are offered as alternative therapies for type I diabetes mellitus and are increasingly being offered for type II diabetes [86]. The number of pancreas transplantations performed in the United States has declined since 2004 [114]. While SCS provides adequate preservation for DBD grafts [7], DCD pancreases have a higher risk of early graft loss from thrombosis [158]. A number of preservation technologies beyond SCS have been tested in pancreas preservation, including the two-layer method, oxygen persufflation, and MP. Historical and recent attempts at HMP and NMP of the pancreas will be reviewed.

2.7.1 History of Pancreas Perfusion

Babkin and Starling performed the earliest isolated, in situ pancreas perfusions in 1926 using a heart-lung preparation to oxygenate and pump normothermic blood to the organ [6]. After this point, the development of pancreas perfusion models mirrored kidney perfusion. Early HMP used plasma to sustain pancreases for 24 h [216, 235]. Albumin-based solutions [36] and silica-filtered plasma [8] were also attempted. With the advent of SCS, attention shifted away from HMP in favor of the simpler and equally effective method [70]. Preservation with UW solution and HTK became the standard of care [197].

2.7.2 *Pancreas HMP*

The pancreas is particularly susceptible to edema; strict pressure and flow restriction are required to minimize swelling [70, 116]. This is a double-edged sword. When solid organ pancreas transplant is desired, edema is detrimental, as it increases the risk of thrombosis and graft failure [241]. However, in the case of beta-islet cell retrieval, edema can be beneficial. Taylor et al. found that HMP and its associated edema increased the yield and purity of retrieved islet cells compared to SCS [214]. The only reported instance of clinical pancreatic HMP was perfused with UW solution following a period of SCS, resulting in satisfactory islet isolation [128].

2.7.3 *Pancreas NMP*

The pancreas has been the subject of numerous NMP investigations since the 1960s, but these experiments were typically performed for physiologic investigations rather than organ preservation. Perfusion solutions varied but often contained whole blood or RBCs supplemented with electrolyte solutions [145] and colloid agents [64, 65]. Isolated pancreas perfusion did not extend beyond 5 h [64]. To supply adequate nutrients and oxygen, pancreases were perfused at physiologic pressures resulting in edema [170].

Recently, there has been renewed interest in pancreas NMP. Building off their kidney NMP conditioning system, Barlow et al. reported NMP in discarded human pancreases [7]. To assess organ quality, 2-h perfusions were performed with RBCs in a colloid solution. Over the course of brief but stable perfusions, pancreatic enzymes were released. Since these enzymes continuously recirculate, they can potentiate cellular damage. This problem must be dealt with before this early perfusion system can translate into clinical applications.

2.8 Machine Perfusion of Heart Grafts

With the donor heart utilization rate at only 32% [118], the donor heart shortage is a major problem limiting the availability of heart transplantation. Expansion into marginal quality grafts is limited, since the major predictors of primary graft failure are donor age and ischemic time [194, 244]. Moreover, the relatively short acceptable ischemic time (less than 4–6 h) of SCS limits the range of organ allocation and creates a significant time restriction for recipients and surgical teams.

2.8.1 History of Heart MP: The Langendorff Heart Perfusion Model

A small animal isolated heart perfusion model was established in 1897 by Oscar Langendorff and later modified. This model consists of a heart, reservoir, pump, oxygenator, and heater [10, 203]. In this model, retrograde flow from the aorta travels to the coronary arteries, and outflow from the coronary sinus is collected and returned to the reservoir. The standard perfusate is Krebs-Henseleit solution, but this limits the oxygen-carrying capacity. Furthermore, its low oncotic pressure causes edema. Some researchers have compensated for this by using whole blood or supplementing Krebs-Henseleit solution with RBCs and/or oncotic agents like albumin or dextran. Maintaining temperature at 36 °C prevents a change in heart rate and cardiac contractile function inherent to lower temperatures. The Langendorff model has been an important platform for cardiovascular and pharmacological research, and it has been particularly valuable for heart transplantation, as much has been learned about cardiac SCS and IRI with this model.

2.8.2 HMP of the Heart

The first four cases of clinical cardiac transplantation following donor heart HMP were reported in 1984 [236]. Following cardioplegia at the donor hospital, the aorta was cannulated, and the heart was continuously perfused with an oxygenated, hyperosmolar extracellular electrolyte solution containing glucose, sucrose, and glycerol. Heterotopic transplant followed 6–15 h of perfusion. The results were mixed. One patient experienced irreversible acute rejection leading to donor heart excision after 5 days, but the three remaining patients survived for 6–16 months after transplant.

Like most other organs, SCS became the standard preservation strategy for the heart, but recently interest in HMP has returned. For instance, HMP was performed in canine orthotopic transplant model [69]. Hearts that underwent 24-h HMP with B-MPS were compared to 4-h SCS. Despite the organ weight increasing by 27.4% in the HMP group, there was no significant difference between the two groups in terms of survival, inotropic support, and cardiac output.

2.8.3 NMP of the Heart

The Organ Care System (OCS) Heart (TransMedics, Andover, MA) is a portable normothermic heart perfusion machine. Its portable design is based on the concept that initiating perfusion at the donor hospital minimizes cold ischemia and that continuous perfusion with oxygen and nutrient-rich, blood-based perfusate maintains viability during transportation [92]. The system is primed with donor blood, and maintenance solution is added during perfusion. Following cardioplegia, the aorta and pulmonary artery are cannulated, and the superior and inferior vena cava are ligated. Retrograde flow passes through the aorta to perfuse the coronary arteries at 650–860 ml/min. Outflow from the coronary sinus drains to the reservoir via the right heart to a cannulated pulmonary artery.

Four clinical trials have been performed using OCS Heart: PROTECT I, PROTECT II, PROCEED I, and PROCEED II. The most recent trial, PROCEED II, was a prospective, multicenter, randomized non-inferiority trial in the United States and Europe [4]. Patients were assigned to Organ Care System NMP ($n = 67$) or SCS ($n = 63$), and the donor population was limited to standard criteria hearts. Thirty-day patient and graft survival rates were 94% in the OCS and 97% in the standard cold storage group. Cardiac-related serious adverse events occurred in eight OCS patients (13%) and nine SCS patients (14%). Perfusate lactate slightly increased throughout NMP (start 2.2 mmol/L, end 2.4 mmol/L). Of five hearts that were perfused but declined for the study, four demonstrated an increase in lactate during perfusion.

Dhital et al. recently described three successful cases of DCD heart transplant using OCS heart preservation [58]. Donors were Maastricht category III with less than 30 min of warm ischemia and younger than 40 years old. The three recipients were discharged on days 21–28 and survived through 77–171 days of follow-up. At this point, the OCS heart is not approved for use in the United Sates

2.9 Machine Perfusion of the Lungs

Donor lungs are susceptible to injury during intensive care unit management and may be rejected for a variety of reasons including pulmonary edema, atelectasis, ventilation-induced lung injury, pneumonia, lung contusion, and aspiration. Consequently, donor lung utilization rate is only 20% [225]. While marginal and standard donor lungs have similar outcomes [211], a dramatic increase in lung supply has not occurred. Ex vivo lung perfusion (EVLP) is a novel technology to preserve, evaluate, and treat initially rejected donor lungs, salvaging them for clinical use and thereby increasing supply.

2.9.1 History of Lung Perfusion Research

2.9.1.1 Small Animal Models of Isolated Lung Perfusion

The isolated perfused lung (IPL) model, a controlled perfusion and ventilation system for small animals, was developed in the 1980s [164] to provide an array of physiological data for pathological and pharmacological studies [54, 72, 73]. It consisted of a pump, organ chamber, ventilator, deoxygenator, and reservoir. Temperature was kept at 37 °C with a water-jacketed organ chamber. Perfusates were cellular (heparinized autologous whole blood or washed red blood cell with albumin and Krebs-Henseleit solution) or acellular (albumin and Krebs-Henseleit solution), and perfusion was pressure or flow controlled. Pulmonary edema limited perfusion duration, possibly due to suboptimal perfusates or perfusion protocols. This model was expanded to discarded human lungs [127, 131].

2.9.1.2 HMP of the Lungs

Lung HMP is not performed clinically, and there is only one recent report of experimental HMP. Nakajima et al. reported a 1-h perfusion of rat lungs that resulted in IRI compared to SCS [160]. A historical report of lung HMP demonstrated the deterioration of pulmonary function and lung edema in a canine model [148]. High vascular resistance and low pulmonary compliance at low temperature may be technical impediments to hypothermic lung perfusion [18, 106].

2.9.2 Clinical Ex Vivo Lung Perfusion

A research group at Lund University developed the first EVLP system for clinical use in the 1990s. Their perfusate consisted of washed RBCs with STEEN solution, which contains a supraphysiologic concentration of albumin that reduces pulmonary edema through high oncotic pressure [208, 237]. Unlike IPL models, perfusion with 100% cardiac output was stable for 1–2 h and did not produce edema.

Currently, three EVLP systems (Lund, Toronto, and Organ Care Systems) are available for clinical use (Table 2.3). The three systems are similarly structured, with a pump and ventilator for normothermic perfusion. Each has a unique protocol requiring different settings (i.e., perfusate, flow, left atrium cannulation, and ventilation), and consequently, the criteria of transplant suitability are different. The three EVLP systems are reviewed below.

2.9.3 Lund EVLP

Steen and colleagues at University Hospital of Lund developed the first prototype of EVLP and STEEN solution [208]. Their EVLP system consisted of lung chamber, the pulmonary artery cannula, centrifugal pump, reservoir, leukocyte filter, membrane, and heat exchanger. This group reported the first clinical EVLP case from an uncontrolled DCD donor lungs [209]. A pair of donor lungs was perfused in EVLP

Table 2.3 Three clinical lung perfusion systems

EVLP	Lund	Toronto	OCS
RBC	Yes	No	Yes
Perfusate	STEEN solution	STEEN solution	OCS solution
Cannula	PA	PA and LA	PA
Flow	100% CO	40% CO	2.5 L/min
PA pressure	≤20 mmHg	<15 mmHg	≤20 mmHg
Temperature, °C	37	37	37
Pump	Roller	Centrifugal	Piston (pulsatile)

RBC red blood cell, *PA* pulmonary artery, *LA* left atrium, *CO* cardiac output system and transplanted following the EVLP evaluation of lung function using blood gas analysis and other pulmonary parameters. After demonstrating the potential to recondition initially rejected donor lungs [237], they published the first transplant using initially rejected donor lungs [207]. In a series of EVLP cases, lung transplantation outcomes following EVLP were equivalent to standard lung transplantation [105, 231].

2.9.4 *Toronto EVLP*

A group at University of Toronto modified the original Lund EVLP protocol to prolong perfusion time. Their protocol calls for a flow at 40% of cardiac output to reduce shear stress and edema, an acellular perfusate to avoid hemolysis, and left atrial cannulation to create positive pressure in the left atrium [48]. They demonstrated that six porcine lungs and five human lungs could be perfused for 12 h with stable physiological parameters. They also showed the superiority of EVLP to SCS in a porcine model [47]. In the human ex vivo lung perfusion (HELP) study, rejected donor lungs were perfused and transplanted. EVLP outcomes were comparable to a retrospective standard lung transplant group, and the lung recovery rate was 87% [49]. In a report describing their first 50 EVLP cases, donor lung usage increased 20% [50]. A prospective nonrandomized multicenter clinical study (NOVEL lung trial) was conducted using rejected donor lungs with the Toronto EVLP protocol and commercial machine (XPS system). The 30-day mortality of EVLP group was 3% and recovery rate was 57%. There was no significant difference in outcomes between EVLP group and standard lung transplant control group [195], signifying that EVLP can resuscitate donor lungs to increase the supply. With the XPS system now FDA approved, this group established Lung Bioengineering (Baltimore, MD), a dedicated facility that receives donor lungs and performs EVLP for a transplant center.

2.9.5 *Organ Care System EVLP*

The Organ Care System (OCS) Lung (TransMedics, Andover, MA) is a commercial portable lung perfusion machine. There are two major differences between OCS and the Lund/Toronto EVLP systems. In the Lund and Toronto systems, donor lungs are transported, and perfusion is started at the transplant center. In the OCS protocol, perfusion is initiated following cold flush at the donor hospital. Secondly, OCS targets both standard donor lungs and rejected donor lungs, whereas the others focus only on rejected donor lungs. Their central hypothesis is that immediate perfusion with OCS can shorten cold ischemia, thus minimizing IRI and improving postoperative outcomes. The OCS protocol calls for a cellular perfusate of RBCs plus OCS solution, 2.5 L/

min of flow, and an opened left atrium. In a pilot series of SCD lung cases, all 12 patients survived to 30 days [232]. Trials are underway comparing OCS to SCS using SCD lungs (INSPIRE Trial NCT01630434) and initially rejected donor lungs (Lung EXPAND Trial NCT01963780). The OCS lung machine is not FDA approved.

2.9.6 Principle Benefits of EVLP

2.9.6.1 Preservation

The Lund EVLP system has a limited perfusion time of 1–2 h. In a porcine DCD model, Erasmus et al. demonstrated that 4.0 L/min of perfusion for 6 h resulted in pulmonary edema [66]. To reduce the risk of hydrostatic edema, perfusion is limited to 1 h if at that point lungs are acceptable for transplantation [105, 231]. In the Toronto EVLP system, with flow of 40% of cardiac output, clinical data support perfusing rejected donor lungs for 4 h [49, 50]. In an experimental setting, human and porcine lungs were safely preserved for 12 h [48]. The mean perfusion duration of SCD lungs in the OCS system with 2.5 L/min flow was 303 min (range, 188–622) [232]. Although low flow poses a risk of hypoperfusion, the clinical data of the Toronto EVLP and OCS indicates that 40% of cardiac output or 2.5 L/min is enough for lung preservation. This suggests that flow may dictate the limits of preservation time: lower flow rate enables longer preservation.

2.9.6.2 Assessment

Since the primary EVLP donor population is initially rejected donor lungs, accurate assessment is vital for patient safety. EVLP systems provide multiple assessment opportunities: blood gas analysis, airway parameters, vascular function, and visual findings. While blood gas analysis is the standard metric in the Lund protocol [208], it has limited utility in the Toronto system probably due to the absence of RBCs in the perfusate [243]. The Lund protocol uses a PaO_2/FiO_2 ratio of 300 mmHg as a threshold for transplant suitability. It is important to note that the PaO_2/FiO_2 ratio varies at different FiO_2 levels and it is not clear which FiO_2 provides the greatest predictive value to determine transplant suitability [173]. The OCS system uses blood gas analysis, but the ability to differentiate good and poor lungs with this type of assessment remains to be determined.

Airway parameters such as the trend of peak airway pressure and both dynamic and static compliance are useful in the Toronto system. In the Lund system, PaO_2/FiO_2 ratio and airway parameters correlate in porcine lungs [172]. Since the major presentation of IRI during EVLP is airway fluid accumulation, which reduces the PaO_2/FiO_2 ratio, it stands to reason that the deterioration of airway parameters is informative.

Vascular parameters are considered useful in the Lund system. A weight-based target flow is determined for each graft, and when this flow cannot be achieved without surpassing a pulmonary artery pressure threshold of 20 mmHg, a thrombus or other vascular abnormality may be present. The significance of vascular parameters in the Toronto and OCS systems has not been reported. Visual findings are a part of EVLP assessment but limited by their inherent subjectivity.

2.9.6.3 Repair

EVLP repairs damaged lungs mainly by reducing edema and atelectasis. The high oncotic pressure of STEEN solution in the Lund and Toronto systems reduces edema. STEEN solution is not used in the OCS system, and the effect of OCS solution on lung edema is unclear. In all EVLP systems, atelectasis is eliminated by alveolar recruitment. EVLP provides a platform for treatment with medication and genetic therapies prior to transplantation. Cypel et al. demonstrated this by administering a vector expressing IL-10, which promoted anti-inflammatory cytokine expression and improved pulmonary function [46]. In the near future, medications administered during EVLP could be used to reduce IRI.

2.10 Machine Perfusion of Limb Grafts

The major factor limiting the replantation of amputated limbs after injury is warm ischemia time. While irreversible muscle damage occurs within 4–6 h at room temperature, the best functional outcomes are achieved when ischemia time is less than 2 h [210]. Currently, the only method to preserve limbs is SCS, but with ongoing low-level metabolic activity, energy reserves are quickly depleted. Cooling can also directly damage cells, causing swelling and cell death. These effects are compounded after replantation/transplantation and reperfusion of the limb by the production of reactive oxygen species and inflammation [51, 83].

Unlike visceral organs, extremities contain a wide range of tissues with different resistance to ischemia. While the skin can be grafted after weeks of storage at 4 °C [33], the muscle will irreversibly deteriorate within hours [22]. Sustaining physiologic limb metabolism through MP may offer a solution by supporting the survival of all the different types of tissues present in a whole limb (Fig. 2.4).

2.10.1 History of Limb Perfusion

The first successful human extremity perfusion at physiologic temperature was reported in 1964 [55]. The limb was perfused with whole blood at a physiologic arterial pressure. Normal electrical properties of nerve and muscle were lost after

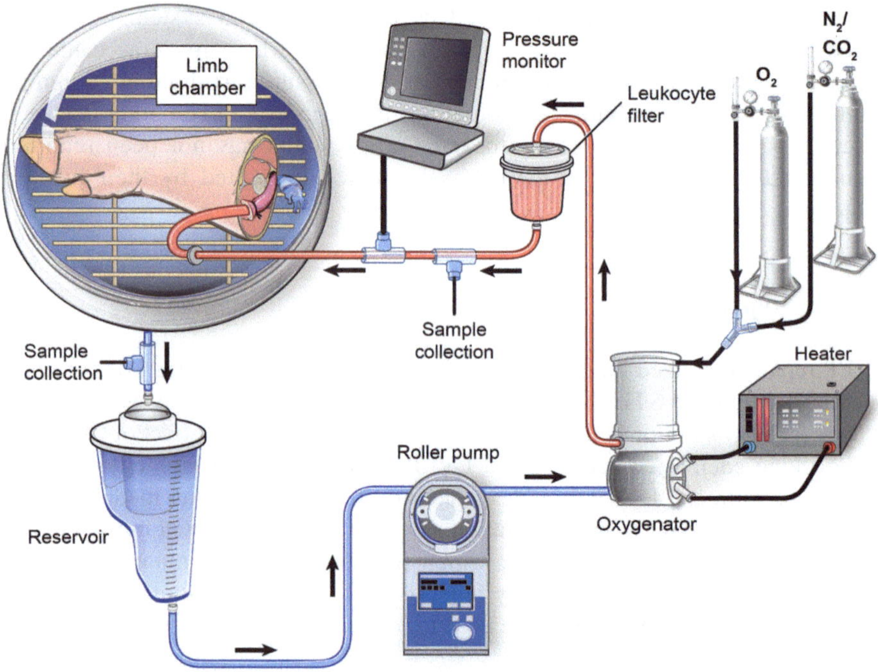

Fig. 2.4 Schematic of a normothermic perfusion system for the limb. (Reprinted with permission, Cleveland Clinic Center for Medical Art & Photography © 2014-2016. All Rights Reserved)

the first hour of circulatory arrest but were restored with normothermic oxygenated perfusion. In 1976, O'Donovan et al. perfused amputated human limbs for 7 h with whole blood and dextrose at physiologic pressures and temperature and registered muscle contractility within 30% of normal function [169].

The first cases of extremity replantation in the early 1960s prompted further study to improve the preservation of the amputated limbs [139]. It became evident that microcirculatory changes and the development of edema were two major barriers to the reestablishment of flow following replantation [94, 143]. The use of whole blood was abandoned, and extremities were perfused with crystalloids and colloid solutions [94, 143] including standard organ preservation solutions such as UW solution [82, 165] and Euro-Collins solution [3]. To increase the oxygenation of the perfused extremity, artificial oxygen carriers like Fluosol and fluorocarbon were tested [192, 222]. In 1991 canine hind limbs were maintained with 24-h HMP using a perfusate containing whole blood, dextran, mannitol, and lactated Ringer's solution [59]. Periodic perfusate exchange reduced acidosis during perfusion, and the extremities were successfully replanted. Wagner et al. reintroduced porcine limb NMP with a blood-dialysate and glucose mixture [230]. Physiologic flow rates were associated with lower arterial resistance and lactate production over 6-h perfusion.

2.10.2 Recent Work in Limb Perfusion and Future Directions

The advent of hand transplantation in the late 1990s renewed interest in the development of limb perfusion systems to allow procurement and transportation of the limbs from distant locations. Constantinescu et al. successfully performed SMP (32 °C) of porcine limbs with heparinized whole blood for 12 h [45]. In preliminary studies, the researchers were unable to maintain NMP at physiologic arterial pressure without experiencing significant weight gain. However, after perfusion at a mean pressure of 34 mmHg, no increase in limb weight or compartment pressure was observed, and the motor response of the muscles was stable throughout perfusion. In a subsequent study, the same group replanted limbs following 6 h of ischemia and 12 h of SMP [156]. The study found that prolonged perfusion was not associated with increased IRI. Ozer et al. performed an 11-h SMP of porcine limbs using a perfusate of plasma with RBCs and glucose, followed by limb transplantation [179]. Limb weight was increased after perfusion, but single fiber contractility was stable during the perfusion. They subsequently perfused 4 limbs for 24 h and transplanted them. Limb weight increased by 20%, and muscle contractility was maintained throughout the perfusion [180].

Although the previous studies showed that ex vivo perfusion can prolong the viability of major limb segments, many factors must still be determined including the ideal perfusate composition, temperature, pressure, and pump design. A few studies that performed the transplantation of the limbs following perfusion did not evaluate long-term limb function, in particular bone healing, nerve regeneration, and muscle function restoration. The limb-specific criteria defining the viability and replantability of the limb need further investigation. Defining these criteria and specific monitoring parameters is important to decide if a limb can be replanted or transplanted following perfusion or should be discarded (Table 2.4).

Table 2.4 Common and limb-specific perfusion parameters

Common perfusion monitoring parameters	Limb-specific monitoring parameters
Perfusion pressure	Muscle contraction after nerve stimulation (motor conduction velocities)
Temperature of perfusate	Tissue O_2 saturation (Vioptix©)
Perfusate gasometry (pO_2)	Peripheral perfusion (indocyanine green angiography, SPY© Elite System)
Perfusate electrolytes (Na, K, Ca)	Temperature uniformity (thermography, Fluke TiS thermal imaging scanner)
Glucose	Compartment pressure (Stryker intracompartmental pressure monitor)
Lactate	Perfusate creatine kinase
Histology (H&E, electron microscopy)	Perfusate myoglobin

2.11 Conclusion

Across the spectrum of organ transplantation, machine perfusion is being tested as a potential solution to the organ shortage crisis. Through improved preservation, assessment, and resuscitation, MP can reduce the chance that a potentially viable organ will be discarded unnecessarily and increase the likelihood that transplanted grafts will have prolonged function in recipients. The last two decades have been notable for the introduction, and in some cases reintroduction, of MP in the clinical setting for the heart, lung, and liver, along with the clinical testing and technical optimization of kidney perfusion. At the same time, MP is approaching the clinical threshold for the pancreas and intestines and is being innovatively explored as a promising option for limb transplantation and replantation.

Increased complexity and cost could be initial barriers to the implementation of MP, as was the case back when this technology was first explored in the mid-twentieth century. However, as renal HMP has shown, reducing morbidity and mortality and avoiding their associated costs to the healthcare system may outweigh these potential barriers. As MP is optimized for each organ and clinical scenario, perhaps the day is approaching when organs will be banked for future use and access to transplantation will be available to all who need it.

References

1. Abouna, G. M. (2008). Organ shortage crisis: Problems and possible solutions. *Transplantation Proceedings, 40*, 34–38.
2. Abu-Elmagd, K., Reyes, J., Todo, S., et al. (1998). Clinical intestinal transplantation: New perspectives and immunologic considerations. *Journal of the American College of Surgeons, 186*, 512–527.
3. Arai, K., Hotokebuchi, T., Miyahara, H., et al. (1993). Successful long-term storage of rat limbs. The use of simple immersion in Euro-Collins solution. *International Orthopaedics, 17*, 389–396.
4. Ardehali, A., Esmailian, F., Deng, M., et al. (2015). Ex-vivo perfusion of donor hearts for human heart transplantation (PROCEED II): A prospective, open-label, multicentre, randomised non-inferiority trial. *Lancet, 385*, 2577–2584.
5. Arthur, P. G., Niu, X.-W., Huang, W.-H., et al. (2013). Desferrioxamine in warm reperfusion media decreases liver injury aggravated by cold storage. *World Journal of Gastroenterology, 19*, 673–681.
6. Babkin, B. P., & Starling, E. H. (1926). A method for the study of the perfused pancreas. *The Journal of Physiology, 61*, 245–247.
7. Barlow, A. D., Hosgood, S. A., & Nicholson, M. L. (2013). Current state of pancreas preservation and implications for DCD pancreas transplantation. *Transplant Journal, 95*, 1419–1424.
8. Baumgartner, D., Sutherland, D. E., & Najarian, J. S. (1980). Studies on segmental pancreas autotransplants in dogs: Technique and preservation. *Transplantation Proceedings, 12*, 163–171.
9. Beecher, H. K., Adams, R. D., Barger, C., et al. (1968). A definition of irreversible coma. Report of the Ad Hoc Committee of the Harvard Medical School to Examine the Definition of Brain Death. *JAMA, 205*, 337–340.

10. Bell, R. M., Mocanu, M. M., & Yellon, D. M. (2011). Retrograde heart perfusion: The Langendorff technique of isolated heart perfusion. *Journal of Molecular and Cellular Cardiology, 50*, 940–950.
11. Bellomo, R., Suzuki, S., Marino, B., et al. (2012). Normothermic extracorporeal perfusion of isolated porcine liver after warm ischaemia: A preliminary report. *Critical Care and Resuscitation, 14*, 173–176.
12. Belzer, F. O. (1991). Organ preservation: A personal perspective. In P. I. Terasaki (Ed.), *History of transplantation: Thirty-five recollections* (pp. 595–613). Los Angeles: UCLA Tissue Typing Laboratory.
13. Belzer, F. O., Ashby, B. S., & Dunphy, J. E. (1967). 24-hour and 72-hour preservation of canine kidneys. *Lancet (London, England), 2*, 536–538.
14. Belzer, F. O., Ashby, B. S., Gulyassy, P. F., & Powell, M. (1968a). Successful seventeen-hour preservation and transplantation of human-cadaver kidney. *The New England Journal of Medicine, 278*, 608–610.
15. Belzer, F. O., Ashby, B. S., Huang, J. S., & Dunphy, J. E. (1968b). Etiology of rising perfusion pressure in isolated organ perfusion. *Annals of Surgery, 168*, 382–391.
16. Belzer, F. O., D'Alessandro, A. M., Hoffmann, R. M., et al. (1992). The use of UW solution in clinical transplantation. A 4-year experience. *Annals of Surgery, 215*, 579–585.
17. Belzer, F. O., Glass, N. R., Sollinger, H. W., et al. (1982). A new perfusate for kidney preservation. *Transplantation, 33*, 322–323.
18. Benumof, J. L., & Wahrenbrock, E. A. (1977). Dependency of hypoxic pulmonary vasoconstriction on temperature. *Journal of Applied Physiology, 42*, 56–58.
19. Bessems, M., Doorschodt, B. M., Kolkert, J. L. P., et al. (2007). Preservation of steatotic livers: A comparison between cold storage and machine perfusion preservation. *Liver Transplantation, 13*, 497–504.
20. Bessems, M., Doorschodt, B. M., van Vliet, A. K., & van Gulik, T. M. (2005). Improved rat liver preservation by hypothermic continuous machine perfusion using polysol, a new, enriched preservation solution. *Liver Transplantation, 11*, 539–546.
21. Bhangoo, R. S., Hall, I. E., Reese, P. P., & Parikh, C. R. (2012). Deceased-donor kidney perfusate and urine biomarkers for kidney allograft outcomes: A systematic review. *Nephrology, Dialysis, Transplantation, 27*, 3305–3314.
22. Blaisdell, F. W. (2002). The pathophysiology of skeletal muscle ischemia and the reperfusion syndrome: A review. *Cardiovascular Surgery, 10*, 620–630.
23. Boehnert, M. U., Yeung, J. C., Bazerbachi, F., et al. (2013). Normothermic acellular ex vivo liver perfusion reduces liver and bile duct injury of pig livers retrieved after cardiac death. *American Journal of Transplantation, 13*, 1441–1449.
24. Brasile, L., Buelow, R., Stubenitsky, B. M., & Kootstra, G. (2003a). Induction of heme oxygenase-1 in kidneys during ex vivo warm perfusion. *Transplantation, 76*, 1145–1149.
25. Brasile, L., Glowacki, P., Castracane, J., & Stubenitsky, B. M. (2010). Pretransplant kidney-specific treatment to eliminate the need for systemic immunosuppression. *Transplantation, 90*, 1294–1298.
26. Brasile, L., Green, E., & Haisch, C. (1997). Oxygen consumption in warm-preserved renal allografts. *Transplantation Proceedings, 29*, 1322–1323.
27. Brasile, L., Stubenitsky, B., Haisch, C. E., et al. (2005). Potential of repairing ischemically damaged kidneys ex vivo. *Transplantation Proceedings, 37*, 375–376.
28. Brasile, L., Stubenitsky, B. M., Booster, M. H., et al. (2001). Hypothermia--a limiting factor in using warm ischemically damaged kidneys. *American Journal of Transplantation, 1*, 316–320.
29. Brasile, L., Stubenitsky, B. M., Booster, M. H., et al. (2002a). Transfection and transgene expression in a human kidney during ex vivo warm perfusion. *Transplantation Proceedings, 34*, 2624.
30. Brasile, L., Stubenitsky, B. M., Booster, M. H., et al. (2002b). Overcoming severe renal ischemia: The role of ex vivo warm perfusion. *Transplantation, 73*, 897–901.

31. Brasile, L., Stubenitsky, B. M., Booster, M. H., et al. (2003b). NOS: The underlying mechanism preserving vascular integrity and during ex vivo warm kidney perfusion. *American Journal of Transplantation, 3*, 674–679.
32. Braun, F., Schütz, E., Laabs, S., et al. (1998). Development of a porcine small bowel ex vivo perfusion model. *Transplantation Proceedings, 30*, 2613–2615.
33. Bravo, D., Rigley, T. H., Gibran, N., et al. (2000). Effect of storage and preservation methods on viability in transplantable human skin allografts. *Burns, 26*, 367–378.
34. Brockmann, J., Reddy, S., Coussios, C., et al. (2009). Normothermic perfusion: A new paradigm for organ preservation. *Annals of Surgery, 250*, 1–6.
35. Bruinsma, B. G., Yeh, H., Ozer, S., et al. (2014). Subnormothermic machine perfusion for ex vivo preservation and recovery of the human liver for transplantation. *American Journal of Transplantation, 14*, 1400–1409.
36. Brynger, H., & Claes, G. (1975). Behaviour of the duct-ligated canine pancreas during hypothermic albumin perfusion. *Eur Surg Res Eur Chir Forschung Rech Chir Eur, 7*, 287–296.
37. Butler, A. J., Rees, M. A., Wight, D. G. D., et al. (2002). Successful extracorporeal porcine liver perfusion for 72 hr. *Transplantation, 73*, 1212–1218.
38. Candinas, D., Largiadèr, F., Binswanger, U., et al. (1996). A novel dextran 40-based preservation solution. *Transplant International, 9*, 32–37.
39. Cannon, R. M., Brock, G. N., Garrison, R. N., et al. (2013). To pump or not to pump: A comparison of machine perfusion vs cold storage for deceased donor kidney transplantation. *Journal of the American College of Surgeons, 216*, 625–633.
40. Changani, K. K., Fuller, B. J., Bryant, D. J., et al. (1997). Non-invasive assessment of ATP regeneration potential of the preserved donor liver. A 31P MRS study in pig liver. *Journal of Hepatology, 26*, 336–342.
41. Cho, Y. W., Terasaki, P. I., Cecka, J. M., & Gjertson, D. W. (1998). Transplantation of kidneys from donors whose hearts have stopped beating. *The New England Journal of Medicine, 338*, 221–225.
42. Claes, G., Aurell, M., Blohmé, I., & Pettersson, S. (1972). Experimental and clinical results of continuous hypothermic albumin perfusion. *Proceedings of the European Dialysis and Transplant Association, 9*, 484–490.
43. Collins, G. M., Bravo-Shugarman, M., & Terasaki, P. I. (1969). Kidney preservation for transportation. Initial perfusion and 30 hours' ice storage. *Lancet, 2*, 1219–1222.
44. Colvin-Adams, M., Smith, J. M., Heubner, B. M., et al. (2015). OPTN/SRTR 2013 annual data report: Heart. *American Journal of Transplantation, 15*(Suppl 2), 1–28.
45. Constantinescu, M. A., Knall, E., Xu, X., et al. (2011). Preservation of amputated extremities by extracorporeal blood perfusion; a feasibility study in a porcine model. *The Journal of Surgical Research, 171*, 291–299.
46. Cypel, M., Liu, M., Rubacha, M., et al. (2009a). Functional repair of human donor lungs by IL-10 gene therapy. *Science Translational Medicine, 1*, 4ra9.
47. Cypel, M., Rubacha, M., Yeung, J., et al. (2009b). Normothermic ex vivo perfusion prevents lung injury compared to extended cold preservation for transplantation. *American Journal of Transplantation, 9*, 2262–2269.
48. Cypel, M., Yeung, J. C., Hirayama, S., et al. (2008). Technique for prolonged normothermic ex vivo lung perfusion. *The Journal of Heart and Lung Transplantation, 27*, 1319–1325.
49. Cypel, M., Yeung, J. C., Liu, M., et al. (2011). Normothermic ex vivo lung perfusion in clinical lung transplantation. *The New England Journal of Medicine, 364*, 1431–1440.
50. Cypel, M., Yeung, J. C., Machuca, T., et al. (2012). Experience with the first 50 ex vivo lung perfusions in clinical transplantation. *The Journal of Thoracic and Cardiovascular Surgery, 144*, 1200–1206.
51. de Groot, H., & Rauen, U. (2007). Ischemia-reperfusion injury: Processes in pathogenetic networks: A review. *Transplantation Proceedings, 39*, 481–484.
52. de Rougemont, O., Breitenstein, S., Leskosek, B., et al. (2009). One hour hypothermic oxygenated perfusion (HOPE) protects nonviable liver allografts donated after cardiac death. *Annals of Surgery, 250*, 674–683.

53. de Vries, E. E., Hoogland, E. R. P., Winkens, B., et al. (2011). Renovascular resistance of machine-perfused DCD kidneys is associated with primary nonfunction. *American Journal of Transplantation, 11*, 2685–2691.
54. DeCampos, K. N., Keshavjee, S., Slutsky, A. S., & Liu, M. (1999). Alveolar recruitment prevents rapid-reperfusion-induced injury of lung transplants. *The Journal of Heart and Lung Transplantation, 18*, 1096–1102.
55. Delorme, T. L., Shaw, R. S., & Austen, W. G. (1964). A method of studying "normal" function in the amputated human limb using perfusion. *The Journal of Bone and Joint Surgery, 46*, 161–164.
56. Demacker, P. N. M., Beijers, A. M., van Daal, H., et al. (2009). Plasma citrulline measurement using UPLC tandem mass-spectrometry to determine small intestinal enterocyte pathology. *Journal of Chromatography. B, Analytical Technologies in the Biomedical and Life Sciences, 877*, 387–392.
57. Desai, T. R., Sisley, A. C., Brown, S., & Gewertz, B. L. (1996). Defining the critical limit of oxygen extraction in the human small intestine. *Journal of Vascular Surgery, 23*, 832–838.
58. Dhital, K. K., Iyer, A., Connellan, M., et al. (2015). Adult heart transplantation with distant procurement and ex-vivo preservation of donor hearts after circulatory death: A case series. *Lancet, 385*, 2585–2591.
59. Domingo-Pech, J., Garriga, J. M., Toran, N., et al. (1991). Preservation of the amputated canine hind limb by extracorporeal perfusion. *International Orthopaedics, 15*, 289–291.
60. Durand, F., Renz, J. F., Alkofer, B., et al. (2008). Report of the Paris consensus meeting on expanded criteria donors in liver transplantation. *Liver Transplantation, 14*, 1694–1707.
61. Dutkowski, P., Furrer, K., Tian, Y., et al. (2006). Novel short-term hypothermic oxygenated perfusion (HOPE) system prevents injury in rat liver graft from non-heart beating donor. *Annals of Surgery, 244*, 968–976.
62. Dutkowski, P., Polak, W. G., Muiesan, P., et al. (2015). First comparison of hypothermic oxygenated perfusion versus static cold storage of human donation after cardiac death liver transplants: An international-matched case analysis. *Annals of Surgery, 262*, 764–771.
63. Dutkowski, P., Schlegel, A., de Oliveira, M., et al. (2014). HOPE for human liver grafts obtained from donors after cardiac death. *Journal of Hepatology, 60*, 765–772.
64. Eckhauser, F., Knol, J. A., Porter-Fink, V., et al. (1981). Ex vivo normothermic hemoperfusion of the canine pancreas: Applications and limitations of a modified experimental preparation. *The Journal of Surgical Research, 31*, 22–37.
65. Eloy, M. R., Kachelhoffer, J., Pousse, A., et al. (1974). Ex vivo vascular perfusion of the isolated canine pancreas. Experimental procedure, haemodynamic data and experimental applications. *Eur Surg Res Eur Chir Forschung Rech Chir Eur, 6*, 341–353.
66. Erasmus, M. E., Fernhout, M. H., Elstrodt, J. M., & Rakhorst, G. (2006). Normothermic ex vivo lung perfusion of non-heart-beating donor lungs in pigs: From pretransplant function analysis towards a 6-h machine preservation. *Transplant International, 19*, 589–593.
67. Feng, L., Zhao, N., Yao, X., et al. (2007). Histidine-tryptophan-ketoglutarate solution vs. University of Wisconsin solution for liver transplantation: A systematic review. *Liver Transplantation, 13*, 1125–1136.
68. Ferrigno, A., Rizzo, V., Boncompagni, E., et al. (2011). Machine perfusion at 20°C reduces preservation damage to livers from non-heart beating donors. *Cryobiology, 62*, 152–158.
69. Fitton, T. P., Barreiro, C. J., Bonde, P. N., et al. (2005). Attenuation of DNA damage in canine hearts preserved by continuous hypothermic perfusion. *The Annals of Thoracic Surgery, 80*, 1812–1820.
70. Florack, G., Sutherland, D. E., Heil, J., et al. (1983). Preservation of canine segmental pancreatic autografts: Cold storage versus pulsatile machine perfusion. *The Journal of Surgical Research, 34*, 493–504.
71. Fontes, P., Lopez, R., van der Plaats, A., et al. (2015). Liver preservation with machine perfusion and a newly developed cell-free oxygen carrier solution under subnormothermic conditions. *American Journal of Transplantation, 15*, 381–394.

72. Fujinaga, T., Nakamura, T., Fukuse, T., et al. (2006). Isoflurane inhalation after circulatory arrest protects against warm ischemia reperfusion injury of the lungs. *Transplantation, 82*, 1168–1174.
73. Fukuse, T., Albes, J. M., Takahashi, Y., et al. (1995). Influence of red blood cells on lung function in an ex vivo rat heart-lung model. *The Journal of Surgical Research, 59*, 399–404.
74. Fuller, B. J., & Lee, C. Y. (2007). Hypothermic perfusion preservation: The future of organ preservation revisited? *Cryobiology, 54*, 129–145.
75. Gage, F., Leeser, D. B., Porterfield, N. K., et al. (2009). Room temperature pulsatile perfusion of renal allografts with Lifor compared with hypothermic machine pump solution. *Transplantation Proceedings, 41*, 3571–3574.
76. Gallinat, A., Efferz, P., Paul, A., & Minor, T. (2014). One or 4 h of "in-house" reconditioning by machine perfusion after cold storage improve reperfusion parameters in porcine kidneys. *Transplant International, 27*, 1214–1219.
77. Gallinat, A., Moers, C., Smits, J. M., et al. (2013). Machine perfusion versus static cold storage in expanded criteria donor kidney transplantation: 3-year follow-up data. *Transplant International, 26*, E52–E53.
78. Gallinat, A., Moers, C., Treckmann, J., et al. (2012a). Machine perfusion versus cold storage for the preservation of kidneys from donors ≥ 65 years allocated in the Eurotransplant Senior Programme. *Nephrology, Dialysis, Transplantation, 27*, 4458–4463.
79. Gallinat, A., Paul, A., Efferz, P., et al. (2012b). Role of oxygenation in hypothermic machine perfusion of kidneys from heart beating donors. *Transplantation, 94*, 809–813.
80. Gallinat, A., Paul, A., Efferz, P., et al. (2012c). Hypothermic reconditioning of porcine kidney grafts by short-term preimplantation machine perfusion. *Transplantation, 93*, 787–793.
81. Ginzel, K. H. (1960). A method for measurement of peristaltic activity and for perfusion of the vascular system and lumen of an isolated loop of small intestine in the guinea pig. *Naunyn-Schmiedebergs Archiv für Experimentelle Pathologie und Pharmakologie, 238*, 231–232.
82. Gordon, L., Levinsohn, D. G., Borowsky, C. D., et al. (1992). Improved preservation of skeletal muscle in amputated limbs using pulsatile hypothermic perfusion with University of Wisconsin solution. A preliminary study. *The Journal of Bone and Joint Surgery American, 74*, 1358–1366.
83. Greaney, P. J., Cordisco, M., Rodriguez, D., et al. (2010). Use of an extracorporeal membrane oxygenation circuit as a bridge to salvage a major upper-extremity replant in a critically ill patient. *Journal of Reconstructive Microsurgery, 26*, 517–522.
84. Gringeri, E., Bonsignore, P., Bassi, D., et al. (2012). Subnormothermic machine perfusion for non-heart-beating donor liver grafts preservation in a Swine model: A new strategy to increase the donor pool? *Transplantation Proceedings, 44*, 2026–2028.
85. Groen, H., Moers, C., Smits, J. M., et al. (2012). Cost-effectiveness of hypothermic machine preservation versus static cold storage in renal transplantation. *American Journal of Transplantation, 12*, 1824–1830.
86. Gruessner, A. C. (2011). 2011 update on pancreas transplantation: Comprehensive trend analysis of 25,000 cases followed up over the course of twenty-four years at the international pancreas transplant registry (IPTR). *The Review of Diabetic Studies, 8*, 6–16.
87. Grundmann, R., Raab, M., Meusel, E., et al. (1975). Analysis of the optimal perfusion pressure and flow rate of the renal vascular resistance and oxygen consumption in the hypothermic perfused kidney. *Surgery, 77*, 451–461.
88. Guarrera, J. V., Henry, S. D., Samstein, B., et al. (2015). Hypothermic machine preservation facilitates successful transplantation of "orphan" extended criteria donor livers. *American Journal of Transplantation, 15*, 161–169.
89. Guarrera, J. V., Henry, S. D., Samstein, B., et al. (2010). Hypothermic machine preservation in human liver transplantation: The first clinical series. *American Journal of Transplantation, 10*, 372–381.
90. Guibert, E. E., Petrenko, A. Y., Balaban, C. L., et al. (2011). Organ preservation: Current concepts and new strategies for the next decade. *Transfusion Medicine and Hemotherapy, 38*, 125–142.

91. Hamilton, D. (2013). Kidney transplantation: A history. In P. Morris & S. J. Knechtle (Eds.), *Kidney transplantation: Principles and practice* (7th ed., pp. 1–9). Oxford: Saunders.
92. Hassanein, W. H., Zellos, L., Tyrrell, T. A., et al. (1998). Continuous perfusion of donor hearts in the beating state extends preservation time and improves recovery of function. *The Journal of Thoracic and Cardiovascular Surgery, 116*, 821–830.
93. Henry, S. D., Nachber, E., Tulipan, J., et al. (2012). Hypothermic machine preservation reduces molecular markers of ischemia/reperfusion injury in human liver transplantation. *American Journal of Transplantation, 12*, 2477–2486.
94. Hicks, T. E., Boswick, J. A., & Solomons, C. C. (1980). The effects of perfusion on an amputated extremity. *The Journal of Trauma, 20*, 632–648.
95. Hoffmann, R. M., Southard, J. H., Lutz, M., et al. (1983). Synthetic perfusate for kidney preservation. Its use in 72-hour preservation of dog kidneys. *Archives of Surgery, 118*, 919–921.
96. Hohenleitner, F. J., & Senior, J. R. (1969). Metabolism of canine small intestine vascularly perfused in vitro. *Journal of Applied Physiology, 26*, 119–128.
97. Hoogland, E. R. P., de Vries, E. E., Christiaans, M. H. L., et al. (2013). The value of machine perfusion biomarker concentration in DCD kidney transplantations. *Transplantation, 95*, 603–610.
98. Hosgood, S. A., Bagul, A., Kaushik, M., et al. (2008a). Application of nitric oxide and carbon monoxide in a model of renal preservation. *The British Journal of Surgery, 95*, 1060–1067.
99. Hosgood, S. A., Bagul, A., Yang, B., & Nicholson, M. L. (2008b). The relative effects of warm and cold ischemic injury in an experimental model of nonheartbeating donor kidneys. *Transplantation, 85*, 88–92.
100. Hoyer, D. P., Gallinat, A., Swoboda, S., et al. (2014a). Subnormothermic machine perfusion for preservation of porcine kidneys in a donation after circulatory death model. *Transplant International, 27*, 1097–1106.
101. Hoyer, D. P., Gallinat, A., Swoboda, S., et al. (2014b). Influence of oxygen concentration during hypothermic machine perfusion on porcine kidneys from donation after circulatory death. *Transplantation, 98*, 944–950.
102. Huang, H., He, Z., Roberts, L. J., & Salahudeen, A. K. (2003). Deferoxamine reduces cold-ischemic renal injury in a syngeneic kidney transplant model. *American Journal of Transplantation, 3*, 1531–1537.
103. Humphries, A. L., Russell, R., Gregory, J., et al. (1964). Hypothermic perfusion of the canine kidney for 48 hours followed by Reimplantation. *The American Surgeon, 30*, 748–752.
104. Humphries, A. L., Russell, R., Stoddard, L. D., & Moretz, W. H. (1968). Three-day kidney preservation: Perfusion of kidneys with hypothermic, diluted blood of plasma. *Surgery, 63*, 646–652.
105. Ingemansson, R., Eyjolfsson, A., Mared, L., et al. (2009). Clinical transplantation of initially rejected donor lungs after reconditioning ex vivo. *The Annals of Thoracic Surgery, 87*, 255–260.
106. Inoue, H., Inoue, C., & Hildebrandt, J. (1982). Temperature effects on lung mechanics in air- and liquid-filled rabbit lungs. *Journal of Applied Physiology, 53*, 567–575.
107. Jamieson, R. W., Zilvetti, M., Roy, D., et al. (2011). Hepatic steatosis and normothermic perfusion-preliminary experiments in a porcine model. *Transplantation, 92*, 289–295.
108. Jiao, B., Liu, S., Liu, H., et al. (2013). Hypothermic machine perfusion reduces delayed graft function and improves one-year graft survival of kidneys from expanded criteria donors: A meta-analysis. *PloS One, 8*, e81826.
109. Jochmans, I., Moers, C., Ploeg, R., & Pirenne, J. (2011a). To perfuse or not to perfuse kidneys donated after cardiac death. *American Journal of Transplantation, 11*, 409–410.
110. Jochmans, I., Moers, C., Smits, J. M., et al. (2011b). The prognostic value of renal resistance during hypothermic machine perfusion of deceased donor kidneys. *American Journal of Transplantation, 11*, 2214–2220.
111. Jochmans, I., Moers, C., Smits, J. M., et al. (2010). Machine perfusion versus cold storage for the preservation of kidneys donated after cardiac death: A multicenter, randomized, controlled trial. *Annals of Surgery, 252*, 756–764.

112. Jochmans, I., O'Callaghan, J. M., Pirenne, J., & Ploeg, R. J. (2015). Hypothermic machine perfusion of kidneys retrieved from standard and high-risk donors. *Transplant international, 28*(6), 665–676.
113. Kachelhoffer, J., Dauchel, J., Pousse, A., et al. (1976). A simple device to obtain a pulsatile flow. Application to the vascular perfusion of dogs isolated intestinal segments. *Eur Surg Res Eur Chir Forschung Rech Chir Eur, 8*, 461–470.
114. Kandaswamy, R., Skeans, M. A., Gustafson, S. K., et al. (2016). Pancreas. *American Journal of Transplantation, 16*(Suppl 2), 47–68.
115. Karangwa, S. A., Dutkowski, P., Fontes, P., et al. (2016). Machine perfusion of donor livers for transplantation: A proposal for standardized nomenclature and reporting guidelines. *American Journal of Transplantation, 16*(10), 2932–2942.
116. Karcz, M., Cook, H. T., Sibbons, P., et al. (2010). An ex-vivo model for hypothermic pulsatile perfusion of porcine pancreata: Hemodynamic and morphologic characteristics. *Experimental and Clinical Transplantation, 8*, 55–60.
117. Kaths, J. M., Spetzler, V. N., Goldaracena, N., et al. (2015). Normothermic Ex Vivo Kidney Perfusion for the Preservation of Kidney Grafts prior to Transplantation. *Journal of Visualized Experiments, 15*(101), e52909.
118. Khush, K. K., Zaroff, J. G., Nguyen, J., et al. (2015). National decline in donor heart utilization with regional variability: 1995-2010. *American Journal of Transplantation, 15*, 642–649.
119. Kim, W. R., Lake, J. R., Smith, J. M., et al. (2016). Liver. *American Journal of Transplantation, 16*(Suppl 2), 69–98.
120. Klein, A. S., Messersmith, E. E., Ratner, L. E., et al. (2010). Organ donation and utilization in the United States, 1999-2008. *American Journal of Transplantation, 10*, 973–986.
121. Knaak, J. M., Spetzler, V. N., Goldaracena, N., et al. (2014). Subnormothermic ex vivo liver perfusion reduces endothelial cell and bile duct injury after donation after cardiac death pig liver transplantation. *Liver Transplantation, 20*, 1296–1305.
122. Kosieradzki, M., & Rowiński, W. (2008). Ischemia/reperfusion injury in kidney transplantation: Mechanisms and prevention. *Transplantation Proceedings, 40*, 3279–3288.
123. Kwiatkowski, A., Wszoła, M., Kosieradzki, M., et al. (2009). The early and long term function and survival of kidney allografts stored before transplantation by hypothermic pulsatile perfusion. A prospective randomized study. *Annals of Transplantation, 14*, 14–17.
124. Lam, V. W. T., Laurence, J. M., Richardson, A. J., et al. (2013). Hypothermic machine perfusion in deceased donor kidney transplantation: A systematic review. *The Journal of Surgical Research, 180*, 176–182.
125. Lauschke, H., Olschewski, P., Tolba, R., et al. (2003). Oxygenated machine perfusion mitigates surface antigen expression and improves preservation of predamaged donor livers. *Cryobiology, 46*, 53–60.
126. Lee, C. Y., & Mangino, M. J. (2009). Preservation methods for kidney and liver. *Organogenesis, 5*, 105–112.
127. Lee, J. W., Fang, X., Gupta, N., et al. (2009). Allogeneic human mesenchymal stem cells for treatment of E. coli endotoxin-induced acute lung injury in the ex vivo perfused human lung. *Proceedings of the National Academy of Sciences of the United States of America, 106*, 16357–16362.
128. Leeser, D. B., Bingaman, A. W., Poliakova, L., et al. (2004). Pulsatile pump perfusion of pancreata before human islet cell isolation. *Transplantation Proceedings, 36*, 1050–1051.
129. Lindell, S. L., Compagnon, P., Mangino, M. J., & Southard, J. H. (2005). UW solution for hypothermic machine perfusion of warm ischemic kidneys. *Transplantation, 79*, 1358–1361.
130. Lindell, S. L., Muir, H., Brassil, J., & Mangino, M. J. (2013). Hypothermic machine perfusion preservation of the DCD kidney: Machine effects. *Journal of Transplantation, 2013*, 802618.
131. Linder, A., Friedel, G., Fritz, P., et al. (1996). The ex-vivo isolated, perfused human lung model: Description and potential applications. *The Thoracic and Cardiovascular Surgeon, 44*, 140–146.

132. Liu, Q., Berendsen, T., Izamis, M.-L., et al. (2013a). Perfusion defatting at subnormothermic temperatures in steatotic rat livers. *Transplantation Proceedings, 45*, 3209–3213.
133. Liu, Q., Izamis, M.-L., Xu, H., et al. (2013b). Strategies to rescue steatotic livers before transplantation in clinical and experimental studies. *World Journal of Gastroenterology, 19*, 4638–4650.
134. Liu, Q., Nassar, A., Farias, K., et al. (2015). Comparing normothermic machine perfusion preservation with different perfusates on porcine livers from donors after circulatory death. *American Journal of Transplantation, 16*, 794-807.
135. Liu, Q., Nassar, A., Farias, K., et al. (2014a). Sanguineous normothermic machine perfusion improves hemodynamics and biliary epithelial regeneration in donation after cardiac death porcine livers. *Liver Transplantation, 20*, 987–999.
136. Liu, Q., Vekemans, K., Iania, L., et al. (2014b). Assessing warm ischemic injury of pig livers at hypothermic machine perfusion. *The Journal of Surgical Research, 186*, 379–389.
137. Liu, Q., Vekemans, K., van Pelt, J., et al. (2009). Discriminate liver warm ischemic injury during hypothermic machine perfusion by proton magnetic resonance spectroscopy: A study in a porcine model. *Transplantation Proceedings, 41*, 3383–3386.
138. Lockett, C. J., Fuller, B. J., Busza, A. L., & Proctor, E. (1995). Hypothermic perfusion preservation of liver: The role of phosphate in stimulating ATP synthesis studied by 31P NMR. *Transplant International, 8*, 440–445.
139. Malt, R., & McKhann, C. (1964). Replantation of severed arms. *JAMA, 189*, 716–722.
140. Mangino, M. (2011). Hypothermic machine perfusion of kidneys. In K. Uygun & C. Y. Lee (Eds.), *Methods in bioengineering: Organ preservation and reengineering* (pp. 35–57). Boston: Artech House.
141. Mangus, R. S., Fridell, J. A., Vianna, R. M., et al. (2008). Comparison of histidine-tryptophan-ketoglutarate solution and University of Wisconsin solution in extended criteria liver donors. *Liver Transplantation, 14*, 365–373.
142. Matas, A. J., Smith, J. M., Skeans, M. A., et al. (2015). OPTN/SRTR 2013 Annual data report: Kidney. *American Journal of Transplantation, 15*(Suppl 2), 1–34.
143. Mehl, R., Paul, H., Shorey, W., et al. (1964). Patency of the microcirculation in the traumatically amputated limb--a comparison of common perfusates. *The Journal of Trauma, 4*, 495–505.
144. Mercer, D. F., Vargas, L., Sun, Y., et al. (2011). Stool calprotectin monitoring after small intestine transplantation. *Transplantation, 91*, 1166–1171.
145. Meyer, W., Castelfranchi, P. L., Schulz, L. S., et al. (1973). Physiologic studies during perfusion of the isolated canine pancreas. The endocrine and exocrine behavior. *Eur Surg Res Eur Chir Forschung Rech Chir Eur, 5*, 105–115.
146. Minor, T., Manekeller, S., Sioutis, M., & Dombrowski, F. (2006). Endoplasmic and vascular surface activation during organ preservation: Refining upon the benefits of machine perfusion. *American Journal of Transplantation, 6*, 1355–1366.
147. Miyazaki, K., Sunada, K., Iseki, K., & Arita, T. (1986). Simultaneous vascular and luminal perfusion of rat small intestine. *Chemical & Pharmaceutical Bulletin (Tokyo), 34*, 3830–3835.
148. Modry, D. L., Jirsch, D. W., Boehme, G., et al. (1973). Hypothermic perfusion preservation of the isolated dog lung. *The Annals of Thoracic Surgery, 16*, 583–597.
149. Moers, C., Pirenne, J., Paul, A., & Ploeg, R. J. (2012). Machine perfusion or cold storage in deceased-donor kidney transplantation. *The New England Journal of Medicine, 366*, 770–771.
150. Moers, M., Smits, J.M., Maathuis, M.J., et al. (2009). Machine Perfusion or Cold Storage in Deceased-Donor Kidney Transplantation. *The New England Journal of Medicine, 360*, 7–19.
151. Moers, C., Varnav, O. C., van Heurn, E., et al. (2010). The value of machine perfusion perfusate biomarkers for predicting kidney transplant outcome. *Transplantation, 90*, 966–973.
152. Monbaliu, D., Liu, Q., Vekemans, K., & Pirenne, J. (2009). History of organ perfusion in organ transplantation. In D. Talbot & A. M. D'Alessandro (Eds.), *Organ donation and transplantation after cardiac death* (pp. 31–49). New York: Oxford University Press.

153. Monbaliu, D., Pirenne, J., & Talbot, D. (2012). Liver transplantation using donation after cardiac death donors. *Journal of Hepatology, 56*, 474–485.
154. Moustafellos, P., Hadjianastassiou, V., Roy, D., et al. (2007). The influence of pulsatile preservation in kidney transplantation from non-heart-beating donors. *Transplantation Proceedings, 39*, 1323–1325.
155. Mozes, M. F., Skolek, R. B., & Korf, B. C. (2005). Use of perfusion parameters in predicting outcomes of machine-preserved kidneys. *Transplantation Proceedings, 37*, 350–351.
156. Müller, S., Constantinescu, M. A., Kiermeir, D. M., et al. (2013). Ischemia/reperfusion injury of porcine limbs after extracorporeal perfusion. *The Journal of Surgical Research, 181*, 170–182.
157. Muñoz-Abraham, A. S., Patrón-Lozano, R., Narayan, R. R., et al. (2015). Extracorporeal hypothermic perfusion device for intestinal graft preservation to decrease ischemic injury during transportation. *Journal of Gastrointestinal Surgery, 20*, 313–21.
158. Muthusamy, A. S. R., Mumford, L., Hudson, A., et al. (2012). Pancreas transplantation from donors after circulatory death from the United Kingdom. *American Journal of Transplantation, 12*, 2150–2156.
159. Nagrath, D., Xu, H., Tanimura, Y., et al. (2009). Metabolic preconditioning of donor organs: Defatting fatty livers by normothermic perfusion ex vivo. *Metabolic Engineering, 11*, 274–283.
160. Nakajima, D., Chen, F., Yamada, T., et al. (2011). Hypothermic machine perfusion ameliorates ischemia-reperfusion injury in rat lungs from non-heart-beating donors. *Transplantation, 92*, 858–863.
161. Nassar, A., Liu, Q., Farias, K., et al. (2014). Role of vasodilation during normothermic machine perfusion of DCD porcine livers. *The International Journal of Artificial Organs, 37*, 165–172.
162. Nassar, A., Liu, Q., Farias, K., et al. (2016). Impact of temperature on porcine liver machine perfusion from donors after cardiac death. *Artificial Organs, 40*, 999–1008.
163. Nicholson, M. L., & Hosgood, S. A. (2013). Renal transplantation after ex vivo normothermic perfusion: The first clinical study. *American Journal of Transplantation, 13*, 1246–1252.
164. Niemeier, R. W. (1984). The isolated perfused lung. *Environmental Health Perspectives, 56*, 35–41.
165. Norden, M. A., Rao, V. K., & Southard, J. H. (1997). Improved preservation of rat hindlimbs with the University of Wisconsin solution and butanedione monoxime. *Plastic and Reconstructive Surgery, 100*, 957–965.
166. Nyrén, O., Blank, M. A., & Jaffe, B. M. (1992). Evaluation of a rat model for the study of local regulation of intestinal blood flow: Ex vivo asanguineous perfusion of the ileal vascular bed. *The Journal of Surgical Research, 53*, 455–463.
167. O'Callaghan, J., Leuvenink, H. G. D., Friend, P. J., & Ploeg, R. J. (2013a). Kidney preservation. In P. Morris & S. J. Knechtle (Eds.), *Kidney transplantation: Principles and practice* (7th ed., pp. 130–141). New York: Saunders.
168. O'Callaghan, J. M., Morgan, R. D., Knight, S. R., & Morris, P. J. (2013b). Systematic review and meta-analysis of hypothermic machine perfusion versus static cold storage of kidney allografts on transplant outcomes. *The British Journal of Surgery, 100*, 991–1001.
169. O'Donovan, M. J., Rowlerson, A., & Taylor, A. (1976). Proceedings: Contraction characteristics and histochemistry of motor units studied in perfused human limb muscles. *The Journal of Physiology, 257*, 24P–25P.
170. O'Malley, V. P., Keyes, D. M., & Postier, R. G. (1986). The fluosol-perfused isolated canine pancreas: A model for the study of blood component effects in acute pancreatitis. *The Journal of Surgical Research, 40*, 210–215.
171. Okada, N., Mizuta, K., Oshima, M., et al. (2015). A novel split liver protocol using the subnormothermic oxygenated circuit system in a porcine model of a marginal donor procedure. *Transplantation Proceedings, 47*, 419–426.
172. Okamoto, T., Wheeler, D., Liu, Q., et al. (2016). Correlation between PaO2/FiO2 and airway and vascular parameters in the assessment of cellular ex vivo lung perfusion system. *The Journal of Heart and Lung Transplantation, 35*(11), 1330–1336.

173. Okamoto, T., Wheeler, D., Liu, Q., et al. (2015). Variability in pressure of arterial oxygen to fractional inspired oxygen concentration ratio during cellular ex vivo lung perfusion: Implication for decision making. *Transplantation, 99*, 2504–2513.
174. Olschewski, P., Gass, P., Ariyakhagorn, V., et al. (2010). The influence of storage temperature during machine perfusion on preservation quality of marginal donor livers. *Cryobiology, 60*, 337–343.
175. Oltean, M., & Olausson, M. (2010). The Chiu/park scale for grading intestinal ischemia-reperfusion: If it ain't broke don't fix it! *Intensive Care Medicine, 36*, 1095. author reply 1096.
176. op den Dries, S., Karimian, N., Sutton, M. E., et al. (2013). Ex vivo normothermic machine perfusion and viability testing of discarded human donor livers. *American Journal of Transplantation, 13*, 1327–1335.
177. Op den Dries, S., Sutton, M. E., Karimian, N., et al. (2014). Hypothermic oxygenated machine perfusion prevents arteriolonecrosis of the peribiliary plexus in pig livers donated after circulatory death. *PloS One, 9*, e88521.
178. Opelz, G., & Terasaki, P. I. (1982). Advantage of cold storage over machine perfusion for preservation of cadaver kidneys. *Transplantation, 33*, 64–68.
179. Ozer, K., Rojas-Pena, A., Mendias, C. L., et al. (2015). Ex situ limb perfusion system to extend vascularized composite tissue allograft survival in Swine. *Transplantation, 99*, 2095–2101.
180. Ozer, K., Rojas-Pena, A., Mendias, C. L., et al. (2016). The effect of ex situ perfusion in a Swine limb vascularized composite tissue allograft on survival up to 24 hours. *Journal of Hand Surgery American, 41*, 3–12.
181. Parrish, D., Lindell, S. L., Reichstetter, H., et al. (2015). Cell Impermeant-based low-volume resuscitation in hemorrhagic shock: A biological basis for injury involving cell swelling. *Annals of Surgery, 263*, 565–572.
182. Pascual, J., Zamora, J., & Pirsch, J. D. (2008). A systematic review of kidney transplantation from expanded criteria donors. *American Journal of Kidney Diseases, 52*, 553–586.
183. Patel, S. K., Pankewycz, O. G., Nader, N. D., et al. (2012). Prognostic utility of hypothermic machine perfusion in deceased donor renal transplantation. *Transplantation Proceedings, 44*, 2207–2212.
184. Pegg, D. E., & Green, C. J. (1976). Renal preservation by hypothermic perfusion. III. The lack of influence of pulsatile flow. *Cryobiology, 13*, 161–167.
185. Ploeg, R. J., Goossens, D., McAnulty, J. F., et al. (1988). Successful 72-hour cold storage of dog kidneys with UW solution. *Transplantation, 46*, 191–196.
186. Poitras, P., Trudel, L., Miller, P., & Gu, C. M. (1997). Regulation of motilin release: Studies with ex vivo perfused canine jejunum. *The American Journal of Physiology, 272*, G4–G9.
187. Polyak, M. M., Arrington, B. O., Stubenbord, W. T., et al. (2000). The influence of pulsatile preservation on renal transplantation in the 1990s. *Transplantation, 69*, 249–258.
188. Ravikumar, R., Jassem, W., Mergental, H., et al. (2016). Liver transplantation after ex vivo normothermic machine preservation: A phase 1 (first-in-man) clinical trial. *American Journal of Transplantation, 16*(6), 1779–1787.
189. Reznik, O. N., Bagnenko, S. F., Loginov, I. V., et al. (2008). Machine perfusion as a tool to select kidneys recovered from uncontrolled donors after cardiac death. *Transplantation Proceedings, 40*, 1023–1026.
190. Rijkmans, B. G., Buurman, W. A., & Kootstra, G. (1984). Six-day canine kidney preservation. Hypothermic perfusion combined with isolated blood perfusion. *Transplantation, 37*, 130–134.
191. Rogers, J. W., Sellers, E. A., & Gornall, A. G. (1947). Intestinal perfusion in the treatment of uremia. *Science, 106*, 108.
192. Rosen, H. M., Slivjak, M. J., & McBrearty, F. X. (1987). The role of perfusion washout in limb revascularization procedures. *Plastic and Reconstructive Surgery, 80*, 595–605.
193. Ruiz, J. O., Schultz, L. S., Hendrickx, J., et al. (1971). Isolated intestinal perfusion: A method for assessing preservation methods and viability before transplantation. *Transactions American Society for Artificial Internal Organs, 17*, 42–48.

194. Russo, M. J., Chen, J. M., Sorabella, R. A., et al. (2007). The effect of ischemic time on survival after heart transplantation varies by donor age: An analysis of the United Network for Organ Sharing database. *The Journal of Thoracic and Cardiovascular Surgery, 133*, 554–559.
195. Sanchez, P. G., Davis, R. D., D'Ovidio, F., et al. (2014). The NOVEL lung trial one-year outcomes. *The Journal of Heart and Lung Transplantation, 33*, S71–S72.
196. Schlegel, A., Kron, P., Graf, R., et al. (2014). Warm vs. cold perfusion techniques to rescue rodent liver grafts. *Journal of Hepatology, 61*, 1267–1275.
197. Schneeberger, S., Biebl, M., Steurer, W., et al. (2009). A prospective randomized multicenter trial comparing histidine-tryptophan-ketoglutarate versus University of Wisconsin perfusion solution in clinical pancreas transplantation. *Transplant International, 22*, 217–224.
198. Schold, J. D., Kaplan, B., Howard, R. J., et al. (2005). Are we frozen in time? Analysis of the utilization and efficacy of pulsatile perfusion in renal transplantation. *American Journal of Transplantation, 5*, 1681–1688.
199. Schön, M. R., Kollmar, O., Wolf, S., et al. (2001). Liver transplantation after organ preservation with normothermic extracorporeal perfusion. *Annals of Surgery, 233*, 114–123.
200. Seal, J. B., & Gewertz, B. L. (2005). Vascular dysfunction in ischemia-reperfusion injury. *Annals of Vascular Surgery, 19*, 572–584.
201. Sgourakis, G., Papapanagiotou, A., Kontovounisios, C., et al. (2013). The value of plasma neurotensin and cytokine measurement for the detection of bowel ischaemia in clinically doubtful cases: A prospective study. *Experimental Biology and Medicine (Maywood, N.J.), 238*, 874–880.
202. Shepherd, A. P., & Riedel, G. L. (1982). Effect of pulsatile pressure and metabolic rate on intestinal autoregulation. *The American Journal of Physiology, 242*, H769–H775.
203. Skrzypiec-Spring, M., Grotthus, B., Szelag, A., & Schulz, R. (2007). Isolated heart perfusion according to Langendorff – still viable in the new millennium. *Journal of Pharmacological and Toxicological Methods, 55*, 113–126.
204. Southard, J. H., van Gulik, T. M., Ametani, M. S., et al. (1990). Important components of the UW solution. *Transplantation, 49*, 251–257.
205. Stangl, M. J., Krapp, J., Theodorou, D., et al. (2000). Computer-assisted ex vivo, normothermic small bowel perfusion. *Eur Surg Res Eur Chir Forschung Rech Chir Eur, 32*, 100–106.
206. Starzl, T. E., Groth, C. G., Brettschneider, L., et al. (1968). Orthotopic homotransplantation of the human liver. *Annals of Surgery, 168*, 392–415.
207. Steen, S., Ingemansson, R., Eriksson, L., et al. (2007). First human transplantation of a non-acceptable donor lung after reconditioning ex vivo. *The Annals of Thoracic Surgery, 83*, 2191–2194.
208. Steen, S., Liao, Q., Wierup, P. N., et al. (2003). Transplantation of lungs from non-heart-beating donors after functional assessment ex vivo. *The Annals of Thoracic Surgery, 76*, 244–252.
209. Steen, S., Sjöberg, T., Pierre, L., et al. (2001). Transplantation of lungs from a non-heart-beating donor. *Lancet, 357*, 825–829.
210. Stevanovic, M., & Sharpe, F. (2014). Functional free muscle transfer for upper extremity reconstruction. *Plastic and Reconstructive Surgery, 134*, 257e–274e.
211. Straznicka, M., Follette, D. M., Eisner, M. D., et al. (2002). Aggressive management of lung donors classified as unacceptable: Excellent recipient survival one year after transplantation. *The Journal of Thoracic and Cardiovascular Surgery, 124*, 250–258.
212. Sudan, D., Vargas, L., Sun, Y., et al. (2007). Calprotectin: A novel noninvasive marker for intestinal allograft monitoring. *Annals of Surgery, 246*, 311–315.
213. Sutton, M. E., op den Dries, S., Karimian, N., et al. (2014). Criteria for viability assessment of discarded human donor livers during ex vivo normothermic machine perfusion. *PloS One, 9*, e110642.
214. Taylor, M. J., Baicu, S., Leman, B., et al. (2008). Twenty-four hour hypothermic machine perfusion preservation of porcine pancreas facilitates processing for islet isolation. *Transplantation Proceedings, 40*, 480–482.

215. Taylor, M. J., & Baicu, S. C. (2010). Current state of hypothermic machine perfusion preservation of organs: The clinical perspective. *Cryobiology, 60*, S20–S35.
216. Tersigni, R., Toledo-Pereyra, L. H., Pinkham, J., & Najarian, J. S. (1975). Pancreaticoduodenal preservation by hypothermic pulsatile perfusion for twenty-four hours. *Annals of Surgery, 182*, 743–748.
217. Thuillier, R., Allain, G., Celhay, O., et al. (2013). Benefits of active oxygenation during hypothermic machine perfusion of kidneys in a preclinical model of deceased after cardiac death donors. *The Journal of Surgical Research, 184*, 1174–1181.
218. Tolboom, H., Izamis, M.-L., Sharma, N., et al. (2012). Subnormothermic machine perfusion at both 20°C and 30°C recovers ischemic rat livers for successful transplantation. *The Journal of Surgical Research, 175*, 149–156.
219. Toledo-Pereyra, L. H., Condie, R. M., Malmberg, R., et al. (1974). A fibrinogen-free plasma perfusate for preservation of kidneys for one hundred and twenty hours. *Surgery, Gynecology & Obstetrics, 138*, 901–905.
220. Treckmann, J., Moers, C., Smits, J. M., et al. (2011). Machine perfusion versus cold storage for preservation of kidneys from expanded criteria donors after brain death. *Transplant International, 24*, 548–554.
221. Uematsu, T., Asano, T., Enomoto, K., et al. (1987). Predictable viability assay of isolated canine liver using hypothermic continuous machine perfusion. *Transplantation Proceedings, 19*, 1321–1323.
222. Usui, M., Sakata, H., & Ishii, S. (1985). Effect of fluorocarbon perfusion upon the preservation of amputated limbs. An experimental study. *Journal of Bone and Joint Surgery British, 67*, 473–477.
223. Vairetti, M., Ferrigno, A., Carlucci, F., et al. (2009). Subnormothermic machine perfusion protects steatotic livers against preservation injury: A potential for donor pool increase? *Liver Transplantation, 15*, 20–29.
224. Vairetti, M., Ferrigno, A., Rizzo, V., et al. (2008). Correlation between the liver temperature employed during machine perfusion and reperfusion damage: Role of Ca2+. *Liver Transplantation, 14*, 494–503.
225. Valapour, M., Paulson, K., Smith, J. M., et al. (2013). OPTN/SRTR 2011 annual data report: Lung. *American Journal of Transplantation, 13*(Suppl 1), 149–177.
226. Valapour, M., Skeans, M. A., Heubner, B. M., et al. (2015). OPTN/SRTR 2013 annual data report: Lung. *American Journal of Transplantation, 15*(Suppl 2), 1–28.
227. Van Raemdonck, D., Neyrinck, A., Rega, F., et al. (2013). Machine perfusion in organ transplantation: A tool for ex-vivo graft conditioning with mesenchymal stem cells? *Current Opinion in Organ Transplantation, 18*, 24–33.
228. Vercaemst, L. (2008). Hemolysis in cardiac surgery patients undergoing cardiopulmonary bypass: A review in search of a treatment algorithm. *The Journal of Extra-Corporeal Technology, 40*, 257–267.
229. Vogel, T., Brockmann, J. G., Coussios, C., & Friend, P. J. (2012). The role of normothermic extracorporeal perfusion in minimizing ischemia reperfusion injury. *Transplantation Reviews (Orlando, Fla.), 26*, 156–162.
230. Wagner, S. M., Nogueira, A. C., Paul, M., et al. (2003). The isolated normothermic hemoperfused porcine forelimb as a test system for transdermal absorption studies. *Journal of Artificial Organs, 6*, 183–191.
231. Wallinder, A., Ricksten, S.-E., Hansson, C., et al. (2012). Transplantation of initially rejected donor lungs after ex vivo lung perfusion. *The Journal of Thoracic and Cardiovascular Surgery, 144*, 1222–1228.
232. Warnecke, G., Moradiellos, J., Tudorache, I., et al. (2012). Normothermic perfusion of donor lungs for preservation and assessment with the organ care system lung before bilateral transplantation: A pilot study of 12 patients. *Lancet, 380*, 1851–1858.
233. Watson, C. J. E., Kosmoliaptsis, V., Randle, L. V., et al. (2016). Preimplant normothermic liver perfusion of a suboptimal liver donated after circulatory death. *American Journal of Transplantation, 16*, 353–357.

234. Watson, C. J. E., Wells, A. C., Roberts, R. J., et al. (2010). Cold machine perfusion versus static cold storage of kidneys donated after cardiac death: A UK multicenter randomized controlled trial. *American Journal of Transplantation, 10*, 1991–1999.
235. Westbroek, D. L., De Gruyl, J., Dijkhuis, C. M., et al. (1974). Twenty-four-hour hypothermic preservation perfusion and storage of the duct-ligated canine pancreas with transplantation. *Transplantation Proceedings, 6*, 319–322.
236. Wicomb, W. N., Cooper, D. K., Novitzky, D., & Barnard, C. N. (1984). Cardiac transplantation following storage of the donor heart by a portable hypothermic perfusion system. *The Annals of Thoracic Surgery, 37*, 243–248.
237. Wierup, P., Haraldsson, A., Nilsson, F., et al. (2006). Ex vivo evaluation of nonacceptable donor lungs. *The Annals of Thoracic Surgery, 81*, 460–466.
238. Wight, J., Chilcott, J., Holmes, M., & Brewer, N. (2003a). The clinical and cost-effectiveness of pulsatile machine perfusion versus cold storage of kidneys for transplantation retrieved from heart-beating and non-heart-beating donors. *Health Technology Assessment, 7*, 1–94.
239. Wight, J. P., Chilcott, J. B., Holmes, M. W., & Brewer, N. (2003b). Pulsatile machine perfusion vs. cold storage of kidneys for transplantation: A rapid and systematic review. *Clinical Transplantation, 17*, 293–307.
240. Wilson, C. (2009). Perfusate development for the NHBD. In D. Talbot & A. M. D'Alessandro (Eds.), *Organ donation and transplantation after cardiac death* (pp. 67–102). New York: Oxford University Press.
241. Wright, F. H., Wright, C., Ames, S. A., et al. (1990). Pancreatic allograft thrombosis: Donor and retrieval factors and early postperfusion graft function. *Transplantation Proceedings, 22*, 439–441.
242. Xu, H., Berendsen, T., Kim, K., et al. (2012). Excorporeal normothermic machine perfusion resuscitates pig DCD livers with extended warm ischemia. *The Journal of Surgical Research, 173*, e83–e88.
243. Yeung, J. C., Cypel, M., Machuca, T. N., et al. (2012). Physiologic assessment of the ex vivo donor lung for transplantation. *The Journal of Heart and Lung Transplantation, 31*, 1120–1126.
244. Young, J. B., Naftel, D. C., Bourge, R. C., et al. (1994). Matching the heart donor and heart transplant recipient. Clues for successful expansion of the donor pool: A multivariable, multiinstitutional report. The cardiac transplant research database group. *The Journal of Heart and Lung Transplantation, 13*, 353–365.
245. Zhu, J. Z. J., Castillo, E. G., Salehi, P., et al. (2003). A novel technique of hypothermic luminal perfusion for small bowel preservation. *Transplantation, 76*, 71–76.

Chapter 3
Mobile Health Technology in Transplantation

James N. Fleming, John W. McGillicuddy, and Frank A. Treiber

Abstract Mobile health (mHealth) technology encompasses a wide range of potential clinical tools, with end-users ranging from the individual patient to national healthcare systems. This chapter focuses on mHealth with the patient as the end-user. It describes mHealth applications studied within solid organ transplant, as well as applications that have been used in non-transplant patients with chronic diseases common in transplant. It also provides an overview of fundamental necessities for successful mHealth systems and provides a roadmap to addressing current barriers to widespread study and implementation of mHealth.

3.1 Introduction

Optimal treatment of existing diseases requires the involvement of both patients and practitioners in managing multiple aspects of their disease processes. There continues to be multiple barriers to optimal care in the current healthcare model, including overcrowding of clinics and costs related to treatments, travel, and clinic visits.

Patients with chronic diseases are frequent users of complex and costly healthcare services, which are increasingly delivered in the outpatient setting. Consequently, there has been increased emphasis placed on improving the efficiency and equitability of health systems' delivery of care to these patients. Suggested areas of focus include equity, effectiveness, safety, responsiveness, continuity of care, efficiency, and accessibility [1]. Many of the barriers to these areas of focus are related to demographic disparity, accessibility of the clinic (distance, parking, and transportation), wait times, appointment scheduling, affordability, inability to meet medical fees, delays in availability, and limited support of self-care practices [1]. Telemedicine, more particularly mobile health (mHealth), offers a promising approach to overcoming these significant barriers to better outcomes.

Although no standardized definition of mHealth has been established, the Global Observatory for eHealth defined it as medical and public health practice supported by mobile devices, such as mobile phones, patient monitoring devices, personal

J.N. Fleming • J.W. McGillicuddy (✉) • F.A. Treiber
Medical University of South Carolina, Charleston, SC, USA
e-mail: mcgillij@musc.edu

S.N. Nadig, J.A. Wertheim (eds.), *Technological Advances in Organ Transplantation*, DOI 10.1007/978-3-319-62142-5_3

digital assistants, and other wireless devices [2]. The appeal of mHealth as a medium for telemedicine lies in its accessibility and mobility for both patients and practitioners. There are several different modalities of mHealth, each of which can be individualized, including remote monitoring, store and forward of data, and interactive telemedicine. As most telemedicine modalities are, or soon will be, deployable with mobile devices, this chapter includes studies that involve either telemedicine or mHealth.

3.2 Telemedicine in Solid Organ Transplant

While telemedicine has existed in one form or another for more than 40 years, its application in solid organ transplantation has lagged behind other disciplines. In a 2015 Cochrane review, there were 93 randomized controlled trials evaluating the effectiveness, acceptability, and costs of telemedicine in addition to, or as an alternative to, usual care; while only one of the included trials was performed in solid organ transplant recipients [3], there have been several less robust studies performed in the transplant population. Select studies related to solid organ transplant are described in Table 3.1. For the sake of completeness, studies related to telemedicine (vs mHealth) and protocol-only manuscripts are also included.

3.3 Potential for Benefits

Although there is a dearth of quality analyses of mHealth in solid organ transplant, the vast majority of recipients suffer from at least one chronic disease, many of which have been targeted by telemedicine and mHealth studies. Examples include nicotine addiction, diabetes, coronary heart disease, hypertension, and heart failure (Table 3.2). Cardiovascular death is the leading cause of death and graft loss in kidney transplant recipients and is driven by high rates of hypertension, dyslipidemia, and diabetes. These same risk factors, driven in part by the rising incidence of NAFLD and the aging of the transplant population, are also highly prevalent in liver and heart transplant recipients [12]. Nonadherence to medical therapy for these comorbid conditions is associated with significant increases in cardiovascular morbidity [13–15]. In addition, nonadherence to immunosuppressants, a very common problem in transplant recipients, is a well-known risk factor for poor outcomes and is directly responsible for up to 47% of late rejections and 36% of late graft losses in the kidney transplant population [13]. mHealth programs offer an opportunity to better identify and combat medication nonadherence through monitoring of prescription refills and the proportion of days covered (PDC) and self-reporting of medication administration, Medication Event Monitoring System (MEMS), ingestible sensors, and other forms of electronic medication monitoring [9, 16–18]. Additionally, they may be useful in education to increase living organ donation in populations with low-living donation rates [19].

Table 3.1 Trials investigating telemedicine technology in solid organ transplant

Study	Patients	Population	Telehealth	Comparator	Outcomes	Results
Kidney transplant						
Foucher et al. [4]	250 patient goal	Functional kidney at 1-year posttransplant, high-speed Internet access	iPad and USB flash drive for teleconference program	Usual care	Difference in major complications at 2-year posttransplant (death, rejection, graft loss, malignancy, decrease in eGFR by ≥25%)	Registered on ClinicalTrials.gov in June 2012, no information on current status will be considerably underpowered based on power analysis
Winsett et al. [5]	21	Abdominal transplants following up in transplant clinic	Remote telemedicine clinic sites with distance RNs, digitized stethoscope, otoscope, handheld camera	None, pilot study	Patient and provider satisfaction	Almost all patients reported the service at least as good as conventional visits. Providers (NPs) were uncomfortable with relying on expertise for physical exam with distance RNs. Scheduling between multiple distance sites was inefficient
Kreuzer et al. [6]	114 patient goal	Kidney transplant recipients between age 16 and 22 years old	Smartphone apps to support documentation and transfer of clinical data, facilitate appointment scheduling, and communicate with case managers	Usual care	Variability in immunosuppressant levels	No data available yet

(continued)

Table 3.1 (continued)

Study	Patients	Population	Telehealth	Comparator	Outcomes	Results
McKenzie et al. [7]	33	Liver transplant recipients between 12 and 21 years old	Text message reminders for labwork	Usual care	Adherence to laboratory testing	Significant improvement in adherence to laboratory testing (58% vs 78%, $p < 0.001$)
Yoon et al. [8]	246	Adult lung transplants that took part in the University of Minnesota Lung Transplant Home Monitoring Program between 1992 and 2002	Daily home spirometry, symptoms, and vital signs stored electronically and submitted to center through phone lines	Compared patient adherent to >75% of home monitoring to those less adherent	Survival	Trend toward better survival in adherent group, no statistical difference
McGillicuddy et al. [9]	19	Kidney transplant recipients identified as nonadherent based on screening using electronic medication tray	Electronic medication tray with reminders, bluetooth BP monitor, smartphone for transmitting data and send reminders to check BP (SMASK System)	Usual care along with electronic medication tray (no reminders)	3-month BP, medication adherence, satisfaction	Medication adherence improved significantly more in mHealth group at each month after baseline. SBP decreased from 138 mmHg to 122 mmHg in mHealth vs 135 to 138 mmHg in control ($p < 0.05$ at month 3). High overall satisfaction with mHealth system (4.8/5 Likert scale)

McGillicuddy et al. [10]	18	Retrospective review of kidney transplant recipients studied for 3 months using mHealth system [9] and BP control 9 months after trial ended	mHealth from months 0 to 3 listed in previous study, usual care from months 3 to 12	Usual care with electronic medication tray (no reminders) from months 0 to 3, usual care with no device from months 3 to 12	Clinic-measured BP	Prior mHealth group sustained lower BP at 12 months compared to usual care group (131 vs 155 mmHg, $p = 0.004$)
Ertel AE et al. [11]	20 patients in preliminary results	Liver transplant recipients immediately postsurgery	Tablet containing automated text messages with health questions, videoconferencing ability, connectivity with remote devices to track temperature, blood pressure, glucose, weight	None reported	30- and 90-day readmissions, health service utilization, health metrics, healthcare professional satisfaction, patient feasibility	No results available yet

Table 3.2 Trials investigating telemedicine technology in chronic diseases

Study	Patients	Population	Telehealth	Comparator	Outcomes	Results
Hypertension						
McManus et al. [20]	552	≥35 with history of stroke, CHD, DM, or CKD with SBP >145 mmHg; UK	Microlife WatchBP Home self-monitoring and titration	Usual care (PCP management)	Difference in SBP at 12 months	Decrease in SBP of 9.2 mmHg (95% CI 5.7–12.7) more in mHealth group vs control and decrease by 14.9 mmHg overall
Magid et al. [21]	348	18–79 years old with HTN and last two office BP values over goal, three or fewer antihypertensives, computer and Internet access; USA	Omron HEM-790IT BP cuff with uploaded BP via AHA Heart360 web application; weekly averages sent to clinical pharmacy specialist for adjustment	Usual care (PCP management)	Proportion of patients that met BP goal at 6 months	More patients met goal in intervention group (54.1 vs 35.4%, $p < 0.001$), mean reduction in SBP of 12 mmHg (95% CI 16–9); higher satisfaction in care with intervention group (58 vs 42%, $p < 0.001$)
Margollis et al. [22]	450	Adult patients with HTN at the last two office visits	A&D Medical 767PC automated oscillometric BP monitor + pharmacist education and titration of antihypertensives	Usual care (PCP +/− pharmacist consultation)	Proportion of patients with controlled BP at 6- and 12-month visits	Higher proportion of patients with BP control in telemonitoring group (48.5 vs 25.1%, $p = 0.001$); higher decrease in BP in intervention group (−9.7 mmHg, 95% CI −13.4 to −6, $p < 0.001$)

Davidson et al. [23]	38	21–65-year-old African American and Hispanic patients with uncontrolled hypertension	SMASH program included Droid X smartphone, Maya MedMinder wireless GSM electronic medication tray, A&D UA-767PlusBT BP monitor, self-determination theory-guided SMS text for encouragementtexts	Usual care	Proportion of patients with SBP <140 mmHg and proportion of patients with DBP <90 mmHg,	Significantly more patients with SBP at goal in mHealth group at 1 month (70.6 vs 15.8%, $p < 0.001$) and at 6 months (94.4 vs 41.2%, $p < 0.003$). More mHealth patients with DBP at goal at 1 month (100 vs 68.4%, $p = 0.02$) and 6 months (94.4 vs 76.5%, $p = 0.04$)
Diabetes						
Ramachandran et al. [24]	537	35–55-year-old Indian men with impaired glucose tolerance; India	Frequent text messaging to promote exercise and dietary habits	Usual care	Incidence of type 2 DM	Cumulative incidence of type 2 DM was lower in mHealth group (18 vs 27%, HR 0.64, 95% CI 0.45–0.92, $p = 0.015$)
Quinn et al. [25]	30	Type 2 DM from three community practices; USA	WellDoc; cell phone-based system that tracks BS and medication regimens and reported them to MDs along with suggested treatment plans	Usual care	Impact of mHealth on HgA1c; MD adherence to guidelines and adoption of technology	Significantly improved HgA1c in mHealth group (−2.03 vs −0.68%, $p < 0.02$, one-tailed). More mHealth patients had meds titrated vs controls (84 vs 23#, $p = 0.002$)

(continued)

Table 3.2 (continued)

Study	Patients	Population	Telehealth	Comparator	Outcomes	Results
Holmen et al. [26]	151 (three arms)	18 years or older type 2 DM with HgA1c ≥7.1%; Norway	Few Touch Application phone-based self-management system; automatic transfer of BS, diet manual, physical activity registration, personal goals (intervention arm 1) ± health counseling from diabetes nurse x 4 months (intervention arm 2)	Usual care	HgA1c improvement, self-management, health-related QOL at 1 year	HgA1c decreased similarly between groups
Heart failure						
Koehler et al. [27]	710	NYHA II or III with EF ≤35% and decompensation within the past 2 years or EF ≤25%; Germany	Remote cell phone-transmitted data from portable EKG, BP monitor, and scale; physician medical support 24/7	Usual care	Death; composite of CV death and HF hospitalization; at least 12 months of follow-up	No difference in all-cause mortality (HR 0.97, 95% CI 0.67–1.41, $p = 087$) or composite of CV death and HF hospitalization; 81% of mHealth group at least 70% compliant to data transfers
Weintraub et al. [28]	188	Symptomatic HF patients with hospitalization within 2 weeks	Automated health monitoring (AHM) technology; measured weight, BP, HR, and subjective patient assessments via standard telephone line + SPAN-HF disease management program	SPAN-HF nurse-directed disease management program	HF hospitalizations at 90 days	AHM group had decreased HF hospitalizations compared to controls (HR 0.5, 95% CI 0.25–0.99, $p = 0.05$)

Bekelman et al. [29]	384	Symptomatic HF patients from VA	Multidisciplinary collaborative team, screening and treatment of depression, telemonitoring with patient self-care support	Usual care	Patient-reported HF-specific health status	Significant improvement in Kansas City Cardiomyopathy Questionnaire in both groups, no difference between groups (13.5 point mean improvement in both groups, $p = 0.97$)
Chronic diseases						
Steventon et al. [30]	3230	Diabetes, COPD, CHF	Varied among sites, but all COPD had a pulse oximeter, all DM had glucometer, all CHF had scales	Usual care	Admissions to the hospital at 12 months; mortality	Telehealth group had fewer 12-month admissions (42.9 vs 48.2%, OR 0.82 (95% CI 0.70–0.97, $p = 0.017$) and mortality (4.6 vs 8.3%, OR 0.54, 95% CI 0.39–0.75, $p < 0.001$)
Greving et al. [31]	330	Recent manifestations of atherosclerosis and with ≥2 treatable risk factors not at goal	Personalized website with progress toward risk factor goals, communication with NP via website for 12 months in addition to usual care	Usual care	Societal costs, QALY, cost-effectiveness	Societal costs trended lower, no significant difference in QALY, reduction in Framingham Risk Score by −2.1 (95% CI −3.8 to −0.3), a 14% (95% CI −25,−2%) decrease in 10-year risk for CHD

(continued)

Table 3.2 (continued)

Study	Patients	Population	Telehealth	Comparator	Outcomes	Results
Coronary heart disease						
Chow et al. [32]	710	CHD, prior history of MI, or angiographically proven	Four semipersonal text messages per week for 6 months providing advice, motivational reminders, and support to change lifestyle behaviors	Usual care	LDL at 6 months, SBP, BMI, physical activity, smoking status	Patients in the telehealth group had lower mean differences in LDL (−5 mg/dL (95% CI −9 to 0,$p = 0.04$)), SBP (−8 mmHg (95% CI −10 to −5, $p < 0.001$)), BMI (−1.3(95% CI −1.6 to −0.9, $p < 0.001$)), more physically active minutes per week (345 (95% CI 195–495, $p < 0.001$)), and a lower percentage of current smokers (0.61(95% CI 0.48–0.76, $p < 0.001$))
Smoking cessation						
Free et al. [33].	5800	Smokers aged 16 or older willing to make an attempt to quit smoking	Txt2stop program; 5 text messages/day for 5 weeks, then 3 text messages daily for 26 weeks as well as information on smoking cessation helpline numbers	Information on smoking cessation helpline numbers	Abstinence by salivary cotinine testing or carbon monoxide testing	Increased percentage of patients abstinent from smoking (10.7 vs 4.9%, RR 2.20 (95% CI 1.80–2.68, $p < 0.0001$))

3.4 Utilization in Resource-Limited Areas

A central promise of mHealth is the opportunity to provide cost-effective and comprehensive care for patients with limited means or patients living in rural or remote areas, a demographic with the worst outcomes after solid organ transplantation [34–38].

In the United States, 64% of the population owns a smartphone, up dramatically from only 35% in 2011, a trend that is closely mirrored in the kidney transplant population [39, 40]. Approximately 20% of American smartphone owners report that their phone serves as their only or primary means of accessing the Internet [39]. Both of these trends suggest an increasingly smartphone-dependent population for whom mHealth will be a useful, and natural, extension of their mobile lives. The demographic drivers of these trends are young adults (aged 18–29), those with low household incomes and low levels of educational attainment, and nonwhites [39]. These same demographic groups have poor access to, and poor outcomes after, solid organ transplant. It is important to note that this same population is most likely to have to cancel or suspend their service due to financial constraints (48% vs 21%) or occasionally reach the maximum data allowed on their smartphone plan (51% vs 35%) [39]. For this reason mHealth options with small amounts of data transfer and the option of store and forward may be preferable.

This high penetration of mobile phones, even in resource-poor settings, offers an excellent opportunity to utilize mHealth to improve care delivery. A retrospective analysis in southern India indicated that their telemedicine program for type 2 DM may increase costs to patients. Patients received an average of 17 telemedicine follow-ups that would not have otherwise occurred; yet improved their HbA1c by 2.2% [41]. In this circumstance, a cost-effectiveness study could help clarify whether the intervention was beneficial or not. In a more clear-cut analysis, a 15-year retrospective review of telemedicine services in a rural, underserved province of Korea, telemedicine services for multiple chronic illnesses, mostly in patients over 60 years old, improved satisfaction, compliance, HbA1c, and blood pressure control and decreased healthcare costs by 50% per patient [42]. Although not in a predominantly resource-poor area, a retrospective survey study in British Columbia of self-monitored blood glucose compared to usual care in 200 patients showed a trend toward decrease estimated cost ($131.26 vs 210.89, $p = 0.128$) associated with physician visits [43]. Unfortunately, the results were inconclusive and only the costs of one aspect of care were considered. Regrettably, there have been only a few well-done studies investigating the cost-effectiveness of telemedicine, and there are limitations to these existing economic evaluations including disparate estimation methods, lack of randomized control trials, lack of long-term evaluation studies, small sample sizes, and absence of quality data and appropriate measures [44].

The current state of the literature indicates a real need for high-quality cost-effectiveness or cost-minimization studies so that the benefits of mHealth can be better realized and applied in a more systematic fashion.

3.5 Types of mHealth Modalities

mHealth's major strength is the ubiquity of the technology and the rapid expansion in the number of different ways the technology can be used to further remote patient care. mHealth can be as simple as text messaging or as complex as whole-system mHealth programs. Simple text messaging has been shown to be effective as a reminder for laboratories but has largely failed to demonstrate efficacy in most prospective studies in patients with chronic illnesses [7, 45]. Text messaging seems to be most effective when used in combination with other modalities. The National Organ Retrieval Imaging System (NORIS), studied in the UK since 2004, is a physician-focused mHealth system used in liver transplantation. In an effort to increase liver allograft utilization, donor liver information and images are taken by the retrieval team, and the accepting surgeon is notified via text message or page to review the information. The accepting surgeon is then able to communicate with the retrieval team and request additional information or images. The system has been piloted in order to demonstrate the technical feasibility of such a system to increase the utilization of marginal liver grafts and split liver allografts [46, 47]. To date, only pilot studies have been published, and data on how such a system might improve the utilization of high-risk allografts is not yet available.

At the other end of the spectrum are patient-centered modalities that largely aim to improve patient self-care and adherence to their medical regimen. Simple text messaging reminders have been used in attempts to improve, with mixed results, the management of congestive heart failure and diabetes mellitus [48, 49]. Whole-system programs seem to be much more effective. The SMASH program, which uses patient-centered, self-determination theory-guided, iterative SMS messages to support adherence, along with SMS reminders, a bluetooth-enabled medication device with reminders, and a bluetooth-enabled BP monitor, has been very successful at reducing BP and achieving BP goals, while improving medication adherence in resource-limited populations [23, 50, 51].

There are a large number of mHealth applications (apps) available for free or inexpensive download to most smartphones. Although there is an enormous breadth of functionality, the most useful apps allow patients to track medications, sync/export/print regimens, track missed and taken doses, generate reminders to take medications, and allow providers to access and adjust medication regimens and complex instructions. Unfortunately, although the last comprehensive review of such apps was just performed and published in 2016, it will almost certainly be outdated by the time this chapter reaches print [52]. Nonetheless, many of the apps evaluated are likely to remain active, as some have been in use since the first review from this research group in 2003 [53].

To improve usefulness for both patients and clinicians, some apps have increased their functionality to allow for integration with other information systems, most often patient electronic health records. While most proprietary apps that include this type of interconnectivity are available for free or at low cost to the patient, they only provide full functionality with healthcare electronic health records at substantial cost to the hosting healthcare system.

3.5.1 *Smartphone Connected Devices*

The number of smartphone connected devices is rising rapidly and, in contrast to most mHealth apps, is regulated by the FDA as medical devices. The requirement for FDA approval means that there is more complete published data on the outcomes associated with their use.

Just a few examples of the available devices include wireless EKG monitors, continuous BP monitoring devices, continuous glucose monitoring devices, and pulse oximeter monitoring devices [54–57]. The future of healthcare may change dramatically in the face of our newfound ability to monitor patients in their home environment. mHealth will let providers extend their reach to patients in remote areas using handheld diagnostic devices without the usual costs that are associated with the more traditional modes of care [58]. Not only will clinicians have access to health information that was not previously available, but they will be able to better tailor therapy to the individual based on patient-specific data, behaviors, and preferences. mHealth offers an opportunity to create a truly collaborative approach to medical care that allows for the full investment of both the patient and the care providers and works to increase the patient's quality of life and the amount of time they spend out of the clinic and out of the hospital.

Numerous miniature sensor-based technologies have been developed for home use. There are implantable devices that remotely monitor patient hemodynamics and allow for improved management and a reduction in the utilization of the healthcare system for high-utilization chronic diseases such as chronic heart failure [59]. Miniaturized ingestible sensors are being used to combat medication nonadherence by providing extremely accurate data on medication-taking behavior [18]. It is imperative that, when faced with this rapid expansion of mHealth capacity, the scientific community is committed to rigorously evaluating the benefits and costs to insure that the justification exists to implement the most efficacious and cost-effective modalities [60, 61].

3.6 Patient Acceptance

Critical to the widespread adoption of mHealth will be patient willingness and acceptance. In a kidney transplant population, high rates of willingness and acceptance were seen when recipients were offered the opportunity to employ an mHealth program that included a smartphone, an electronic medication tray, and a bluetooth-enabled blood pressure monitor [40]. It was found that nearly 90% of kidney transplant patients with hypertension were receptive to incorporating the program into their medical care [40]. In the subsequent proof-of-concept randomized controlled trial that followed, 75% of the potential subjects that were approached agreed to participate and reported very high rates of satisfaction at the completion of the 3-month study [9]. Similarly, in a survey in heart failure patients and clinicians, both patients and providers reported being comfortable and willing to use an mHealth system, provided the system was easy to use, had clear tangible benefits, maintained good patient-provider communication, and did not increase clinical workload [62].

3.6.1 Design Factors Important for Acceptance

mHealth requires both patients and providers to accept and "buy in" to the technology. Essential to acceptance for both parties is perceived utility. Perceived utility is the individual's perception that a product, service, or technique has benefits at the personal or organizational level over the current standard. Other chief design factors are perceived ease of use, reliability, and security. From a functionality perspective, the mHealth app must have interoperability, or integration with other systems, such as EHR, and work within the providers' work flow. From a health system perspective, it must be cost-effective, improve outcomes, save time, or reduce errors [63].

3.6.2 Necessities to Produce Positive Behavioral Change: Self-Care and Shared Decision-Making

The long-term success of mHealth depends on its capacity to produce durable behavioral changes in patients that improve self-care and shared decision-making. Although device-related factors are certainly important, patient selection and motivation toward self-management are equally important. Bhavnani et al. theorize four distinct categories of patients in regard to mHealth adoption (Table 3.3) [64]. High-efficiency users will modify their behaviors with improved knowledge regardless of whether or not an mHealth program is a part of their care plan. While they might enjoy the convenience and trackability of an mHealth program, it is largely a bystander to their positive behavioral change. Initial adopters will be highly accepting and motivated initially, but then their motivation and device use decline rapidly. For example, Mattila et al. initially saw high utilization of an mHealth program designed to increase activity, improve cholesterol, and aid with weight loss, but

Table 3.3 Categories of patients who engage in mHealth

Category	Patient	Device
High-efficiency users	Predetermined to modify their behaviors	Bystander in positive behavioral change
Initial adopters	Rapidly decline in willingness	Do not retain use
Nonadopters	Do not adopt changes	Do not adopt use
Gradual believers	Condition improves as they modify behaviors	Use device and understand benefit for change

Adapted from Bhavnani et al. [64] with permission from Oxford University Press persistent utilization was seen in fewer than 30% of the subjects [65]. Nonadopters are unlikely to adopt any behavior change and see the mHealth programs as a waste of time. Gradual believers will attempt positive behavioral change with utilization of the mHealth program and will, on seeing tangible improvements, become long-term adopters of the positive behavioral change as enabled by the mHealth program. The goal is to move patients from initial adopters and nonadopters into the gradual believers category [64].

3.7 Challenges of Implementation

With the availability of so many mHealth applications, why is it not more widespread? Part of the challenge is creating a comprehensive mHealth application that meets all of the patient, provider, and institution needs. With most mHealth programs created by separate companies from the EHR, there are always challenges with the programs working together in a seamless arrangement. In addition, the focus of the institution (typically the purchaser) is frequently divergent in some ways from the end users (providers and patients), leading to the risk of a product selection that doesn't meet all needs or is primarily focused on the institutional goals. In addition, practically speaking, a conversion to mHealth is expensive. While EHR adoption is widespread among hospitals now, it was only when the Affordable Care Act (ACA) provided $30 billion to support EHR adoption that these changes began to be widespread. And, unfortunately, it has not come without its complications. EHR have been shown to undermine the therapeutic bond, reduce provider satisfaction, and increase documentation requirements [66–68].

3.7.1 Fitting mHealth into Clinical Practice

One of the most important design factors of mHealth applications, as described by providers, is the efficiency and ease of use. It must be able to simplify or speed up tasks that were previously arduous. If the mHealth program duplicates work that already must be done in a different fashion (double-work), it is doomed from the beginning. In addition, there is concern about the massive amounts of potentially irrelevant data leading to clinicians' missing important data points, which may have medicolegal repercussions [69]. Two proposed methods to reduce this concern are electronic and human, although the method that may be the most successful is a combination of the two. The electronic method would have rules built in to alert clinicians based off of perceived singular outcomes or groupings of information (i.e., one critical value laboratory marker or a significant increase in a laboratory marker that wouldn't normally signal an alarm, being within the perceived "normal" range). The human method to manage the data overload includes ancillary staff playing critical roles in processing and reviewing the data so that clinicians are not overburdened [69]. Another option that has been suggested is a new healthcare position called "digitalists" [70]. This position does not exist yet, but the argument is that it may be a new clinician necessary to meet the digital needs, similar to the way hospitalists were created in the late 1990s. This new clinician would be trained in mobile health and would monitor, interpret, and act upon remote patient data. They would have interactions with the patient, another healthcare practitioner, or both in order to resolve an issue that is identified [70].

3.7.2 Resistance to Changing Responsibilities

Providers have been shown to be resistant to changing their tasks and responsibilities without proven benefit and adequate training. Frequently, clinicians have noted that a critical barrier in mHealth adoption has been a lack of effective training on optimal use [63]. This leads to frustration and inefficiency, with an underestimation of the time required to perform tasks [69]. In order to maximize the mHealth system, the end users (providers, patients, administrators) must all be involved in the choice or development of a system [69].

3.7.3 Building, Coordinating, and Sustaining mHealth Services in Addition to the Existing Healthcare Delivery System

There are many challenges to generating data to validate mHealth use in healthcare in order to drive the desire to build a complete mHealth system that works in conjunction with the existing healthcare delivery system. It is a major challenge to study and generate data on mHealth interventions in a rapid fashion before the technology is outdated [64]. In addition, the applications where mHealth can be used to show benefit are limited to chronic conditions that have easily measurable physiologic parameters that frequently change in a short period of time. Examples of these conditions are diabetes, hypertension, chronic heart failure, and asthma. These chronic conditions must be targeted for adequate control to reduce the risk of end-result conditions, such as myocardial infarctions and stroke, where acute changes develop suddenly after a prolonged period of clinical stability [64]. One potential exception to this rule is nonadherence. Nonadherence is a complex problem that, unfortunately, is not easily measurable and may not respond significantly to singular changes. mHealth may have the potential to help monitor for nonadherence real time through adequate control of chronic conditions such as hypertension and diabetes, as well as real-time tracking of medication administration (electronic medication boxes, self-report, MMES, or ingestible sensors) and the proportion of days covered (PDC) based on refill prescription monitoring. Indeed, mHealth may have the potential to finally accurately describe the context of the nonadherence epidemic and help identify combinations of methods to combat it.

Once adopted, a system that integrates mHealth into a two-way communication with the existing EHR is a necessity. The data from the mobile device must be able to integrate seamlessly into the already large volume of existing patient data in the EHR. Data is useless if it is not able to be viewed in one field, versus having to compare between different panes of an electronic system. It is also critical for the data to be stored in the same fashion in the laboratory database maintained by health systems. These are often used for EHR, lab systems, financial systems, administration, billing and payment processing, and quality improvement projects.

3.8 Regulations and Reimbursement

3.8.1 FDA Regulations

The FDA regulates all medical devices, including those that are attached to smartphones or other mobile devices, as described previously. However, the FDA does not intend, at this time, to regulate applications unless they are classified as medical devices or whose functionality could pose a risk to a patient's safety if the app were to not function as intended. They state in a guidance document "In general, if a mobile app is intended for use in performing a medical device function (i.e. for diagnosis of disease or other conditions, or the cure, mitigation, treatment, or prevention of a disease) it is a medical device, regardless of the platform on which it is run" [71]. In the absence of that designation, there is no regulation of medical apps used by patients or clinicians. This places the accuracy, reliability, and data security squarely on the shoulders of any institution that decides to implement widespread utilization of medical apps in an mHealth program. This concept may be another reason for the slow adoption of mHealth programs by companies and institutions.

3.8.2 Medicare Reimbursement of Telemedicine and Mobile Health

Additionally, Medicare only reimburses clinicians and institutions for limited types of Part B services, and the details of the communication must meet Medicare requirements (Table 3.4) [72]. First of all, the Medicare beneficiary must be in an area that has limited access to healthcare, and they must be present in a healthcare facility that is authorized by law. The telemedicine service details and practitioner are also limited based on CMS requirements. These requirements limit mHealth applications and must be provided using real-time audio and video. Although any

Table 3.4 DHHS CMS telehealth services reimbursement requirements (CY 2016)

1. Geographical originating site
Rural Health Professional Shortage Area (HPSA) located either: Outside of a metropolitan statistical area (MSA) or in a rural census tract A county outside of an MSA
2. Originating site detail
Office of a physician or practitioner Hospital Critical Access Hospital (CAH) Rural Health Clinic Federally Qualified Health Center Hospital-based or CAH-based renal dialysis center Skilled nursing facility Community Mental Health Center
3. Distant site practitioners
Physician Nurse practitioner Physician assistant Nurse midwife Clinical nurse specialist Certified registered nurse anesthetist Clinical psychologist Clinical social worker Registered dietitians
4. Telemedicine requirements
Interactive audio and video telecommunication system Real-time communication
5. Services covered
Updated regularly at https://www.cms.gov/

coverage of telemedicine is encouraging, the current reimbursement profile for Medicare beneficiaries limits forward movement of technology and still requires transportation to healthcare facilities. Additionally, it limits the type of clinicians that can have services reimbursed. Noticeably absent from the approved list are pharmacists, who have demonstrated the ability to improve outcomes in chronic conditions at decreased costs [73–75].

3.9 Conclusion

In general, there appear to be three main paths to translating mHealth into the real world. First of all, we must identify new methods for patient engagement that results in beneficial and measurable behavioral changes. Secondly, we must develop the necessary tools to streamline the clinical integration and data analytics between the mHealth program and the main EHR. And finally, we must work with federal authorities to outline regulatory factors that promote the most effective and robust technologies for clinical use.

The solid organ transplant patient population is ideal for studying and realizing the potential health and cost benefits of mHealth, with the high concentration of chronic disease and the close follow-up by clinicians. Patients and clinicians in this specialty have demonstrated a willingness to use mHealth, and a global benefit can be seen through better management of chronic diseases and nonadherence.

References

1. Fradgley, E. A., Paul, C. L., & Bryant, J. (2015). A systematic review of barriers to optimal outpatient specialist services for individuals with prevalent chronic diseases: What are the unique and common barriers experienced by patients in high income countries? *International Journal for Equity in Health, 14*, 52.
2. World Health Organization. (2011). *mHealth: New horizons for health through mobile technologies – Based on the findings of the second global survey on eHealth* (Global Observatory for eHealth series). Geneva, Switzerland.
3. Flodgren, G., Rachas, A., Farmer, A. J., Inzitari, M., & Shepperd, S. (2015). Interactive telemedicine: effects on professional practice and health care outcomes. *Cochrane Database of Systematic Reviews*, Issue 9. Art. No.: CD002098.
4. Foucher, Y., Meurette, A., Daguin, P., et al. (2015). A personalized follow-up of kidney transplant recipients using video conferencing based on a 1-year scoring system predictive of long term graft failure (TELEGRAFT study): Protocol for a randomized controlled trial. *BMC Nephrology, 16*, 6.
5. Winsett, R. P., Thompson, D., Gower, G., & Gaber, A. O. (2006). Introducing telemedicine into a transplant outpatient clinic. *Journal of Telemedicine and Telecare, 12*(4), 215–216.
6. Kreuzer, M., Prufe, J., Bethe, D., et al. (2014). The TRANSNephro-study examining a new transition model for post-kidney transplant adolescents and an analysis of the present health care: Study protocol for a randomized controlled trial. *Trials, 15*, 505.
7. McKenzie, R. B., Berquist, W. E., Foley, M. A., Park, K. T., Windsheimer, J. E., & Litt, I. F. (2015). Text messaging improves participation in laboratory testing in adolescent liver transplant patients. *Journal of Participatory Medicine, 7*, 1–15.
8. Yoon, H. J., Guo, H., Hertz, M., & Finkelstein S. (2008). Adherence to home-monitoring and its impact on survival in post-lung transplantation patients. *American Medical Informatics Association Annual Symposium Proceedings* (pp. 835–838).
9. McGillicuddy, J. W., Gregoski, M. J., Weiland, A. K., et al. (2013). Mobile health medication adherence and blood pressure control in renal transplant recipients: A proof-of-concept randomized controlled trial. *JMIR Res Protoc., 2*(2), e32. doi:10.2196/resprot.2633.
10. McGillicuddy, J. W., Taber, D. J., Mueller, M., et al. (2015). Sustainability of improvements in medication adherence through a mobile health intervention. *Progress in Transplantation, 25*(3), 217–223.
11. Ertel, A. E., Kaiser, T., & Shah, S. A. (2015). Using telehealth to enable patient-centered care for liver transplantation. *JAMA Surgery, 150*(7), 674–675.
12. Gillis, K. A., Patel, R. K., & Jardine, A. G. (2014). Cardiovascular complications after transplantation: Treatment options in solid organ recipients. *Transplantation Reviews (Orlando, Fla.), 28*(2), 47–55.
13. Bulter, J. A., Roderick, P., Mullee, M., Mason, J. C., & Peveler, R. C. (2004). Frequency and impact of nonadherence to immunosuppressants after renal transplantation: A systematic review. *Transplantation, 77*(5), 769–776.
14. Mazzaglia, G., Ambrosioni, E., Alacqua, M., et al. (2009). Persistence to antihypertensive medications and cardiovascular morbidity among newly diagnosed hypertensive patients. *Circulation, 120*, 1598–1605.

15. Bohm, M., Schumacher, H., Laufs, U., et al. (2013). Effects of nonpersistence with medication on outcomes in high-risk patients with cardiovascular disease. *American Heart Journal, 166*(2), 306–314.
16. Heldenbrand, S., Martin, B. C., Gubbins, P. O., et al. (2016). Assessment of medication adherence app features, functionality, and health literacy level and the creation of a searchable web-based adherence app resource for health care professionals and patients. *Journal of the American Pharmaceutical Association*, 2016, 56(3), 29–302.
17. Lelubre, M., Kamal, S., Genre, N., et al. (2015). Interdisciplinary medication adherence program: The example of a university community pharmacy in Switzerland. *BioMed Research International, 2015*, 103546.
18. Belknap, R., Weis, S., Brookens, A., et al. (2013). Feasibility of an ingestible sensor-based system for monitoring adherence to tuberculosis therapy. *PLoS One, 8*(1), e53373.
19. Sieverdes, J. C., Nemeth, L. S., Magwood, G. S., et al. (2015). Patient-centered mHealth living donor transplant education program for African Americans: Development and analysis. *JMIR Res Protoc., 4*(3), e84.
20. McManus, R. J., Mant, J., Haque, M. S., et al. (2014). Effect of self-monitoring and medication self-titration on systolic blood pressure in hypertensive patients at high risk of cardiovascular disease: The TASMIN-SR randomized clinical trial. *Journal of the American Medical Association, 312*(8), 799–808.
21. Magid, D. J., Olson, K. L., Billups, S. J., Wagner, N. M., Lyons, E. E., & Kroner, B. A. (2013). A pharmacist-led, American Heart Association Heart360 Web-enabled home blood pressure monitoring program. *Circulation. Cardiovascular Quality and Outcomes, 6*(2), 157–163.
22. Margollis, K. L., Asche, S. E., Bergdall, A. R., et al. (2013). Effect of home blood pressure telemonitoring and pharmacist management on blood pressure control: A cluster randomized clinical trial. *Journal of the American Medical Association, 310*(1), 46–56.
23. Davidson, T. M., McGillicuddy, J., Mueller, M., et al. (2015). Evaluation of an mHealth medication regimen self-management program for African American and Hispanic uncontrolled hypertensives. *J Pers Med., 5*(4), 389–405.
24. Ramachandran, A., Snehalatha, C., Ram, J., et al. (2013). Effectiveness of mobile phone messaging in prevention of type 2 diabetes by lifestyle modification in men in India: A prospective, parallel-group, randomized controlled trial. *The Lancet Diabetes and Endocrinology, 1*(3), 191–198.
25. Quinn, C. C., Clough, S. S., Minor, J. M., Lender, D., Okafor, M. C., & Gruber-Baldini, A. (2008). WellDoc mobile diabetes management randomized controlled trial: Change in clinical and behavioral outcomes and patient and physician satisfaction. *Diabetes Technology & Therapeutics, 10*(3), 160–168.
26. Holmen, H., Torbjornsen, A., Wahl, A. K., et al. (2014). A mobile health intervention for self-management and lifestyle change for persons with type 2 diabetes, part 2: One-year results from the Norwegian randomized controlled trial RENEWING HEALTH. *JMIR Mhealth Uhealth, 2*(4), e57.
27. Koehler, F., Winkler, S., Schieber, M., et al. (2011). Impact of remote telemedical management on mortality and hospitalizations in ambulatory patients with chronic heart failure: The telemedical interventional monitoring in heart failure study. *Circulation, 123*(17), 1873–1880.
28. Weintraub, A., Gregory, D., Patel, A. R., et al. (2010). A multicenter randomized controlled evaluation of automated home monitoring and telephonic disease management in patients recently hospitalized for congestive heart failure: The SPAN-CHF II trial. *Journal of Cardiac Failure, 16*(4), 285–292.
29. Bekelman, D. B., Plomondon, M. E., Carey, E. P., et al. (2015). Primary results of the patient-centered disease management (PCDM) for heart failure study: A randomized clinical trial. *JAMA Internal Medicine, 175*(5), 725–732.
30. Steventon, A., Bardsley, M., Billings, J., et al. (2012). Effect of telehealth on use of secondary care and mortality: Findings from the whole system demonstrator cluster randomised trial. *BMJ, 344*, e3874.

31. Greving, J. P., Kaasjager, H. A., Vernooij, J. W., et al. (2015). Cost-effectiveness of a nurse-led internet-based vascular risk factor management programme: Economic evaluation alongside a randomised controlled clinical trial. *BMJ Open, 5*(5), e007128.
32. Chow, C. K., Redfern, J., Hillis, G. S., et al. (2015). Effect of lifestyle-focused text messaging on risk factor modification in patients with coronary heart disease: A randomized clinical trial. *Journal of the American Medical Association, 314*(12), 1255–1263.
33. Free, C., Knight, R., Robertson, S., et al. (2011). Smoking cessation support delivered via mobile phone text messaging (txt2stop): A single-blind, randomised trial. *Lancet, 378*, 49–55.
34. Fan, P. Y., Ashby, V. B., Fuller, D. S., et al. (2010). Access and outcomes among minority transplant patients, 1999–2008, with a focus on determinants in kidney graft survival. *American Journal of Transplantation, 10*, 1090–1107.
35. Andreoni, K. A., Forbes, R., Andreoni, R. M., Phillips, G., Stewart, H., & Ferris, M. (2013). Age-related kidney transplant outcomes: Health disparities amplified in adolescence. *JAMA Internal Medicine, 173*(16), 1524–1532.
36. Davies, R. R., Russo, M. J., Reinhartz, O., et al. (2013). Lower socioeconomic status is associated with worse outcomes after both listing and transplanting children with heart failure. *Pediatric Transplantation, 17*(6), 573–581.
37. Axelrod, D. A., Dzebisashvili, N., Schnitzler, M. A., et al. (2010). The interplay of socioeconomic status, distance to center, and interdonor service area travel on kidney transplant access and outcomes. *Clinical Journal of the American Society of Nephrology, 5*(12), 2276–2288.
38. Barritt, A. S., 4th, Telloni, S. A., Potter, C. W., Gerber, D. A., & Hayashi, P. H. (2013). Local access to subspecialty care influences the chance of receiving a liver transplant. *Liver Transplantation, 19*(4), 377–382.
39. Smith A. U.S. smartphone use in 2015. www.pewinternet.org/2015/04/01/us-smartphone-use-in-2015/. Published April 1, 2015. Accessed 8 Apr 2016.
40. McGillicuddy, J. W., Weiland, A. K., Frezel, R. M., et al. (2013). Patient attitudes toward mobile phone-based health monitoring: Questionnaire study among kidney transplant recipients. *Journal of Medical Internet Research, 15*(1), e6.
41. Kesavadev, J., Shankar, A., Pillai, P. B., Krishnan, G., & Jothydev, S. (2012). Cost-effective use of telemedicine and self-monitoring of blood glucose via diabetes Tele management system (DTMS) to achieve a target glycosylated hemoglobin values without serious symptomatic hypoglycemia in 1,000 subjects with type 2 diabetes mellitus – A retrospective study. *Diabetes Technology & Therapeutics, 14*(9), 772–776.
42. Kim, H. G., Ahn, M. E., Choi, Y. A., et al. (2015). Fifteen-year experience with telemedicine services in Gangwon Province in Korea. *Healthc Inform Res., 21*(4), 283–291.
43. Deng, L., White, A. S., Pawlowska, M., et al. (2015). Cost-benefit analysis of internet therapeutic intervention on patients with diabetes. *Int J Endocrin Metab., 13*(2), e22803.
44. de la Torre-Diez, I., Lopez-Coronado, M., Vaca, C., Aguado, J. S., & de Castro, C. (2015). Cost-utility and cost-effectiveness studies of telemedicine, electronic, and mobile health systems in the literature: A systematic review. *Telemedicine Journal and E-Health, 21*(2), 81–85.
45. de Jongh, T., Gurol-Urganci, I., Vodopivec-Jamsek, V., Car, J, & Atun, R. (2012). Mobile phone messaging for facilitating self-management of long-term illness. *Cochrane Database Systematic Reviews*, (12), CD007459.
46. Reddy, M. S., Bhati, C., Neil, D., Mirza, D. F., & Manas, D. M. (2008). National organ retrieval imaging system: Results of the pilot study. *Transplant International, 21*(11), 1036–1044.
47. Bhati, C. S., Wigmore, S. J., Reddy, S., et al. (2010). Web-based image transmission: A novel approach to aid communication in split liver transplantation. *Clinical Transplantation, 24*(1), 98–103.
48. Nundy, S., Razi, R., Dick, J. J., et al. (2013). A text messaging intervention to improve heart failure self-management after hospital discharge in a largely African American population. *Journal of Medical Internet Research, 15*(3), e53.
49. Dick, J. J., Nundy, S., Solomon, M. C., Bishop, K. N., Chin, M. H., & Peek, M. E. (2011). Feasibility and usability of text message-based program for diabetes self-management in

an urban African American population. *Journal of Diabetes Science and Technology, 5*(5), 1246–1254.
50. Sieverdes, J. C., Gregoski, M., Patel, S., et al. (2013). mHealth medication and blood pressure self-management program in Hispanic hypertensives: A proof of concept trial. *Smart Homecare Technol. TeleHealth., 1*, 1–10.
51. McGillicuddy, J., Gregoski, M., Brunner-Jackson, B., et al. (2012). Facilitating medication adherence and eliminating therapeutic inertia using wireless technology: proof-of-concept findings with uncontrolled hypertensives and kidney transplant recipients. In Proceedings of the Conference on Wireless Health, New York, HY, USA, doi:10.1145/2448096.2448108.
52. Heldenbrand, S., Martin, B. C., Gubbins, P. O., et al. (2016). Assessment of medication adherence app features, functionality, and health literacy level and the creation of a searchable web-based adherence app resource for health care professionals and patients. *Journal of the American Pharmaceutical Association, 56*(3), 293–302.
53. Dayer, L., Heldenbrand, S., Anderson, P., Gubbins, P. O., & Martin, B. C. (2003). Smartphone medication adherence apps: Potential benefits to patients and providers. *Journal of the American Pharmaceutical Association, 53*(2), 172–181.
54. Tarakji, D. G., Wazni, O. M., Callahan, T., et al. (2015). Using a novel wireless system for monitoring patients after the atrial fibrillation ablation procedure: The iTransmit study. *Heart Rhythm, 12*(3), 554–559.
55. Theilade, S., Hansen, T. W., Joergensen, C., Lajer, M., & Rossing, P. (2013). Tonometric devices for central aortic systolic pressure measurements in patients with type 1 diabetes: Comparison of the BPro and SphygmoCor devices. *Blood Pressure Monitoring, 18*(3), 156–160.
56. Nakamura, K., & Balo, A. (2015). The accuracy and efficacy of the Dexcom G4 platinum continuous glucose monitoring system. *Journal of Diabetes Science and Technology, 9*(5), 1021–1026.
57. Diwei, H., Morgan, S. P., Trachanis, D., et al. (2015). A single-chip CMPS pulse oximeter with on-chip lock-in detection. *Sensors, 15*, 17076–17088.
58. Singh, S., Bansal, M., Maheshwari, P., et al. (2013). American Society of Echocardiography: Remote echocardiography with web-based assessments for referrals at a distance (ASE-REWARD) study. *Journal of the American Society of Echocardiography, 26*(3), 221–233.
59. Abraham, W. T., Adamson, P. B., Bourge, R. C., et al. (2011). Wireless pulmonary artery haemodynamic monitoring in chronic heart failure: A randomized controlled trial. *Lancet, 377*(9766), 658–666.
60. Sandhu, A. T., Goldhaber-Fiebert, J. D., Owens, D. K., Turakhia, M. P., Kaiser, D. W., & Heidenriech, P. A. (2016). Cost-effectiveness of implantable pulmonary artery pressure monitoring in chronic heart failure. *JACC Heart Fail., S2213-1779*(15), 00838-0. doi:10.1016/j.jchf.2015.12.015.
61. Au-Yeung, K. Y., & DiCarlo, L. (2012). Cost comparison of wirelessly vs directly observed therapy for adherence confirmation in anti-tuberculosis treatment. *The International Journal of Tuberculosis and Lung Disease, 16*(11), 1498–1504.
62. Seto, E., Leonard, K. J., Masino, C., Cafazzo, J. A., Barnsley, J., & Ross, H. J. (2010). Attitudes of heart failure patients and health care providers towards mobile phone-based remote monitoring. *Journal of Medical Internet Research, 12*(4), e55.
63. Gagnon, M. P., Ngangue, P., Payne-Gagnon, P., & Desmartis, M. (2016). M-health adoption by healthcare professionals: A systematic review. *Journal of the American Medical Informatics Association, 23*, 212–220.
64. Bhavnani, S. P., Narula, J., & Sengupta, P. P. (2016). Mobile technology and the digitization of healthcare. *European Heart Journal*. epublished ahead of print.
65. Mattila, E., Orsama, A. L., Ahtinen, A., Hopsu, L., Leino, T., & Korhonen, I. (2013). Personal health technologies in employee health promotion: Usage activity, usefulness, and health-related outcomes in a 1-year randomized controlled trial. *JMIR Mhealth Uhealth, 1*(2), e16.

66. Sinsky, C. A., Beasley, J. W., Simmons, G. E., et al. (2014). Electronic health records: Design, implementation, and policy for higher-value primary care. *Annals of Internal Medicine, 160*, 727–728.
67. Babbott, S., Manwell, L. B., Brown, R., et al. (2014). Electronic medical records and physician stress in primary care: Results from the MEMO study. *Journal of the American Medical Informatics Association, 21*, e100–e106.
68. Adler-Milstein, J., & Huckman, R. S. (2013). The impact of electronic health record use on physician productivity. *The American Journal of Managed Care, 19*, SP345–SP352.
69. Davis, M. M., Freeman, M., Kaye, J., Vuckovic, N., & Buckley, D. L. (2014). A systematic review of clinician and staff views on the acceptability of incorporating remote monitoring technology into primary care. *Telemedicine and eHealth., 20*(5), 428–438.
70. Spiegel, B. (2015). American Journal of gastroenterology lecture: How digital health will transform gastroenterology. *The American Journal of Gastroenterology.* 2016. epub ahead of print.
71. USDHHS FDA. Mobile Medical Applications: Guidance for industry and food and drug administration staff. http://www.fda.gov/downloads/MedicalDevices/DeviceRegulationandGuidance/GuidanceDocuments/UCM263366.pdf. February 9, 2015. Accessed 20 Apr 2016.
72. DHHS CMS. Telehealth services. https://www.cms.gov/Outreach-and-Education/Medicare-Learning-Network-MLN/MLNProducts/downloads/telehealthsrvcsfctsht.pdf. December, 2015. Accessed 20 Apr 2016.
73. Obreli-Neto, P. R., Marusic, S., Guidoni, C. M., et al. (2015). Economic evaluation of a pharmaceutical care program for elderly and hypertensive patients in primary health care: A 36-month randomized controlled clinical trial. *J Manag Care Spec Pharm., 21*(1), 66–75.
74. Yu, J., Shah, B. M., Ip, E. J., & Chan, J. (2013). A Markov model of the cost-effectiveness of pharmacist care for diabetes in prevention of cardiovascular diseases: Evidence from Kaiser Permanente Northern California. *Journal of Managed Care Pharmacy, 19*(2), 102–114.
75. Pousinho, S., Morgado, M., Falcao, A., & Alves, G. (2016). Pharmacist interventions in the management of type 2 diabetes mellitus: A systematic review of randomized controlled trials. *Journal of Managed Care Pharmacy, 22*(5), 493–515.

Chapter 4
Advances in Vascular Access

Felicitas L. Koller and Kenneth J. Woodside

Abstract Vascular access is the lifeline of the hemodialysis patient. However, many patients lack early nephrology care that results in late referral for vascular access placement and dialysis initiation with a central venous catheter in 80% of the United States incident dialysis population. Despite these hurdles, arteriovenous fistulae (AVF) have doubled, to 65%, in the United States prevalent hemodialysis population over the last decade. The endovascular revolution has resulted in better rescue and maintenance of established AVF and arteriovenous grafts (AVG). Newer conduits are available for those needing an AVG, and a number of devices have been approved or are in trials to address specific issues, such as high-grade central stenosis, deep AVF, slow maturation, need for immediate cannulation, and persistent backwall injury. Finally, new techniques to deal with distal hypoperfusion ischemic syndrome, or "steal syndrome," have been developed.

4.1 Introduction

Chronic kidney disease (CKD) patients require effective vascular access when they progress to end-stage renal disease (ESRD) and require hemodialysis. Arteriovenous fistulae (AVF) are preferred over arteriovenous grafts (AVG), with either considered better than a hemodialysis central venous catheter. AVF have better long-term outcomes, including lower thrombosis rates and lower infection risk, when compared with AVG and central venous catheters [15, 27, 89, 111, 117]. Despite these advantages, historically, AVG and central venous catheter rates were quite high. Two national initiatives, the National Kidney Foundation's Kidney Disease Outcomes Quality Initiative (KDOQI) and the Fistula First Breakthrough Initiative (FFBI),

F.L. Koller, MD
Feinberg School of Medicine, Northwestern University,
676 N. Saint Clair, Chicago, IL 60611, USA
e-mail: felicitas.koller@nm.org

K.J. Woodside, MD (✉)
Department of Surgery, University of Michigan, 2922 Taubman Center,
1500 E. Medical Center Drive, Ann Arbor, MI 48109-5300, USA
e-mail: woodside@umich.edu

S.N. Nadig, J.A. Wertheim (eds.), *Technological Advances in Organ Transplantation*, DOI 10.1007/978-3-319-62142-5_4

stressed the need for AVF, with KDOQI offering guidelines for specific placement-referral timelines and placement strategies [104, 133]. Despite efforts for early referral and placement, the United States Renal Data System (USRDS) reports that 80% of incident dialysis patients initiate hemodialysis with a central venous catheter (Fig. 4.1), with only a quarter having a maturing AVF or AVG in place [124]. However, AVF rates among prevalent US dialysis patients increased from 32% in 2003 to 65% in 2014, likely due to increased emphasis on endovascular and post-placement interventions, which have been used both to aid maturation and rescue failing or thrombosed access.

4.2 Current Practices

While exact practice varies, based on local surgeon training and resources, KDOQI and FFBI have had significant impact on the timeline and standard of care for CKD patients. Early emphasis on vein sparing phlebotomy techniques (use of veins on the dorsum of the hand, rather than at the antecubital space or wrist) is stressed, although frequently not followed, for CKD patients to prevent scarring or thrombosis of veins that may be of use for AVF placement. Ideally, patients are referred 6 months to a year before the need to initiate hemodialysis. However, early referral is more complex than it appears. In the US, 25.3% of patients have no nephrology care before initiation of dialysis, with another 13.2% having less than 6 months of such care [124]. Patients with shorter duration of predialysis nephrology care are unlikely to have adequate time to establish a mature AVF, or even an AVG, as referral for access will be delayed. Furthermore, prediction of the progression of CKD to ESRD (i.e., progression to the need for hemodialysis) is difficult and fraught with uncertainty [68, 86, 91]. In addition, there are practitioner and patient variables: thresholds for dialysis initiation can vary widely [105, 127], and patients are often reluctant to have access placement [18]. Even patients with long-term nephrology care may have personal and medical logistical issues. Kidney transplant patients with failing allografts have low rates of AVF use at hemodialysis initiation [19, 140], with 65% using a catheter on return to dialysis, despite high rates of hospitalized central venous catheter-related infections [110, 137] and preexisting nephrology care for 93.5% of such patients [19]. Similarly, patients with cystic kidney disease, who often have been referred to a nephrologist before significant CKD based on family history, still had 52.5% rate of utilizing a catheter at dialysis initiation [124].

While AVG are considered less desirable, there are certain patients that it may be reasonable or even preferable. In the first year following dialysis initiation, AVG are associated with earlier catheter removal and less catheter time in the first year of dialysis [85]. AVG have historically had less long-term durability than AVF [71], which may result in more lifetime catheter days. However, there may be improvement in AVG life with modern interventional techniques, as AVG have increasingly

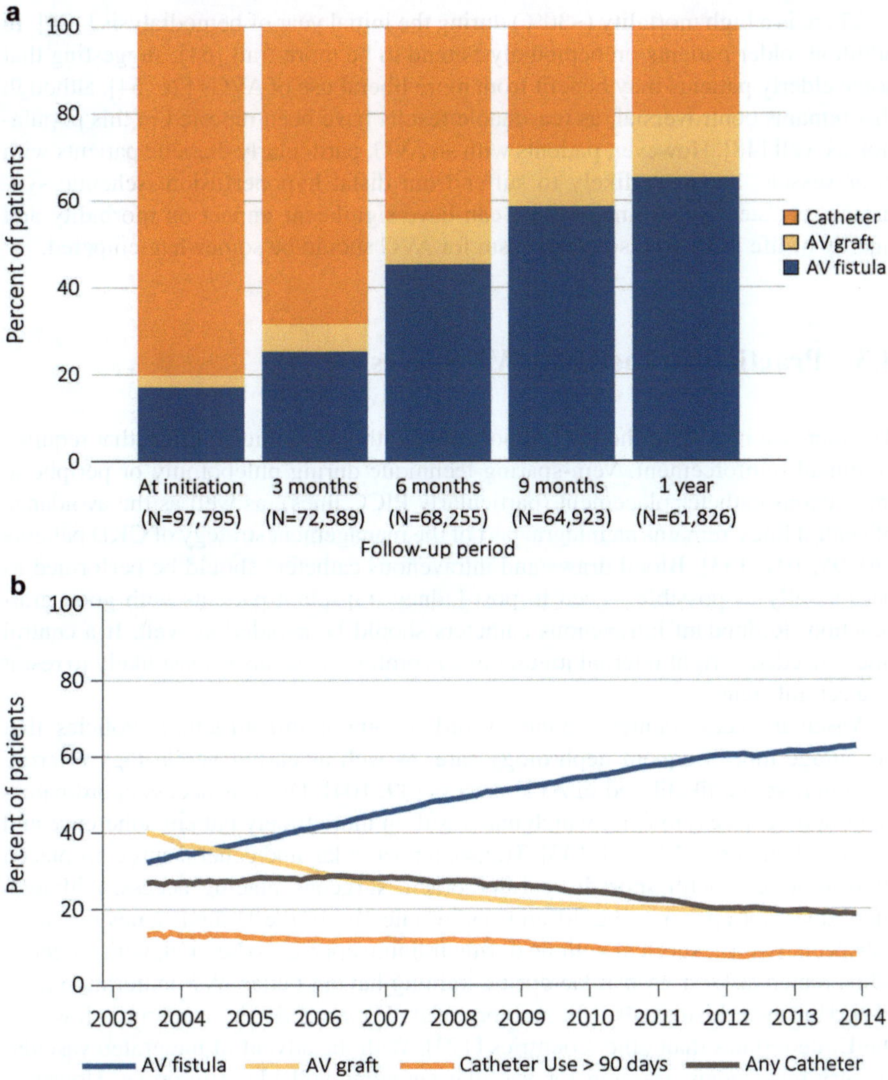

Fig. 4.1 (**a**) Vascular access use during the first year of hemodialysis by time since initiation of ESRD treatment, among patients new to hemodialysis in 2013, from the USRDS special analyses, using ESRD Medical Evidence form (CMS 2728) and CROWNWeb data spanning 2013 and 2014. (**b**) Trends in vascular access type use among ESRD prevalent patients, 2003–2014, from the USRDS special analyses, using USRDS ESRD Database and Fistula First data, with the Fistula First data reported from July 2003 through April 2012 and CROWNWeb data reported from June 2012 through December 2013 (From Saran et al. [124], used with permission)

reasonable secondary patency rates [4]. Femorofemoral AVGs continue to be utilized when traditional options are exhausted but have poorer long-term patency and infectious outcomes [38, 44]. In select candidates, lower extremity AVF using the saphenous vein, or even femoral vein, may be a viable alternative [107].

There is a high mortality (~30%) during the initial year of hemodialysis [124]. In addition, older patients on hemodialysis tend to be more frail [64], suggesting that some elderly patients may benefit from more liberal use of AVGs [26, 54], although this remains controversial, as reasonable results have been reported in this population as well [48]. However, patients with an AVG, particularly diabetic patients with poor vessels, are more likely to suffer from distal hypoperfusion ischemic syndrome, or "steal syndrome," which can have significant impact on morbidity and quality of life [87, 101], so enthusiasm for AVG should be somewhat tempered.

4.3 Practices to Increase AVF Success

The increase in AVF in the US has not come without significant effort that requires continual reinforcement. Vein-sparing technique during phlebotomy or peripheral intravenous catheter placement (particularly PICC lines), as well as the avoidance of central lines, remains an integral part of the management strategy of CKD patients [70, 95, 104, 133]. Blood draws and intravenous catheters should be performed as peripherally as possible—even in post-kidney transplant patients with good graft function. Redundant intravenous catheters should be avoided as well. If a central line is needed, a right internal jugular line is preferred, as this is least likely to result in a central stenosis.

Vascular access centers cannot afford to ignore infrastructure. Policies that encourage more frequent nephrology care, as well as earlier nephrology referral, can improve the likelihood of AVF success [39, 104]. Dialysis access coordinators can improve access to care, which may result in more timely patient adherence and improved outcome [71, 113, 133]. Transplant, vascular, and general surgeons placed dialysis access, with sporadic cardiothoracic surgeons placing access. Although there does not appear to be differences in outcome in the United States based on subspecialty [40], surgical volume during training appears to be vital, with surgeons who created at least 25 fistulae during training having better AVF maturation rates [49]. However, historically, the number of AVF placed during training is lower in the United States than other countries [123]. With the advent of integrated vascular surgery residencies, training for vascular surgeons will likely improve. However, during this time, there has been a significant drop in vascular case numbers for general surgery residents [80], suggesting that the landscape for access providers is rapidly changing. As transplant surgery rotations provide vascular access experience to a set of residents who would otherwise miss the training, institutions need to remain focused on maintaining a training pipeline of both future nephrology and surgery providers [9, 49, 123].

The preoperative ultrasound assessment of the arm and forearm arteries and veins, which has traditionally been called "vein mapping," has long been recommended by the KDOQI guidelines [104], and has even been shown to improve AVF survival in a randomized trial [41]. In addition to age and diabetes [88, 135], arterial calcifications in the radial artery has been associated with worse radiocephalic AVF outcomes [45], likely due to the failure of arterial flow augmentation. As would be expected, smaller arteries and veins (particularly veins 2–3 mm in size) are associated with AVF failure [40]. More advanced arterial testing has yet to become standard of care. However, the Hemodialysis Fistula Maturation Study has found that two measurements of the ability of the brachial artery to dilate, flow-mediated dilation and nitroglycerin-mediated dilation, positively predict 6 week AVF blood flow rate and diameter [6], suggesting that functional testing may be on the horizon.

Intraoperative technique is well established. Although there may be benefit to regional anesthesia for AVF placement in regard to vasodilation, placement, and maturation, the benefit for the AVF itself remains controversial [90, 128]. Anastomoses are typically side to end. Anastomotic size varies by vessel and surgeon [79] but is typically 6–8 mm. Shorter anastomotic size is associated with early thrombosis [40], although caution must also be exercised for overly large anastomoses as well, as these may cause steal syndrome. There is likely benefit to the relatively sharp angle of 30° between the artery and the vein [37], although angles of 45° may also be appropriate [55]. Many surgeons utilize intraoperative heparin. Although the overall benefit remains somewhat controversial, nonsignificant trends favoring weight-based dosing for those who use heparin have been observed. Options for the operative approach for basilic vein transpositions are described in the next section. Interestingly, in the prospective multicenter Hemodialysis Fistula Maturation Study, surgeon perception and intraoperative frustration, as well as intraoperative thrill, were the most consistent predictors of AVF success [40], even compared with more discrete physiological measures.

Perioperative isometric exercise, often with a squeeze ball of some sort, has been controversial over the years, with early studies having mixed results [106, 119]. However, these early studies focused on immediate change soon after exercise. More recent studies have focused on effects after more long-term use, typically of 4–6 weeks of duration. Three recent studies, two randomized to intervention and control group, and one comparing exercise and nonexercise arms found that isometric hand exercise increased (preoperative) cephalic vein size, with one of the studies further showing an increase in brachial artery diameter and flow [82, 84, 131]. Recent data suggest that these exercise-induced increases in vein size may translate to increased maturation [122].

Most typically, patients are seen for a postoperative visit within a couple of weeks of the operation, followed by a maturation assessment at ~6 weeks. The maturation visit assesses for branches requiring ligation, depth, or need for intervention. A duplex ultrasound obtained at this point can improve detection of AVF requiring intervention for maturation, thereby improving successful AVF maturation rates [61, 112]. Approximately half of all AVF will require some sort of intervention to successfully mature, such as balloon angioplasty or collateral ligation, both often fistulogram guided [52, 136]. AVF that require maturation intervention have shorter secondary patency after maturation, even if the intervention was transposition [52, 61]. However, AVF that have detectable lesions but are not intervened upon have significantly lower successful maturation [61]. AVG not only are less likely to require an intervention for maturation (17.7%) but also have a shorter secondary patency if they need such intervention [52].

4.4 Operative Approach to Deep AVF

With the focus on AVF over AVG, basilic vein transpositions (BVT) have become more common, either as a one or two stage operation. There have been a number of small studies examining these two options, but the most recent meta-analysis found that there was essentially no difference for failure rates or patency between the two methods, despite the predilection for the two-stage procedure to be used on smaller veins [24]. The advantage of the one-stage procedure is that the patient only has one planned operation. The two-stage operation allows for the potential use of a smaller basilic vein, as well as a smaller initial incision. Should the fistula fail initial maturation, the patient does not have the associated wound morbidity from the larger incision of the one-stage operation—which becomes increasingly relevant as obesity continues to increase. However, the patient requires a second, transposing (or superficializing) operation, which makes the two-stage operation more logistically complex. Occasionally, the brachial vein is used if a basilic vein is not available, although it is technically more challenging and may have additional morbidity [65].

As obesity has become more prevalent, there has been increasing need to superficialize cephalic vein AVFs, both in the arm and forearm, to get the vein to a depth accessible by the dialysis unit. This superficialization is traditionally performed by incising over the vein, mobilizing it along its course and either making a flap or dividing and tunneling the vein closer to the skin, with reasonable results [16, 97]. Lipectomy-based techniques have been developed as well (Fig 4.2), in which two smaller incisions are made and the fat overlying the vein is removed approximately 4 cm each direction, without mobilizing the AVF [14]. Using a similar two incision approach, a minimally invasive superficialization technique (MIST) can be performed, where the vein can be mobilized and tunneled [59].

4.5 Balloon-Assisted Maturation

Balloon-assisted maturation (BAM) is somewhat controversial, as oversized balloon angioplasty is utilized to induce stenosis by inducing vascular remodeling in animal models [118]. Conceptually, BAM has been applied in two distinct manners: intraoperative balloon dilation and repetitive interventional post-operative dilation, of the outflow vein leading to the central circulation. Intraoperative overwire balloon dilation, which typically occurs after hydrostatic dilation, has been utilized for small veins that are < 2–3 mm, with good short-term success [30, 134], although long-term data are lacking. Postoperative interventional BAM involves serial long-segment angioplasties, often coupled with collateral ligation or coiling [99, 121] . As for the surgical approach, the use of interventional BAM has been somewhat limited, and it remains to be shown if such an approach is biologically feasible in the broader population. In addition, the economic impact of repeated percutaneous angioplasty can be significant.

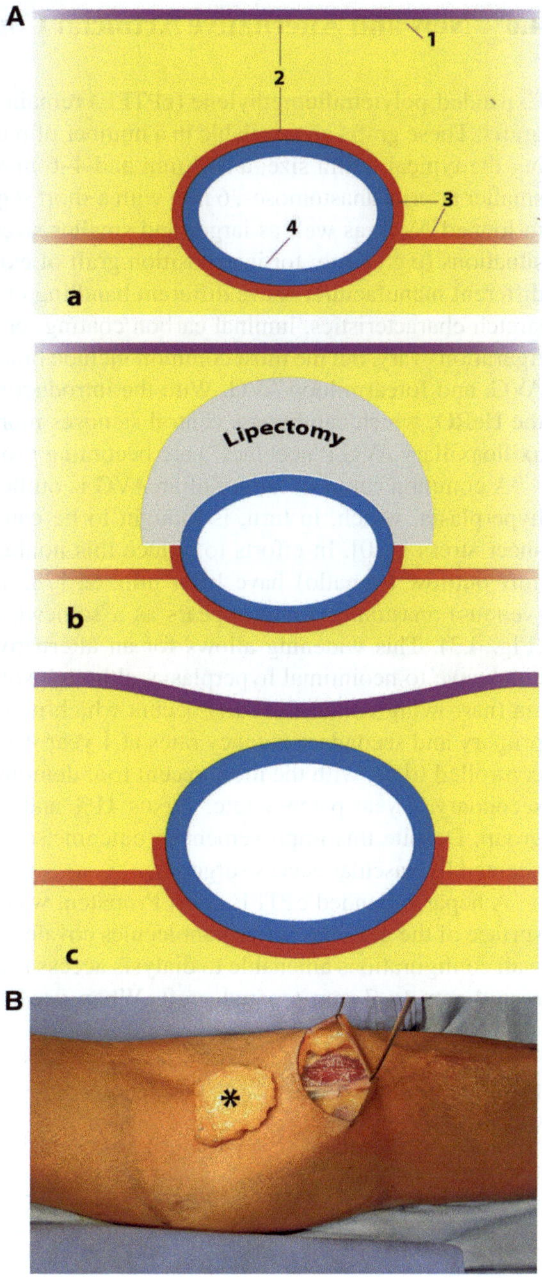

Fig. 4.2 (**A**) Lipectomy technique for superficialization. Operating diagram shows the arm (*a*) before, (*b*) during, and (*c*) after lipectomy. *1* skin, *2* fat, *3* fascia superficialis, *4* vein. (**B**) Operative view shows one incision and the freshly removed fat pad (*) (From Bourquelot et al. [14], used with permission)

4.6 New and Alternative Artificial Conduit

Expanded polytetrafluorethylene (ePTFE) remains widely used for dialysis access grafts. These grafts are available in a number of regular-wall configurations, including the typical 6 mm size, a 4–7 mm and 4–6 mm tapered configurations to allow smaller arterial anastomoses, 6 mm with a short segment of rings to prevent collapse in looped AVG, as well as larger and smaller sizes that may be useful for specific situations (e.g., 8 mm for interposition graft of existing AVF). Of note, PTFE from different manufacturers have different handling properties, as some have additional stretch characteristics, luminal carbon coating, or PTFE wrapping. Operative configurations vary, but the most common include brachioaxillary AVG, femorofemoral AVG, and forearm loop AVG. With the introduction of alternative options, such as the HeRO, which can bypass central stenoses more effectively, brachiojugular and axilloaxillary AVG ("necklace") are becoming progressively more rare.

A common cause of failure of an AVG is outflow stenosis caused by neointimal hyperplasia, which, in turn, is thought to be caused by nonlaminar flow causing sheer stress [120]. In efforts to reduce this nonlaminar flow, ePTFE grafts with a cuff outflow (Venaflo) have been utilized [78, 129]. The cuff is at the outflow (venous) anastomosis and appears as a somewhat bulbous hood for anastomosis (Fig. 4.3). This widening allows for an alternative flow pattern that may be less conducive to neointimal hyperplasia, although significant pseudointimal hyperplasia (narrowing within the cuff) occurs which may abrogate such usage [53]. Both primary and secondary patency rates at 1 year were improved in small randomized controlled trials, with the more recent trial demonstrating a 56% primary, and 91% secondary, 1 year patency rate, versus 41% and 78%, respectively, for the control group. Despite this improvement in outcome, outflow cuffs have not been widely adapted by vascular access surgeons.

A heparin-bonded ePTFE graft, Propaten, was introduced in 2006. The luminal surface of the graft has heparin molecules covalently bonded to the wall. In standard wall configurations amenable to dialysis access usage, there is a 6 mm graft, as well as a 4 mm to 7 mm tapered graft. When used for dialysis access, studies have

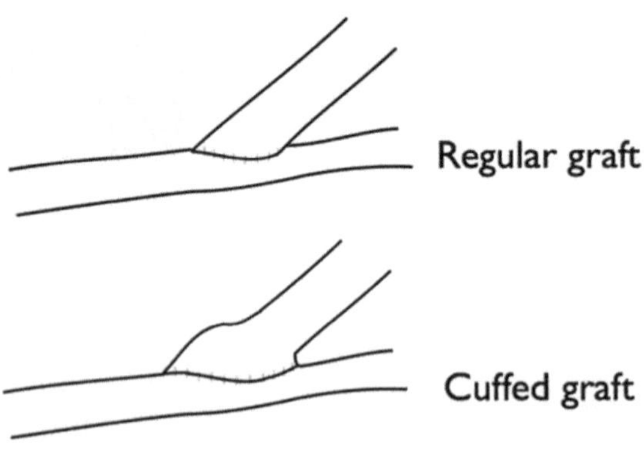

Fig. 4.3 Cuffed AVG. The cuff is on the outflow end (From Ko et al. [78], used with permission)

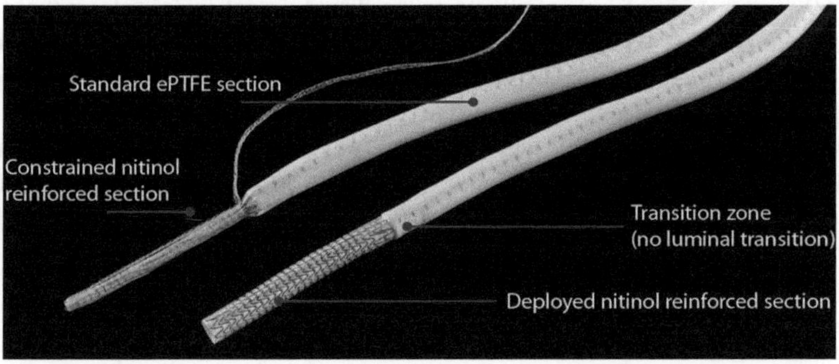

Fig. 4.4 Gore hybrid vascular graft (From Chiesa et al. [22], used with permission)

reported mixed results, with early reports of improved outcome [28], but the overall suggestion that the heparin-bonded ePTFE has similar outcomes to regular stretch ePTFE [3, 29]. In a randomized controlled study of 160 patients, there were significantly lower early thromboses in the Propaten group during the first 5 months [126]. However, although there was a trend toward improved primary and secondary patency in the heparin-bonded group, the difference was not significant, with poor primary patency in both groups and assisted 2-year-assisted primary patency of 41% vs 30% and secondary patency of 83% vs 73% for the heparin-bonded and control groups, respectively. Gore has recently released a hybrid version (Fig. 4.4), with a nitinol-reinforced outflow section that allows for "sutureless" deployment for the outflow limb [1, 22]. As the outflow configuration allows more proximal venous outflow than traditional AVG (e.g., high in the axilla), this device can be useful in selected patients [7, 66].

Early cannulation grafts have been developed, which allow much earlier cannulation, thereby abrogating the need for the central catheter, with their inherent risk of infection and development of central stenosis. Many of these grafts are thicker than typical ePTFE; so, proper patient selection in regard to artery caliber and quality is essential. These grafts are particularly useful in the setting of a prevalent dialysis patient with an AVF or AVG requiring revision, as it allows operative placement or revision without the need for concurrent catheter placement. The Flixene graft was designed to be cannulated within 3 days of placement, although earlier cannulation is reported. It has a trilaminiate composite structure, with an outer microporous layer to promote adherence, a hydrostatic middle layer, and a smooth inner surface. In a study of 46 grafts, primary-assisted patency at 1 year is 56%, with 86% secondary patency at 1 year. Despite early cannulation, infection rates were 4%, although 5 early hematomas were reported [12].

Both the Vectra and Acuseal grafts are designed to be cannulated within 24 h. The Vectra graft is a polyurethane graft. Similar to the Flixene, it has a three-layer design, with microporous inner and outer layers, and a solid middle layer. In a now somewhat older study [69], 3 year infectious rates and pseudoaneurysm formation are similar to PTFE, with significantly improved secondary patency for the Vectra graft (69% versus 57% for the control group). The Acuseal graft is also a three-layer design, with an outer layer of ePTFE, a middle elastomeric membrane layer, and an inner ePTFE layer that is heparin bonded. One year primary unassisted patency is reported as 35%, with a 1 year cumulative graft patency rate of 79% [47]. Other multilayer grafts are available in some countries, such as the Rapidax graft. The Rapidax graft appears to have a similar design to Acuseal, although published data on outcomes with it are more sparse [21, 132].

4.7 Biological Conduits

Biological conduits, including cryopreserved vessel and xenograft vessels, have been utilized for decades [57, 74]. Most frequently, these conduits are utilized when access needs to be placed, preserved, or revised in the setting of infection. Historically, there has been concern for the mechanical strength, aneurysm tendency, and cost for these conduits, although process improvements have somewhat improved upon these concerns. Cryopreserved human vein has been successfully utilized for dialysis access, with patency results are at least as good as expected for AVGs [62, 63, 93]. While there is conflicting data, cryopreserved human vein raises a concern for possible allosensitization for future transplantation [11].

Bovine vessels, including mesenteric vein and carotid artery, are also available. Bovine carotid artery, commercially available as Artegraft, is available in 6–8 mm sizes, with a variety of lengths between 15 cm and 50 cm. In a prospective, randomized trial, bovine carotid artery had similar 1 year secondary patency when compared with ePTFE but improved 1 year primary and assisted primary patency [74]. Infection and pseudoaneurysm rates were similar between the groups, with the bovine carotid artery group having fewer thromboses. Bovine mesenteric vein (ProCol) has shown similar results [72]. Bovine pericardium is also available, and is useful for arterial patching in infected fields [96]. In addition, decellularized extracellular matrix sheets from procine small intestinal submucosa (CorMatrix ECM) that were tubularized at time of surgery to size have been utilized for AVF aneurysm repair [34].

Tissue-engineered grafts are on the horizon. Reports with the Lifeline tissue-engineered vascular graft demonstrated the feasibility of clinical use [94, 138]. This graft was produced by growing monolayers of fibroblasts from a skin biopsy from the proposed recipient. This layer was tubularized and then decellularized. Subsequently, an external layer of fibroblasts was added, and the inner (luminal) surface was endothelialized with endothelial cells cultured from recipient vein (Fig. 4.5). Sixty percent patency was demonstrated at 6 months. More recently, phase-2 trial results have been published for Humacyte's bioengineered human acellular dialysis graft [83, 100]. This conduit was produced by culturing human smooth muscle cells on a polymer scaffold (Fig. 4.6). The conduit was subsequently decellularized. After implantation, there is colonization by host cells, essentially resulting in a population of cells consistent with a vessel undergoing remodeling. In

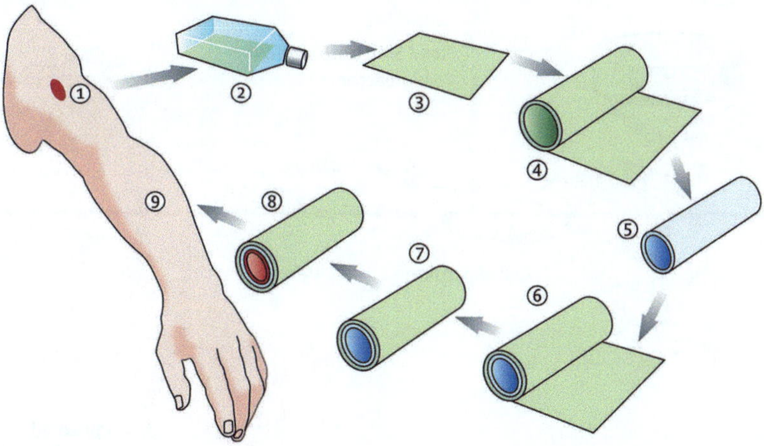

Fig. 4.5 Lifeline vascular-graft system. The key is as follows: *1* skin biopsy, *2* propagation of autologous fibroblasts (*green*) in cell culture, *3* formation of confluent monolayer from autologous fibroblasts (*green*), *4* biofabrication of vascular tube by rolling of first monolayer from autologous fibroblasts (*green*), *5* decellularization and formation of acellular tubular internal membrane (*blue*), *6* rolling of second monolayer (*green*) around acellular internal membrane (*blue*), *7* formation of vascular graft from two concentric acellular and cellular layers, *8* endothelialization (*red*), *9* implantation of endothelialized autologous tissue-engineered vascular graft for hemodialysis access (From Mironov and Kasyanov [100], used with permission)

the combined results of two phase-two studies, with 60 total patients, these conduits had 28% 1 year primary patency, 38% assisted primary patency, and 89% secondary patency, with only three infections. Although there was some aneurysmal formation, it mostly seemed to be limited to cannulation sites.

4.8 Pharmacological Interventions

There are currently a number of investigational adjuvant medical therapeutics designed to increase the patency of AV fistulas or grafts. These therapies can be local therapies delivered at the time of fistula creation or can be systemic medications administered after fistula creation.

Recombinant type-I pancreatic elastase (PRT-201) is an investigational compound which has been studied in phase-1 and phase-2 clinical trials on AV fistulae and AV grafts. The mechanism of action is thought to be related to the elastase's cleavage of amino acid sequences on elastin in adventitial fibers [109]. An investigational group has assessed the effect of PRT-201 applied directly to the adventitia of the inflow and outflow blood vessels of newly created radiocephalic or brachiocephalic AVF [58, 109]. The phase-1 trial compared 45 participants who received a variety of different doses of PRT-201 with 21 participants who received placebo [109]. The phase-2 trial compared 51 participants receiving placebo with 51 participants

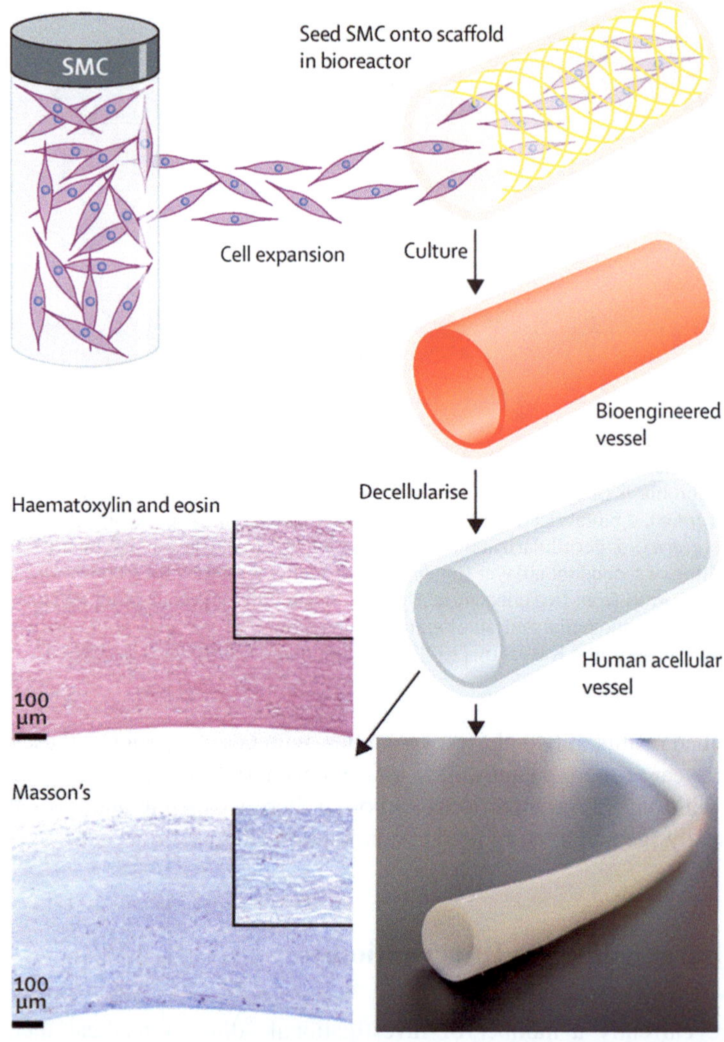

Fig. 4.6 Humacyte's bioengineered human acellular dialysis graft system. Smooth muscle cells (SMC) are seeded onto a biocompatible scaffold within a single-use bioreactor. During culture, a cellular bioengineered vessel is grown, which is then decellularized to produce the human acellular vessel. Gross appearance is an off-white uniform tubular structure, and hematoxylin and eosin stain and Masson's trichrome stain show dense extracellular matrix without cellular or nuclear remnants (From Lawson et al. [83], used with permission)

who received 10 µg, and 49 participants who received 30 µg, of PRT-201 [58]. All participants in both trials were followed up for 12 months. The phase-1 study had as its primary outcome measure safety, and the primary efficacy outcome measure was the proportion with a ≥25% increase in the diameter of the AVF outflow vein or blood flow immediately after treatment. There was no difference between groups for the primary efficacy outcome measure. The phase-2 trial's primary efficacy measure was unassisted primary patency. In that study, PRT-201 appeared safe and was associated with a 33% reduction in the hazard ratio for loss of primary patency in the 30-mg group, but this was not statistically significant [58]. A subgroup of patients undergoing placement of a radiocephalic fistula and treated with elastase did demonstrate a statistically significant increase in primary patency at 3 years [109]. A phase-III trial has begun to enroll patients. The effect of PRT-201 on AV grafts has also been investigated [35]. This study compared 28 participants who received placebo compared with 61 participants who received variable doses of PRT-201 applied directly to the vein-graft anastomosis and the adjacent outflow vein at the time of AV graft construction. Outcomes were reported after 12 months and showed a nonsignificant benefit in favor of treatment.

Another strategy to assist fistula maturation or prevent neointimal hyperplasia is the use of "wraps" or other matrices to provide sustained, local drug release. A recent example is the evaluation of a sirolimus-eluting collagen wrap (Coll-R) in humans [108]. In this study, 12 chronic hemodialysis patients underwent surgical placement of 13 PTFE grafts + Coll-R around the site of the venous and PTFE anastomosis and were followed for up to 24 months. The primary endpoint was safety, and secondary endpoints were pharmacokinetics of sirolimus release, success of Coll-R implantation, and primary unassisted graft patency. There were no technical failures, infections, vascular anastomotic, or wound-healing problems reported, but the lack of controls prevents an analysis of efficacy. A sirolimus-impregnated polymer mesh was recently tested for safety and efficacy for hemodialysis access but was terminated due to an increase in the rate of graft infection in the experimental group [23].

4.9 Systemic Therapies to Facilitate Access Patency

A number of systemic pharmacological agents have been employed in placebo-controlled trials to determine their effectiveness in facilitating access patency including ticlopidine, warfarin, dipyridamole with aspirin, and clopidogrel. Three trials compared ticlopidine with placebo, with a total number of 339 participants undergoing AV fistula formation or graft interposition [42, 50, 51]. These trials showed increased AV fistula patency, but follow-up was only 1 month in duration. Furthermore, ticlopidine has been withdrawn from the market, making the result less meaningful to the dialysis access population. The use of low-intensity warfarin demonstrated no beneficial effect on the group of participants treated with warfarin (versus placebo). This study was stopped prematurely due to an increased risk of bleeding complications [25].

The effects of dipyridamole and aspirin on vascular access have been investigated. A randomized, double-blind, placebo-controlled trial of extended-release dipyridamole (ERDA), at a dose of 200 mg, and aspirin, at a dose of 25 mg, given twice daily after the placement of a new arteriovenous graft until the primary outcome that loss of primary unassisted patency was reached, was performed [33].

The incidence of primary-unassisted patency at 1 year was 23% in the placebo group and 28% in the dipyridamole with aspirin group. Although the ability of dipyridamole with aspirin to inhibit graft failure was significant, the clinical effect of this therapy was modest—a delay of 6 weeks of loss of primary patency. A secondary analysis of the data from this trial examined the relationship of preexisting aspirin use on graft outcomes [32]. As 43% of patients who entered the trial were taking aspirin before randomization (typically 81 mg/d), the study was able to compare outcomes associated with aspirin use across and within randomized groups assigned to either ERDP/aspirin or placebo. After adjustment for unbalanced baseline covariates, there was a relative difference of 17% in favor of patients who were taking aspirin before randomization. The estimated relative risk reduction associated with prerandomization aspirin use was 27% in patients who were randomly assigned to placebo and 10% in patients who were randomly assigned to ERDP/aspirin [32].

Two trials compared the effects of clopidogrel versus placebo. The first compared 46 participants who received 75 mg clopidogrel daily with 47 who received placebo [46]. Treatment was initiated 7–10 days prior to formation of a native arteriovenous fistula and continued for 6 weeks postoperatively. The primary outcome was fistula failure at 8 weeks. In this study, 2 out of the 46 participants in the clopidogrel group compared with 10 out of 47 in the placebo group had AVF failure at 8 weeks. The second study assessed 866 participants for patency failure 6 weeks after native arteriovenous fistula formation as its primary outcome [31]. A total of 436 participants received a loading dose of 300 mg clopidogrel on postoperative day one followed by 75 mg daily; and 430 received matching placebo daily. They reported that 53 out of 436 participants in the treatment group and 84 out of 430 in the placebo arm developed fistula thrombosis at 6 weeks. Clopidogrel in this study reduced the frequency of early thrombosis of new arteriovenous fistulas but did not increase the proportion of fistulas that become suitable for dialysis, which is likely a more relevant clinical outcome.

4.10 Interventional Advancements

Over the last decade, endovascular surgery has blossomed. As would be expected, many of these innovations are useful for the maturation and maintenance of vascular access. Comprehensive surgical care for dialysis patients requires endovascular surgical skills, or at the very least direct coordinated access to an interventionalist. First, there is the obvious outcome advantage of understanding, and being able to provide, comprehensive services to this patient population—as well as their referring nephrologists. Also, in a classic example of “don’t make razors, make razor blades,” access placement is undervalued, whereas effective reimbursement for endovascular access interventions is double that for open primary placements and revisions [5, 92].

For some time, both covered and bare metal stents have been successfully used for dialysis access, particularly for central and other outflow stenoses [8, 17, 116], with pseudoaneurysms of established vascular access particularly well treated by covered stents [125]. However, stents can migrate or fracture, as well as limit future surgical options by preventing usage of the involved vein segment, so long-term view must be entertained before deployment. Often plain angioplasty is performed initially, with avoidance of stenting, until a stenotic lesion is deemed resistant and surgical jump grafting options have been considered. Newer options for angioplasty that avoid stenting, may impact this algorithm, as both cutting balloons and drug eluting (paclitaxel coated) balloons have shown improved outcomes in the treatment of venous (outflow) stenoses [2, 67, 75, 76].

4.11 Distal Hypoperfusion Ischemic Syndrome

The principles of distal hypoperfusion ischemic syndrome, or "steal syndrome," have remained constant. Workup, including measurement of AVF flow rates (i.e., determination of high versus low AVF flow-related distal ischemia) and assessment for anatomically corrected lesions (e.g., upstream arterial lesions) or operative targets remains central to adequate evaluation. Often, patients have associated coronary or peripheral arterial disease [112], which sometimes limits surgical options. High-flow AVF may be amenable to banding. An interventional approach to banding has been developed. The minimally invasive limited ligation endoluminal-assisted revision (MILLER) banding procedure has become more common. The AVF is percutaneously accessed downstream from the anastomosis. Near the anastomosis, the vessel is isolated, and the vessel is banded over a balloon. This approach allows more precise narrowing while also allowing concurrent fistulograms with arterial runoff [98]. For other patients, distal ligation with interval ligation (DRIL) remains an option [13, 77], but proximalization of the arterial inflow (PAI) has become increasingly common (Fig. 4.7). For the PAI procedure, a jump graft is performed from a more proximal location on the brachial artery to the outflow vein close to the anastomosis with a 4 or 5 mm ePTFE graft in an end-to-end fashion, resulting in inflow from a larger vessel and a longer segment of higher resistance conduit. Approximately 85% of patients have resolution of symptoms, with most of the rest having symptom improvement [139].

4.12 Novel Devices

There has been a recent surge in interest in device development for dialysis access. The most commonly recognized device is the Hemodialysis Reliable Outflow (HeRO) vascular access device (Fig. 4.8), which has been on the market for about a

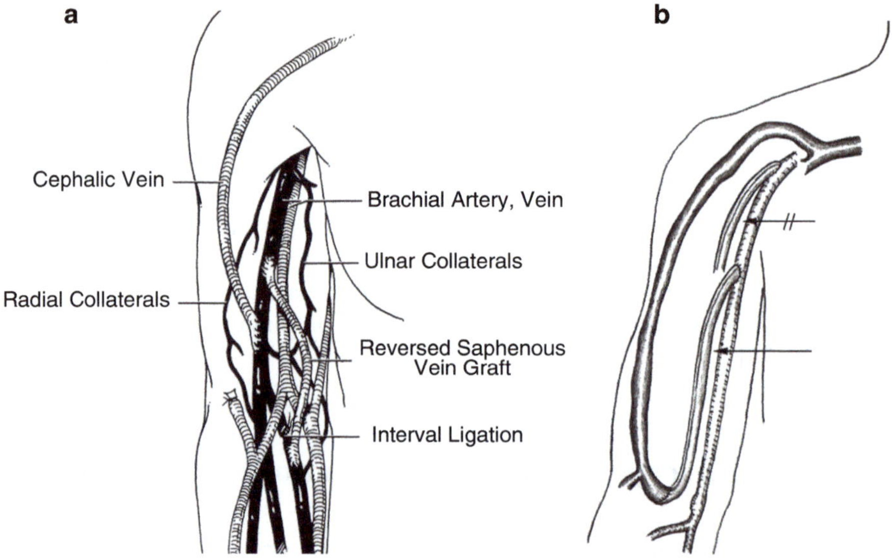

Fig. 4.7 (**a**) DRIL procedure using reversed saphenous vein graft (From Berman et al. [13], used with permission). (**b**) PAI procedure using distal (←) or proximal axillary artery (←//—) (From Zanow et al. [139], used with permission)

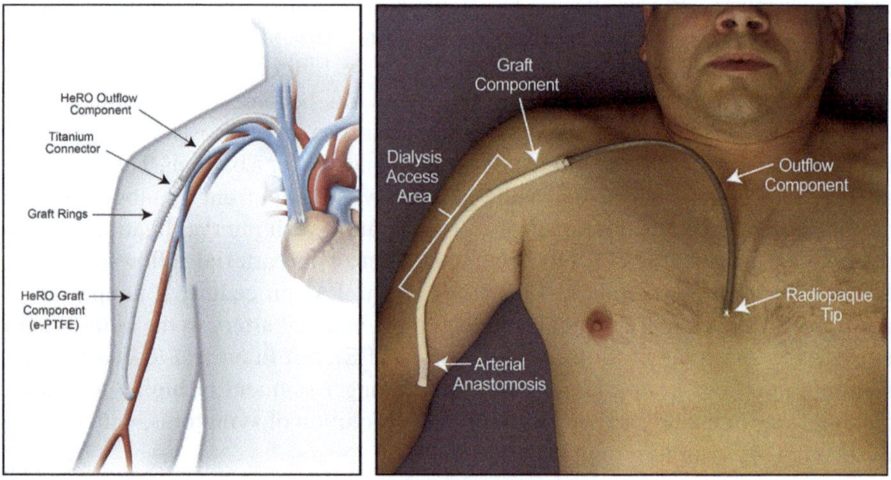

Fig. 4.8 HeRO device (From Katzman et al. [73], used with permission)

decade now. The HeRO has two sections, an AVG component and the outflow (central) extension [73]. The HeRO is designed to bypass severe central stenoses. As its outflow is 20 Fr sized, it requires significant inflow and is not a good option for patients at high risk for steal syndrome. The central portion is placed under fluoroscopy. Stenotic balloon dilation or even caval recanalization is sometimes required to pass this portion of the conduit past the lesion, to the early right atrium. An arterial anastomosis is performed to the AVG portion, with the venous limb of this component connected to the central component by a metal connector, typically at the deltopectoral groove for HeRO placed in the upper extremity. Infectious complications are typically similar to an ePTFE AVG, with similar need for interventions [73, 81, 102, 130].

The Venous Window Needle Guide (VWING) device provides an alternative to superficialization of deep AVF (Fig. 4.9). The VWING is conceptually like a pole vault box; it is a metal guide box that is implanted in one or two locations on fistulae that are too deep. The VWING is effective for a depth range of 6 mm to 15 mm. The device is designed to be palpable through the skin and guides the needle to the designated cannulation sites. In two combined clinical trials, the device maintained continued ability to access the designated sites for 65% of enrolled patients [43]. Only 1 of the 54 patients required removal of the VWING for infection, with another 7 patients having VWINGs removed for cannulation difficulties.

An AVG is under development that is designed to prevent back wall injury to the access. The Bullet Proof vascular access graft is not currently commercially available but under development [60]. This device is an AVG with two cannulation zones (Fig. 4.10). The cannulation zones have a solid back wall that prevents further

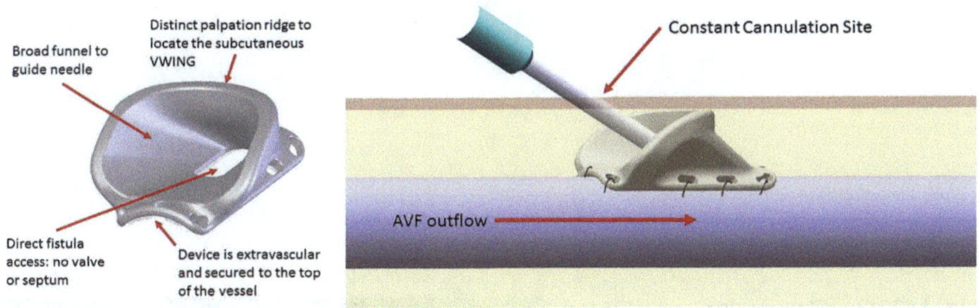

Fig. 4.9 VWING device (From the manufacturer, Vital Access, and Galt et al. [43], used with permission)

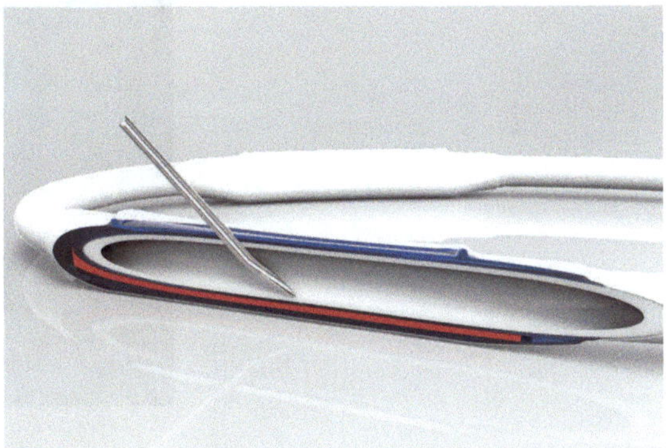

Fig. 4.10 Bullet Proof AVG. The *red area* represents the solid backwall (From the manufacturer, InnAVasc Inc., used with permission)

penetration by the needle. The cannulation zone is larger than the rest of the AVG, allowing needle placement by palpation.

Two investigations devices are in clinical trials that relate to open AVF anastomosis creation. The VasQ is an external support device that slides over the outflow vein, which is then brought over the anastomosis on completion, with the goal of minimizing flow disturbances at the anastomosis (Fig. 4.11). In an uncontrolled study of 20 patients designed to evaluate safety, primary AVF rates at 6 months were 79% with no serious device-related complications [20]. The Optiflow device is an internal insert designed to minimize the need for anastomic suturing (Fig. 4.12). The device has a flange that is deployed in the artery, with a tube that the vein is placed over. The vein is tied in place with an encircling silk suture. In a small trial of patients undergoing brachiocephalic AVF, outcomes were similar to traditional operative placement in matched controls [103].

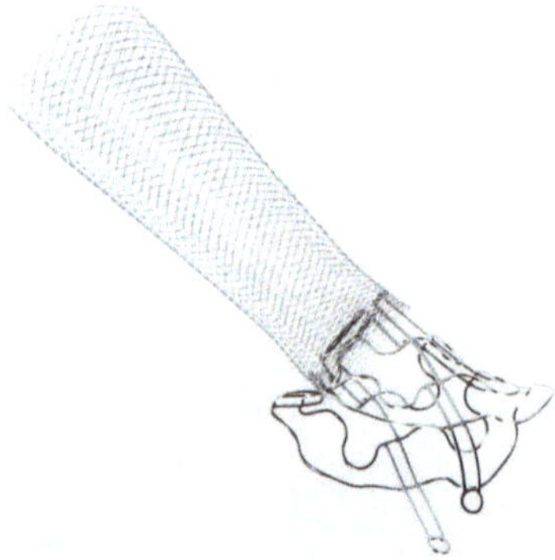

Fig. 4.11 VasQ device (From the manufacturer, Laminate Medical Technologies Ltd., used with permission)

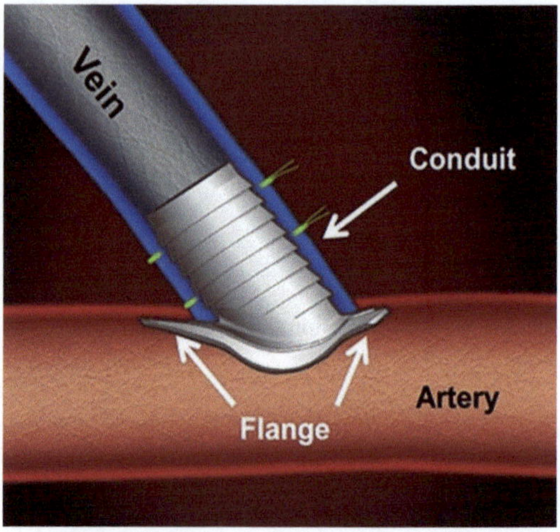

Fig. 4.12 OptiFlow device (From Nikam et al. [103], used with permission)

There are three investigational devices undergoing clinical trials for endovascular AVF anastomosis creation, although none are currently released for general use in the United States. The Ellipsys Vascular Access System, which is considered an investigational device in the United States with the clinical trial "Percutaneous Less Invasive AV Fistula for Vascular Access in ESRD," utilizes a single catheter that is advanced overwire through an appropriate perforating vein into the proximal radial artery. The catheter device is deployed, which sandwiches the vessel wall surfaces of the artery and vein. Utilizing electrocautery, the device creases an elliptical anastomosis. Presumably, collaterals are either coiled or ligated. While results from the larger trial are pending, preliminary international results show 7 of 10 patients obtaining flows > 400 mL/min at 6 weeks or utilizing the AVF for dialysis [10, 55, 56].

The everlinQ system utilizes two catheters with embedded magnets and specifically aligned components to allow radiofrequency cutting between the two catheters [114]. The arterial catheter is advanced into the proximal ulnar artery just distal to the antecubital space, whereas the venous catheter is placed in the nearby ulnar vein (Fig. 4.13).

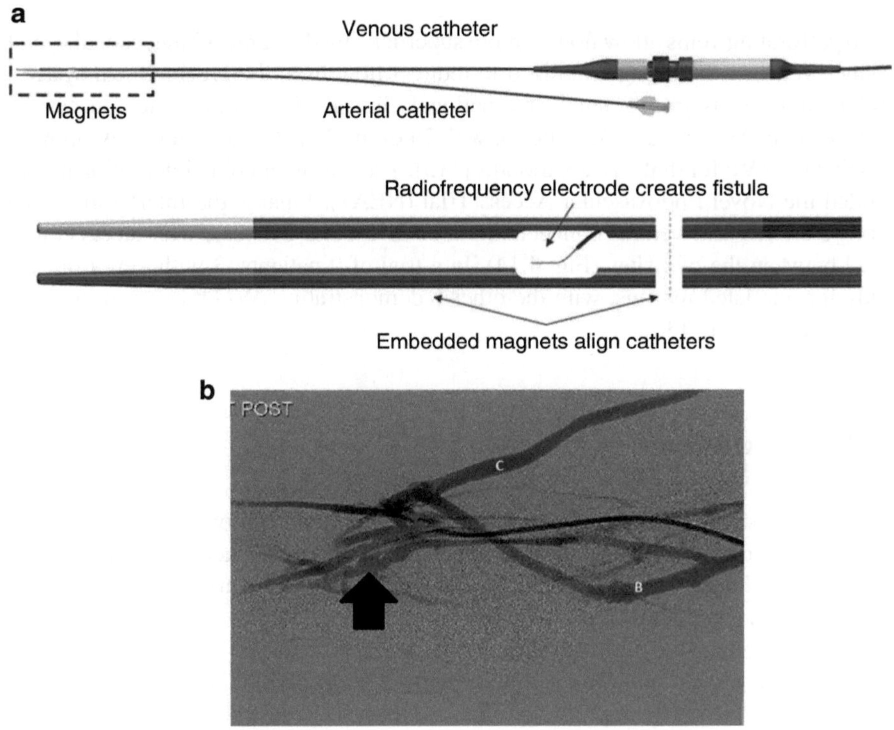

Fig. 4.13 (**a**) Schematic of everlinQ device. (**b**) AVF created using everlinQ between the ulnar vein and ulnar artery (*arrow*). Injection of contrast from the brachial artery demonstrates flow through the fistula filling the cephalic vein (*C*) and the basilic vein (*B*) (From Rajan [114], used with permission)

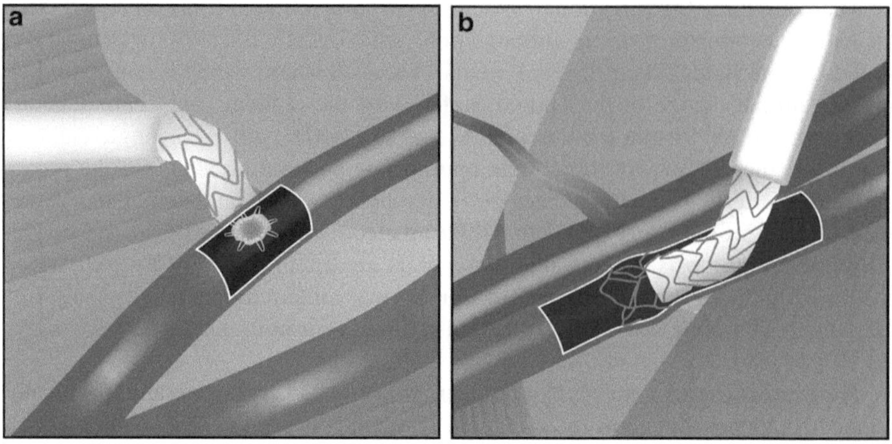

Fig. 4.14 InterGraft device. (**a**) Deployed arterial limb. (**b**) Deployed venous limb (From the manufacturer, Praxis Inc., used with permission)

Deep perforating veins allow flow to more superficial median antecubital, cephalic, and basilica veins, with coiling is utilized to redirect flow from the brachial vein system. Preliminary results from a small international clinical trial demonstrated successful creation of an AVF in 32 of 33 patients, with 24 of the 27 patients with follow-up were utilizing the AVF for dialysis at 6 months [115]. There is an ongoing international trial, entitled the Novel Endovascular Access Trial (NEAT). Finally, the InterGraft device deploys arterial and venous connectors via endovascular delivery, with an AVG tunneled between the two sites (Fig. 4.14). In a trial of 9 patients, 3 withdrew from the study for unrelated reasons, with the other 6 demonstrating AVG patency or assisted patency at 6 months [36].

4.13 Conclusion

Dialysis access remains the lifeline of the ESRD patient. Endovascular techniques have increased the ability to mature and maintain vascular access. A number of conduits and devices, as well as pharmacological agents, that are recently available or on the horizon show promise to significantly impact the field. As with transplant itself, a comprehensive multidisciplinary approach is required, not only to optimize outcomes, but also to maintain a comprehensive program of CKD and ESRD care, which includes not only nephrology care and dialysis access surgical care, but also kidney transplant for appropriate patients.

References

1. Afaneh, C., Aronova, A., Ross, J. R., & Leeser, D. B. (2012). Use of hybrid vascular grafts in failing access for hemodialysis: Report of two cases. *The Journal of Vascular Access, 13*, 513–515.
2. Agarwal, S. K., Nadkarni, G. N., Yacoub, R., Patel, A. A., Jenkins, J. S., Collins, T. J., Annapureddy, N., Kumbala, D., Bodana, S., & Benjo, A. M. (2015). Comparison of cutting balloon angioplasty and percutaneous balloon angioplasty of arteriovenous fistula stenosis: A meta-analysis and systematic review of randomized clinical trials. *Journal of Interventional Cardiology, 28*, 288–295.
3. Allemang, M. T., Schmotzer, B., Wong, V. L., Chang, A., Lakin, R. O., Woodside, K. J., Wang, J., & Kashyap, V. S. (2014a). Heparin bonding does not improve patency of polytetrafluoroethylene arteriovenous grafts. *Annals of Vascular Surgery, 28*, 28–34.
4. Allemang, M. T., Schmotzer, B., Wong, V. L., Lakin, R. O., Woodside, K. J., Schulak, J. A., Wang, J., & Kashyap, V. S. (2014b). Arteriovenous grafts have higher secondary patency in the short term compared with autologous fistulae. *American Journal of Surgery, 208*, 800–805.
5. Allon, M., Dinwiddie, L., Lacson, E., Jr., Latos, D. L., Lok, C. E., Steinman, T., & Weiner, D. E. (2011). Medicare reimbursement policies and hemodialysis vascular access outcomes: A need for change. *Journal of the American Society of Nephrology, 22*, 426–430.
6. Allon, M., Greene, T., Dember, L. M., Vita, J. A., Cheung, A. K., Hamburg, N. M., Imrey, P. B., Kaufman, J. S., Robbin, M. L., Shiu, Y. T., Terry, C. M., Umphrey, H. R., Feldman, H. I., & Hemodialysis Fistula Maturation Study, G. (2016). Association between Preoperative Vascular Function and Postoperative Arteriovenous Fistula Development. *Journal of the American Society of Nephrology*. Epublished ahead of print. 27, 3788–3795.
7. Anaya-Ayala, J. E., Davies, M. G., El-Sayed, H. F., Peden, E. K., & Naoum, J. J. (2015). Early experience with a novel hybrid vascular graft for hemodialysis access creation in patients with disadvantaged anatomy. *Journal of Endovascular Therapy, 22*, 778–785.
8. Anaya-Ayala, J. E., Smolock, C. J., Colvard, B. D., Naoum, J. J., Bismuth, J., Lumsden, A. B., Davies, M. G., & Peden, E. K. (2011). Efficacy of covered stent placement for central venous occlusive disease in hemodialysis patients. *Journal of Vascular Surgery, 54*, 754–759.
9. Asif, A., & Beathard, G. A. (2015). We need to train the trainers. *Clinical Journal of the American Society of Nephrology, 10*, 1711–1713.
10. Avenu. 2016. *Ellipsys vascular access system* [Online]. Avenu Medical, Inc. Available: avenumedical.com/ellipsys/. Accessed 6 July 2016.
11. Benedetto, B., Lipkowitz, G., Madden, R., Kurbanov, A., Hull, D., Miller, M., & Bow, L. (2001). Use of cryopreserved cadaveric vein allograft for hemodialysis access precludes kidney transplantation because of allosensitization. *Journal of Vascular Surgery, 34*, 139–142.
12. Berard, X., Ottaviani, N., Brizzi, V., Deglise, S., De Precigout, V., Ducasse, E., Combe, C., & Midy, D. (2015). Use of the Flixene vascular access graft as an early cannulation solution. *Journal of Vascular Surgery, 62*, 128–134.
13. Berman, S. S., Gentile, A. T., Glickman, M. H., Mills, J. L., Hurwitz, R. L., Westerband, A., Marek, J. M., Hunter, G. C., Mcenroe, C. S., Fogle, M. A., & Stokes, G. K. (1997). Distal revascularization-interval ligation for limb salvage and maintenance of dialysis access in ischemic steal syndrome. *Journal of Vascular Surgery, 26*, 393–402.
14. Bourquelot, P., Tawakol, J. B., Gaudric, J., Natario, A., Franco, G., Turmel-Rodrigues, L., Van Laere, O., & Raynaud, A. (2009). Lipectomy as a new approach to secondary procedure superficialization of direct autogenous forearm radial-cephalic arteriovenous accesses for hemodialysis. *Journal of Vascular Surgery, 50*, 369–374.
15. Bradbury, B. D., Fissell, R. B., Albert, J. M., Anthony, M. S., Critchlow, C. W., Pisoni, R. L., Port, F. K., & Gillespie, B. W. (2007). Predictors of early mortality among incident US hemodialysis patients in the Dialysis Outcomes and Practice Patterns Study (DOPPS). *Clinical Journal of the American Society of Nephrology, 2*, 89–99.

16. Bronder, C. M., Cull, D. L., Kuper, S. G., Carsten, C. G., Kalbaugh, C. A., Cass, A., Watkins, T., & Taylor, S. M. (2008). Fistula elevation procedure: Experience with 295 consecutive cases during a 7-year period. *Journal of the American College of Surgeons, 206*, 1076–1081.
17. Carmona, J., Rits, Y., Jones, B., Dowers, L., Bednarski, D., & Rubin, J. R. (2016). Patency of the Viabahn stent graft for the treatment of outflow stenosis in hemodialysis grafts. *American Journal of Surgery, 211*, 551–554.
18. Casey, J. R., Hanson, C. S., Winkelmayer, W. C., Craig, J. C., Palmer, S., Strippoli, G. F., & Tong, A. (2014). Patients' perspectives on hemodialysis vascular access: A systematic review of qualitative studies. *American Journal of Kidney Diseases, 64*, 937–953.
19. Chan, M. R., Oza-Gajera, B., Chapla, K., Djamali, A. X., Muth, B. L., Turk, J., Wakeen, M., Yevzlin, A. S., & Astor, B. C. (2014). Initial vascular access type in patients with a failed renal transplant. *Clinical Journal of the American Society of Nephrology, 9*, 1225–1231.
20. Chemla, E., Velazquez, C. C., D'abate, F., Ramachandran, V., & Maytham, G. (2016). Arteriovenous fistula construction with the VasQ external support device: A pilot study. *The Journal of Vascular Access, 17*, 243–248.
21. Chemla, E. S., Nelson, S., & Morsy, M. (2011). Early cannulation grafts in straight axillo-axillary angioaccesses avoid central catheter insertions. *Seminars in Dialysis, 24*, 456–459.
22. Chiesa, R., Kahlberg, A., Mascia, D., Tshomba, Y., Civilini, E., & Melissano, G. (2014). Use of a novel hybrid vascular graft for sutureless revascularization of the renal arteries during open thoracoabdominal aortic aneurysm repair. *Journal of Vascular Surgery, 60*, 622–630.
23. Clinicaltrials.gov. 2016. *Safety study of the vascular wrap paclitaxel-eluting mesh for hemodialysis vascular access* [Online]. ClinicalTrials.gov. Available: clinicaltrials.gov/ct2/show/NCT01033357. Accessed 8 July 2016.
24. Cooper, J., Power, A. H., Derose, G., Forbes, T. L., & Dubois, L. (2015). Similar failure and patency rates when comparing one- and two-stage basilic vein transposition. *Journal of Vascular Surgery, 61*, 809–816.
25. Crowther, M. A., Clase, C. M., Margetts, P. J., Julian, J., Lambert, K., Sneath, D., Nagai, R., Wilson, S., & Ingram, A. J. (2002). Low-intensity warfarin is ineffective for the prevention of PTFE graft failure in patients on hemodialysis: A randomized controlled trial. *Journal of the American Society of Nephrology, 13*, 2331–2337.
26. Cui, J., Steele, D., Wenger, J., Kawai, T., Liu, F., Elias, N., Watkins, M. T., & Irani, Z. (2016). Hemodialysis arteriovenous fistula as first option not necessary in elderly patients. *Journal of Vascular Surgery, 63*, 1326–1332.
27. Dalrymple, L. S., Mu, Y., Nguyen, D. V., Romano, P. S., Chertow, G. M., Grimes, B., Kaysen, G. A., & Johansen, K. L. (2015). Risk Factors for Infection-Related Hospitalization in In-Center Hemodialysis. *Clinical Journal of the American Society of Nephrology, 10*, 2170–2180.
28. Davidson, I., Hackerman, C., Kapadia, A., & Minhajuddib, A. (2009). Heparin bonded hemodialysis e-PTFE grafts result in 20% clot free survival benefit. *The Journal of Vascular Access, 10*, 153–156.
29. Davies, M. G., Anaya-Ayala, J. E., & El-Sayed, H. F. (2016). Equivalent outcomes with standard and heparin-bonded expanded polytetrafluoroethylene grafts used as conduits for hemodialysis access. *Journal of Vascular Surgery, 64*, 715–718.
30. De Marco Garcia, L. P., Davila-Santini, L. R., Feng, Q., Calderin, J., Krishnasastry, K. V., & Panetta, T. F. (2010). Primary balloon angioplasty plus balloon angioplasty maturation to upgrade small-caliber veins (<3 mm) for arteriovenous fistulas. *Journal of Vascular Surgery, 52*, 139–144.
31. Dember, L. M., Beck, G. J., Allon, M., et al. (2008). Effect of clopidogrel on early failure of arteriovenous fistulas for hemodialysis: A randomized controlled trial. *JAMA, 299*, 2164–2171.
32. Dixon, B. S., Beck, G. J., Dember, L. M., Vazquez, M. A., Greenberg, A., Delmez, J. A., Allon, M., Himmelfarb, J., Hu, B., Greene, T., Radeva, M. K., Davidson, I. J., Ikizler, T. A., Braden, G. L., Lawson, J. H., Cotton, J. R., Kusek, J. W., Feldman, H. I., & Group, F. T. D. A.

C. S. (2011). Use of aspirin associates with longer primary patency of hemodialysis grafts. *Journal of the American Society of Nephrology, 22*, 773–781.
33. Dixon, B. S., Beck, G. J., Vazquez, M. A., Greenberg, A., Delmez, J. A., Allon, M., Dember, L. M., Himmelfarb, J., Gassman, J. J., Greene, T., Radeva, M. K., Davidson, I. J., Ikizler, T. A., Braden, G. L., Fenves, A. Z., Kaufman, J. S., Cotton, J. R. J., Martin, K. J., Mcneil, J. W., Rahman, A., Lawson, J. H., Whiting, J. F., Hu, B., Meyers, C. M., Kusek, J. W., & Feldman, H. I. (2009). Effect of dipyridamole plus aspirin on hemodialysis graft patency. *The New England Journal of Medicine, 360*, 2191–2201.
34. Dubose, J. J., Fortuna, G. R., Charlton-Ouw, K. M., Saqib, N., Miller, C. C., 3rd, Estrera, A. L., Safi, H. J., & Azizzadeh, A. (2016). Utility of a tubularized extracellular matrix as an alternative conduit for arteriovenous fistula aneurysm repair. *Journal of Vascular Surgery, 63*, 446–452.
35. Dwivedi, A. J., Roy-Chaudhury, P., Peden, E. K., Browne, B. J., Ladenheim, E. D., Scavo, V. A., Gustafson, P. N., Wong, M. D., Magill, M., Lindow, F., Blair, A. T., Jaff, M. R., Franano, F. N., & Burke, S. K. (2014). Application of human type I pancreatic elastase (PRT-201) to the venous anastomosis of arteriovenous grafts in patients with chronic kidney disease. *The Journal of Vascular Access, 15*, 376–384.
36. Ebner, A., Ross, J. R., Setum, C. M., Kallok, M. J., & Yevzlin, A. S. (2016). Transcatheter anastomosis connector system for vascular access graft placement: Results from a first-in-human pilot study. *The Journal of Vascular Access, 17*, 111–117.
37. Ene-Iordache, B., Cattaneo, L., Dubini, G., & Remuzzi, A. (2013). Effect of anastomosis angle on the localization of disturbed flow in 'side-to-end' fistulae for haemodialysis access. *Nephrology, Dialysis, Transplantation, 28*, 997–1005.
38. Englesbe, M. J., Al-Holou, W. N., Moyer, A. T., Robbins, J., Pelletier, S. J., Magee, J., Sung, R. S., Campbell, D., & Punch, J. D. (2006). Single center review of femoral arteriovenous grafts for hemodialysis. *World Journal of Surgery, 30*, 171–175.
39. Erickson, K. F., Mell, M., Winkelmayer, W. C., Chertow, G. M., & Bhattacharya, J. (2015). Provider visits and early vascular access placement in maintenance hemodialysis. *Journal of the American Society of Nephrology, 26*, 1990–1997.
40. Farber, A., Imrey, P. B., Huber, T. S., Kaufman, J. M., Kraiss, L. W., Larive, B., LI, L., Feldman, H. I., & Group, H. F. M. S. (2016). Multiple preoperative and intraoperative factors predict early fistula thrombosis in the Hemodialysis Fistula Maturation Study. *Journal of Vascular Surgery, 63*, 163–170.
41. Ferring, M., Claridge, M., Smith, S. A., & Wilmink, T. (2010). Routine preoperative vascular ultrasound improves patency and use of arteriovenous fistulas for hemodialysis: A randomized trial. *Clinical Journal of the American Society of Nephrology, 5*, 2236–2244.
42. Fiskerstrand, C. E., Thompson, I. W., Burnet, M. E., Williams, P., & Anderton, J. L. (1985). Double-blind randomized trial of the effect of ticlopidine in arteriovenous fistulas for hemodialysis. *Artificial Organs, 9*, 61–63.
43. Galt, S., Crawford, M., Blebea, J., Ladenheim, E., & Browne, B. (2016). The efficacy and durability of the Venous Window Needle Guide implanted on uncannulatable arteriovenous fistulas. *Journal of Vascular Surgery, 64*, 708–714.
44. Geenen, I. L., Nyilas, L., Stephen, M. S., Makeham, V., White, G. H., & Verran, D. J. (2010). Prosthetic lower extremity hemodialysis access grafts have satisfactory patency despite a high incidence of infection. *Journal of Vascular Surgery, 52*, 1546–1550.
45. Georgiadis, G. S., Georgakarakos, E. I., Antoniou, G. A., Panagoutsos, S., Argyriou, C., Mourvati, E., Passadakis, P., & Lazarides, M. K. (2014). Correlation of pre-existing radial artery macrocalcifications with late patency of primary radiocephalic fistulas in diabetic hemodialysis patients. *Journal of Vascular Surgery, 60*, 462–470.
46. Ghorbani, A., Aalamshah, M., Shahbazian, H., Ehsanpour, A., & Aref, A. (2009). Randomized controlled trial of clopidogrel to prevent primary arteriovenous fistula failure in hemodialysis patients. *Indian Journal of Nephrology, 19*, 57–61.

47. Glickman, M. H., Burgess, J., Cull, D., Roy-Chaudhury, P., & Schanzer, H. (2015). Prospective multicenter study with a 1-year analysis of a new vascular graft used for early cannulation in patients undergoing hemodialysis. *Journal of Vascular Surgery, 62*, 434–441.
48. Goh, M. A., Ali, J. M., Iype, S., & Pettigrew, G. J. (2016). Outcomes of primary arteriovenous fistulas in patients older than 70 years. *Journal of Vascular Surgery, 63*, 1333–1340.
49. Goodkin, D. A., Pisoni, R. L., Locatelli, F., Port, F. K., & Saran, R. (2010). Hemodialysis vascular access training and practices are key to improved access outcomes. *American Journal of Kidney Diseases, 56*, 1032–1042.
50. Gröntoft, K. C., Larsson, R., Mulec, H., Weiss, L. G., & Dickinson, J. P. (1998). Effects of ticlopidine in AV-fistula surgery in uremia. *Scandinavian Journal of Urology and Nephrology, 32*, 276–283.
51. Gröntoft, K. C., Mulec, H., Gutierrez, A., & Olander, R. (1985). Thromboprophylactic effect of ticlopidine in arteriovenous fistulas for haemodialysis. *Scandinavian Journal of Urology and Nephrology, 19*, 55–57.
52. Harms, J. C., Rangarajan, S., Young, C. J., Barker-Finkel, J., & Allon, M. (2016). Outcomes of arteriovenous fistulas and grafts with or without intervention before successful use. *Journal of Vascular Surgery, 64*, 155–162.
53. Heise, M., Husmann, I., Gruneberg, A. K., Knobel, A., Kirschner, P., & Heidenhain, C. (2011). Comparison of straight and Venaflo-type cuffed arteriovenous ePTFE grafts in an animal study. *Journal of Vascular Surgery, 53*, 1661–1667.
54. Hiremath, S., Knoll, G., & Weinstein, M. C. (2011). Should the arteriovenous fistula be created before starting dialysis?: A decision analytic approach. *PLoS One, 6*, e28453.
55. Hull, J. E., Balakin, B. V., Kellerman, B. M., & Wrolstad, D. K. (2013). Computational fluid dynamic evaluation of the side-to-side anastomosis for arteriovenous fistula. *Journal of Vascular Surgery, 58*, 187–193.
56. Hull, J. E., Velez, J. H., & Martinez, J. P. (2014). Percutaneous proximal radial artery arteriovenous fistula creation for hemodialysis using the Ellipsys vascular access system. *Journal of Vascular and Interventional Radiology, 25*, S20.
57. Hunter, G. C., Woodside, K. J., & Naoum, J. J. (2009). Healing characteristics and complications of prosthetic and biological vascular grafts. In J. W. J. Hallett, J. L. Mills, J. J. Earnshaw, J. A. Reekers, & T. W. Rooke (Eds.), *Comprehensive vascular and endovascular surgery* (2nd ed.). Philadelphia: Elsevier Mosby.
58. Hye, R. J., Peden, E. K., O'connor, T. P., Browne, B. J., Dixon, B. S., Schanzer, A. S., Jensik, S. C., Dember, L. M., Jaff, M. R., & Burke, S. K. (2014). Human type I pancreatic elastase treatment of arteriovenous fistulas in patients with chronic kidney disease. *Journal of Vascular Surgery, 60*, 454–461.
59. Inkollu, S., Wellen, J., Beller, Z., Zhang, T., Vachharajani, N., & Shenoy, S. (2016). Successful use of minimal incision superficialization technique for arteriovenous fistula maturation. *Journal of Vascular Surgery, 63*, 1018–1025.
60. Innavasc. 2016. *Bullet proof vascular access graft* [Online]. InnAVasc Medical, Inc. Available: innavasc.com/technologies. Accessed 6 July 2016.
61. Itoga, N. K., Ullery, B. W., Tran, K., Lee, G. K., Aalami, O. O., Bech, F. R., & Zhou, W. (2016). Use of a proactive duplex ultrasound protocol for hemodialysis access. *Journal of Vascular Surgery 64*, 1042–1049.
62. Jadlowiec, C. C., Lavallee, M., Mannion, E. M., & Brown, M. G. (2015). An outcomes comparison of native arteriovenous fistulae, polytetrafluorethylene grafts, and cryopreserved vein allografts. *Annals of Vascular Surgery, 29*, 1642–1647.
63. Jadlowiec, C. C., Lavallee, M., Mannion, E. M., Lally, A., Rochon, C., Sheiner, P. A., & Brown, M. G. (2016). Current outcomes and indications for cryopreserved vein allografts in hemodialysis access surgery. *The Journal of Vascular Access, 17*, 47–54.
64. Jassal, S. V., Karaboyas, A., Comment, L. A., Bieber, B. A., Morgenstern, H., Sen, A., Gillespie, B. W., De Sequera, P., Marshall, M. R., Fukuhara, S., Robinson, B. M., Pisoni, R. L., & Tentori, F. (2016). Functional dependence and mortality in the international dialysis

outcomes and practice patterns study (DOPPS). *American Journal of Kidney Diseases, 67*, 283–292.
65. Jennings, W. C., Sideman, M. J., Taubman, K. E., & Broughan, T. A. (2009). Brachial vein transposition arteriovenous fistulas for hemodialysis access. *Journal of Vascular Surgery, 50*, 1121–1125.
66. Jones, R. G., & Inston, N. G. (2015). Commentary: The hybrid vascular graft: A new frontier in arteriovenous access for dialysis or just another graft? *Journal of Endovascular Therapy, 22*, 786–788.
67. Jones, R. G., Willis, A. P., Jones, C., Mccafferty, I. J., & Riley, P. L. (2011). Long-term results of stent-graft placement to treat central venous stenosis and occlusion in hemodialysis patients with arteriovenous fistulas. *Journal of Vascular and Interventional Radiology, 22*, 1240–1245.
68. Kadatz, M. J., Lee, E. S., & Levin, A. (2016). Predicting progression in CKD: Perspectives and precautions. *American Journal of Kidney Diseases, 67*, 779–786.
69. Kakkos, S. K., Topalidis, D., Haddad, R., Haddad, G. K., & Shepard, A. D. (2011). Long-term complication and patency rates of Vectra and IMPRA Carboflo vascular access grafts with aggressive monitoring, surveillance and endovascular management. *Vascular, 19*, 21–28.
70. Kalloo, S., & Wish, J. B. (2016). Nephrologists versus peripherally inserted central catheters: Are the PICCs winning? *Clinical Journal of the American Society of Nephrology 11*, 1333–1334.
71. Kalman, P. G., Pope, M., Bhola, C., Richardson, R., & Sniderman, K. W. (1999). A practical approach to vascular access for hemodialysis and predictors of success. *Journal of Vascular Surgery, 30*, 727–733.
72. Katzman, H. E., Glickman, M. H., Schild, A. F., Fujitani, R. M., & Lawson, J. H. (2005). Multicenter evaluation of the bovine mesenteric vein bioprostheses for hemodialysis access in patients with an earlier failed prosthetic graft. *Journal of the American College of Surgeons, 201*, 223–230.
73. Katzman, H. E., Mclafferty, R. B., Ross, J. R., Glickman, M. H., Peden, E. K., & Lawson, J. H. (2009). Initial experience and outcome of a new hemodialysis access device for catheter-dependent patients. *Journal of Vascular Surgery, 50*, 600–607.
74. Kennealey, P. T., Elias, N., Hertl, M., Ko, D. S., Saidi, R. F., Markmann, J. F., Smoot, E. E., Schoenfeld, D. A., & Kawai, T. (2011). A prospective, randomized comparison of bovine carotid artery and expanded polytetrafluoroethylene for permanent hemodialysis vascular access. *Journal of Vascular Surgery, 53*, 1640–1648.
75. Kitrou, P. M., Katsanos, K., Spiliopoulos, S., Karnabatidis, D., & Siablis, D. (2015a). Drug-eluting versus plain balloon angioplasty for the treatment of failing dialysis access: Final results and cost-effectiveness analysis from a prospective randomized controlled trial (NCT01174472). *European Journal of Radiology, 84*, 418–423.
76. Kitrou, P. M., Spiliopoulos, S., Katsanos, K., Papachristou, E., Siablis, D., & Karnabatidis, D. (2015b). Paclitaxel-coated versus plain balloon angioplasty for dysfunctional arteriovenous fistulae: One-year results of a prospective randomized controlled trial. *Journal of Vascular and Interventional Radiology, 26*, 348–354.
77. Knox, R. C., Berman, S. S., Hughes, J. D., Gentile, A. T., & Mills, J. L. (2002). Distal revascularization-interval ligation: A durable and effective treatment for ischemic steal syndrome after hemodialysis access. *Journal of Vascular Surgery, 36*, 250–255.
78. Ko, P. J., Liu, Y. H., Hung, Y. N., & Hsieh, H. C. (2009). Patency rates of cuffed and non-cuffed extended polytetrafluoroethylene grafts in dialysis access: A prospective, randomized study. *World Journal of Surgery, 33*, 846–851.
79. Konner, K., Lomonte, C., & Basile, C. (2013). Placing a primary arteriovenous fistula that works – More or less known aspects, new ideas. *Nephrology, Dialysis, Transplantation, 28*, 781–784.
80. Krafcik, B. M., Sachs, T. E., Farber, A., Eslami, M. H., Kalish, J. A., Shah, N. K., Peacock, M. R., & Siracuse, J. J. (2016). Assessment of open operative vascular surgical experience among general surgery residents. *Journal of Vascular Surgery, 63*, 1110–1115.

81. Kudlaty, E. A., Pan, J., Allemang, M. T., Kendrick, D. E., Kashyap, V. S., & Wong, V. L. (2015). The end stage of dialysis access: Femoral graft or HeRO vascular access device. *Annals of Vascular Surgery, 29*, 90–97.
82. Kumar, S., Seward, J., Wilcox, A., & Torella, F. (2010). Influence of muscle training on resting blood flow and forearm vessel diameter in patients with chronic renal failure. *The British Journal of Surgery, 97*, 835–838.
83. Lawson, J. H., Glickman, M. H., Ilzecki, M., Jakimowicz, T., Jaroszynski, A., Peden, E. K., Pilgrim, A. J., Prichard, H. L., Guziewicz, M., Przywara, S., Szmidt, J., Turek, J., Witkiewicz, W., Zapotoczny, N., Zubilewicz, T., & Niklason, L. E. (2016). Bioengineered human acellular vessels for dialysis access in patients with end-stage renal disease: Two phase 2 single-arm trials. *Lancet, 387*, 2026–2034.
84. Leaf, D. A., Macrae, H. S., Grant, E., & Kraut, J. (2003). Isometric exercise increases the size of forearm veins in patients with chronic renal failure. *The American Journal of the Medical Sciences, 325*, 115–119.
85. Leake, A. E., Yuo, T. H., Wu, T., Fish, L., Dillavou, E. D., Chaer, R. A., Leers, S. A., & Makaroun, M. S. (2015). Arteriovenous grafts are associated with earlier catheter removal and fewer catheter days in the United States Renal Data System population. *Journal of Vascular Surgery, 62*, 123–127.
86. Lennartz, C. S., Pickering, J. W., Seiler-Mussler, S., Bauer, L., Untersteller, K., Emrich, I. E., Zawada, A. M., Radermacher, J., Tangri, N., Fliser, D., & Heine, G. H. (2016). External validation of the kidney failure risk equation and re-calibration with addition of ultrasound parameters. *Clinical Journal of the American Society of Nephrology, 11*, 609–615.
87. Leon, C., & Asif, A. (2007). Arteriovenous access and hand pain: The distal hypoperfusion ischemic syndrome. *Clinical Journal of the American Society of Nephrology, 2*, 175–183.
88. Lin, S. L., Huang, C. H., Chen, H. S., Hsu, W. A., Yen, C. J., & Yen, T. S. (1998). Effects of age and diabetes on blood flow rate and primary outcome of newly created hemodialysis arteriovenous fistulas. *American Journal of Nephrology, 18*, 96–100.
89. Malas, M. B., Canner, J. K., Hicks, C. W., Arhuidese, I. J., Zarkowsky, D. S., Qazi, U., Schneider, E. B., Black, J. H., 3rd, Segev, D. L., & Freischlag, J. A. (2015). Trends in incident hemodialysis access and mortality. *JAMA Surgery, 150*, 441–448.
90. Malinzak, E. B., & Gan, T. J. (2009). Regional anesthesia for vascular access surgery. *Anesthesia and Analgesia, 109*, 976–980.
91. Marks, A., Fluck, N., Prescott, G. J., Robertson, L., Simpson, W. G., Cairns Smith, W., & Black, C. (2015). Looking to the future: Predicting renal replacement outcomes in a large community cohort with chronic kidney disease. *Nephrology, Dialysis, Transplantation, 30*, 1507–1517.
92. Martin, J. D., Warble, P. B., Hupp, J. A., Mapes, J. E., Stanziale, S. F., Weiss, L. L., Schiller, T. B., & Hanson, L. A. (2010). A real world analysis of payment per unit time in a Maryland Vascular Practice. *Journal of Vascular Surgery, 52*, 1094–1098.
93. Matsuura, J. H., Johansen, K. H., Rosenthal, D., Clark, M. D., Clarke, K. A., & Kirby, L. B. (2000). Cryopreserved femoral vein grafts for difficult hemodialysis access. *Annals of Vascular Surgery, 14*, 50–55.
94. Mcallister, T. N., Maruszewski, M., Garrido, S. A., Wystrychowski, W., Dusserre, N., Marini, A., Zagalski, K., Fiorillo, A., Avila, H., Manglano, X., Antonelli, J., Kocher, A., Zembala, M., Cierpka, L., De La Fuente, L. M., & L'heureux, N. (2009). Effectiveness of haemodialysis access with an autologous tissue-engineered vascular graft: A multicentre cohort study. *Lancet, 373*, 1440–1446.
95. Mcgill, R. L., Ruthazer, R., Meyer, K. B., Miskulin, D. C., & Weiner, D. E. (2016). Peripherally inserted central catheters and hemodialysis outcomes. *Clinical Journal of the American Society of Nephrology 11*, 1434–1440.
96. Mcmillan, W. D., Leville, C. D., & Hile, C. N. (2012). Bovine pericardial patch repair in infected fields. *Journal of Vascular Surgery, 55*, 1712–1715.

97. Miles Maliska, C., 3rd, Jennings, W., & Mallios, A. (2015). When arteriovenous fistulas are too deep: Options in obese individuals. *Journal of the American College of Surgeons, 221*, 1067–1072.
98. Miller, G. A., Goel, N., Friedman, A., Khariton, A., Jotwani, M. C., Savransky, Y., Khariton, K., Arnold, W. P., & Preddie, D. C. (2010). The MILLER banding procedure is an effective method for treating dialysis-associated steal syndrome. *Kidney International, 77*, 359–366.
99. Miller, G. A., Goel, N., Khariton, A., Friedman, A., Savransky, Y., Trusov, I., Jotwani, K., Savransky, E., Preddie, D., & Arnold, W. P. (2009). Aggressive approach to salvage non-maturing arteriovenous fistulae: A retrospective study with follow-up. *The Journal of Vascular Access, 10*, 183–191.
100. Mironov, V., & Kasyanov, V. (2009). Emergence of clinical vascular tissue engineering. *Lancet, 373*, 1402–1404.
101. Morsy, A. H., Kulbaski, M., Chen, C., Isiklar, H., & Lumsden, A. B. (1998). Incidence and characteristics of patients with hand ischemia after a hemodialysis access procedure. *The Journal of Surgical Research, 74*, 8–10.
102. Nassar, G. M., Glickman, M. H., Mclafferty, R. B., Croston, J. K., Zarge, J. I., Katzman, H. E., Peden, E. K., Lawson, J. H., Martinez, J. M., & Thackeray, L. (2014). A comparison between the HeRO graft and conventional arteriovenous grafts in hemodialysis patients. *Seminars in Dialysis, 27*, 310–318.
103. Nikam, M., Chemla, E. S., Evans, J., Summers, A., Brenchley, P., Tavakoli, A., Roy-Chaudhury, P., & Mitra, S. (2015). Prospective controlled pilot study of arteriovenous fistula placement using the novel Optiflow device. *Journal of Vascular Surgery, 61*, 1020–1025.
104. Nkf. (2006). National Kidney Foundation's Vascular Access Work Group. KDOQI Clinical practice guidelines for vascular access. *American Journal of Kidney Diseases, 48*(Suppl 1), S176–S247.
105. O'Hare, A. M., Wong, S. P., Yu, M. K., Wynar, B., Perkins, M., Liu, C. F., Lemon, J. M., & Hebert, P. L. (2015). Trends in the timing and clinical context of maintenance dialysis initiation. *Journal of the American Society of Nephrology, 26*, 1975–1981.
106. Oder, T. F., Teodorescu, V., & Uribarri, J. (2003). Effect of exercise on the diameter of arteriovenous fistulae in hemodialysis patients. *ASAIO Journal, 49*, 554–555.
107. Parekh, V. B., Niyyar, V. D., & Vachharajani, T. J. (2016). Lower extremity permanent dialysis vascular access. *Clinical Journal of the American Society of Nephrology 11*, 1693–1702.
108. Paulson, W. D., Kipshidze, N., Kipiani, K., Beridze, N., Devita, M. V., Shenoy, S., & Iyer, S. S. (2012). Safety and efficacy of local periadventitial delivery of sirolimus for improving hemodialysis graft patency: First human experience with a sirolimus-eluting collagen membrane (Coll-R). *Nephrology, Dialysis, Transplantation, 27*, 1219–1224.
109. Peden, E. K., Leeser, D. B., Dixon, B. S., El-Khatib, M. T., Roy-Chaudhury, P., Lawson, J. H., Menard, M. T., Dember, L. M., Glickman, M. H., Gustafson, P. N., Blair, A. T., Magill, M., Franano, F. N., & Burke, S. K. (2013). A multi-center, dose-escalation study of human type I pancreatic elastase (PRT-201) administered after arteriovenous fistula creation. *The Journal of Vascular Access, 14*, 143–151.
110. Perl, J., Zhang, J., Gillespie, B., Wikstrom, B., Fort, J., Hasegawa, T., Fuller, D. S., Pisoni, R. L., Robinson, B. M., & Tentori, F. (2012). Reduced survival and quality of life following return to dialysis after transplant failure: The Dialysis Outcomes and Practice Patterns Study. *Nephrology, Dialysis, Transplantation, 27*, 4464–4472.
111. Pisoni, R. L., Arrington, C. J., Albert, J. M., Ethier, J., Kimata, N., Krishnan, M., Rayner, H. C., Saito, A., Sands, J. J., Saran, R., Gillespie, B., Wolfe, R. A., & Port, F. K. (2009). Facility hemodialysis vascular access use and mortality in countries participating in DOPPS: An instrumental variable analysis. *American Journal of Kidney Diseases, 53*, 475–491.
112. Plato, S. A., 2nd, Kudlaty, E. A., Allemang, M. T., Kendrick, D. E., Wong, V. L., Wang, J. C., & Kashyap, V. S. (2016). Elevated peak systolic velocity and velocity ratio from duplex

ultrasound are associated with hemodynamically significant lesions in arteriovenous access. *Annals of Vascular Surgery 35*, 68–74.

113. Polkinghorne, K. R., Seneviratne, M., & Kerr, P. G. (2009). Effect of a vascular access nurse coordinator to reduce central venous catheter use in incident hemodialysis patients: A quality improvement report. *American Journal of Kidney Diseases, 53*, 99–106.
114. Rajan, D. K. (2016). New approaches to arteriovenous fistula creation. *Seminars in Interventional Radiology, 33*, 6–9.
115. Rajan, D. K., Ebner, A., Desai, S. B., Rios, J. M., & Cohn, W. E. (2015). Percutaneous creation of an arteriovenous fistula for hemodialysis access. *Journal of Vascular and Interventional Radiology, 26*, 484–490.
116. Rajan, D. K., & Falk, A. (2015). A randomized prospective study comparing outcomes of angioplasty versus VIABAHN stent-graft placement for cephalic arch stenosis in dysfunctional hemodialysis accesses. *Journal of Vascular and Interventional Radiology, 26*, 1355–1361.
117. Ravani, P., Palmer, S. C., Oliver, M. J., Quinn, R. R., Macrae, J. M., Tai, D. J., Pannu, N. I., Thomas, C., Hemmelgarn, B. R., Craig, J. C., Manns, B., Tonelli, M., Strippoli, G. F., & James, M. T. (2013). Associations between hemodialysis access type and clinical outcomes: A systematic review. *Journal of the American Society of Nephrology, 24*, 465–473.
118. Ribichini, F., Joner, M., Ferrero, V., Finn, A. V., Crimins, J., Nakazawa, G., Acampado, E., Kolodgie, F. D., Vassanelli, C., & Virmani, R. (2007). Effects of oral prednisone after stenting in a rabbit model of established atherosclerosis. *Journal of the American College of Cardiology, 50*, 176–185.
119. Rodriguez Moran, M., Almazan Enriquez, A., Ramos Boyero, M., Rodriguez Rodriguez, J. M., & Gomez Alonso, A. (1984). Hand exercise effect in maturation and blood flow of dialysis arteriovenous fistulas ultrasound study. *Angiology, 35*, 641–644.
120. Roy-Chaudhury, P., & Kruska, L. (2015). Future directions for vascular access for hemodialysis. *Seminars in Dialysis, 28*, 107–113.
121. Roy-Chaudhury, P., Lee, T., Woodle, B., Wadehra, D., Campos-Naciff, B., & Munda, R. (2012). Balloon-assisted maturation (BAM) of the arteriovenous fistula: The good, the bad, and the ugly. *Seminars in Nephrology, 32*, 558–563.
122. Salimi, F., Majd Nassiri, G., Moradi, M., Keshavarzian, A., Farajzadegan, Z., Saleki, M., Nikpoor, A., & Ghane, M. (2013). Assessment of effects of upper extremity exercise with arm tourniquet on maturity of arteriovenous fistula in hemodialysis patients. *The Journal of Vascular Access, 14*, 239–244.
123. Saran, R., Elder, S. J., Goodkin, D. A., Akiba, T., Ethier, J., Rayner, H. C., Saito, A., Young, E. W., Gillespie, B. W., Merion, R. M., & Pisoni, R. L. (2008). Enhanced training in vascular access creation predicts arteriovenous fistula placement and patency in hemodialysis patients: Results from the Dialysis Outcomes and Practice Patterns Study. *Annals of Surgery, 247*, 885–891.
124. Saran, R., Li, Y., Robinson, B., Abbott, K. C., Agodoa, L. Y., Ayanian, J., Bragg-Gresham, J., Balkrishnan, R., Chen, J. L., Cope, E., Eggers, P. W., Gillen, D., Gipson, D., Hailpern, S. M., Hall, Y. N., He, K., Herman, W., Heung, M., Hirth, R. A., Hutton, D., Jacobsen, S. J., Kalantar-Zadeh, K., Kovesdy, C. P., Lu, Y., Molnar, M. Z., Morgenstern, H., Nallamothu, B., Nguyen, D. V., O'hare, A. M., Plattner, B., Pisoni, R., Port, F. K., Rao, P., Rhee, C. M., Sakhuja, A., Schaubel, D. E., Selewski, D. T., Shahinian, V., Sim, J. J., Song, P., Streja, E., Kurella Tamura, M., Tentori, F., White, S., Woodside, K., & Hirth, R. A. (2016). US Renal Data System 2015 annual data report: Epidemiology of kidney disease in the United States. *American Journal of Kidney Diseases, 67*, A7–S434.
125. Shah, A. S., Valdes, J., Charlton-Ouw, K. M., Chen, Z., Coogan, S. M., Amer, H. M., Estrera, A. L., Safi, H. J., & Azizzadeh, A. (2012). Endovascular treatment of hemodialysis access pseudoaneurysms. *Journal of Vascular Surgery, 55*, 1058–1062.
126. Shemesh, D., Goldin, I., Hijazi, J., Zaghal, I., Berelowitz, D., Verstandig, A., & Olsha, O. (2015). A prospective randomized study of heparin-bonded graft (Propaten) versus standard graft in prosthetic arteriovenous access. *Journal of Vascular Surgery, 62*, 115–122.

127. Slinin, Y., Guo, H., Li, S., Liu, J., Morgan, B., Ensrud, K., Gilbertson, D. T., Collins, A. J., & Ishani, A. (2014). Provider and care characteristics associated with timing of dialysis initiation. *Clinical Journal of the American Society of Nephrology, 9*, 310–317.
128. Son, A., Mannoia, K., Herrera, A., Chizari, M., Hagdoost, M., & Molkara, A. (2016). Dialysis access surgery: Does anesthesia type affect maturation and complication rates? *Annals of Vascular Surgery, 33*, 116–119.
129. Sorom, A. J., Hughes, C. B., Mccarthy, J. T., Jenson, B. M., Prieto, M., Panneton, J. M., Sterioff, S., Stegall, M. D., & Nyberg, S. L. (2002). Prospective, randomized evaluation of a cuffed expanded polytetrafluoroethylene graft for hemodialysis vascular access. *Surgery, 132*, 135–140.
130. Steerman, S. N., Wagner, J., Higgins, J. A., Kim, C., Mirza, A., Pavela, J., Panneton, J. M., & Glickman, M. H. (2013). Outcomes comparison of HeRO and lower extremity arteriovenous grafts in patients with long-standing renal failure. *Journal of Vascular Surgery, 57*, 776–783. discussion 782-3.
131. Uy, A. L., Jindal, R. M., Herndon, T. W., Yuan, C. M., Abbott, K. C., & Hurst, F. P. (2013). Impact of isometric handgrip exercises on cephalic vein diameter in non-AVF candidates, a pilot study. *The Journal of Vascular Access, 14*, 157–163.
132. Vasctek. 2016. *Rapidax II* [Online]. Available: www.vascutek.com/vascutek/products/item/rapidax. Accessed 6 July 2016.
133. Vassalotti, J. A., Jennings, W. C., Beathard, G. A., Neumann, M., Caponi, S., Fox, C. H., Spergel, L. M., & Fistula First Breakthrough Initiative Community Education, C. (2012). Fistula first breakthrough initiative: Targeting catheter last in fistula first. *Seminars in Dialysis, 25*, 303–310.
134. Veroux, P., Giaquinta, A., Tallarita, T., Sinagra, N., Virgilio, C., Zerbo, D., Gloviczki, P., & Veroux, M. (2013). Primary balloon angioplasty of small (</=2 mm) cephalic veins improves primary patency of arteriovenous fistulae and decreases reintervention rates. *Journal of Vascular Surgery, 57*, 131–136.
135. Woodside, K. J., Bell, S., Eckard, A. R., Plattner, B. W., Schaubel, D. E., Dasmunshi, S., Mukhopadhyay, P., Pearson, J., Pisoni, R. L. & Saran, R. 2016a. Arteriovenous fistula maturation in patients on hemodialysis: Results from the United States Renal Data System (USRDS) (Manuscript in preparation).
136. Woodside, K. J., Ratkowiak, K., Mukhopadhyay, P., Dharmarajan, S. H., Plattner, B. W., Schaubel, D. E., Saran, R. & Pisoni, R. 2016b. Procedural burden during arteriovenous fistula maturation following operative placement: An analysis of the United States Renal Data System (Manuscript in preparation).
137. Woodside, K. J., Schirm, Z. W., Noon, K. A., Huml, A. M., Padiyar, A., Sanchez, E. Q., Sarabu, N., Hricik, D. E., Schulak, J. A., & Augustine, J. J. (2014). Fever, infection, and rejection after kidney transplant failure. *Transplantation, 97*, 648–653.
138. Wystrychowski, W., Mcallister, T. N., Zagalski, K., Dusserre, N., Cierpka, L., & L'heureux, N. (2014). First human use of an allogeneic tissue-engineered vascular graft for hemodialysis access. *Journal of Vascular Surgery, 60*, 1353–1357.
139. Zanow, J., Kruger, U., & Scholz, H. (2006). Proximalization of the arterial inflow: A new technique to treat access-related ischemia. *Journal of Vascular Surgery, 43*, 1216–1221.
140. Zhang, J. C., Al-Jaishi, A., Perl, J., Garg, A. X., & Moist, L. M. (2015). Hemodialysis arteriovenous vascular access creation after kidney transplant failure. *American Journal of Kidney Diseases, 66*, 646–654.

Chapter 5
Robotic Surgery in Organ Transplantation

Ivo Tzvetanov, Sandra Garcia Aroz, Mario Spaggiari, and Enrico Benedetti

Abstract The application of minimally invasive surgical technologies has gained widespread adoption. The surgical robotic system has allowed surgeons to perform complex procedures, otherwise unachievable. Robotic systems afford access to minimally invasive surgery to more patients with a variety of pathological conditions. Better surgical dexterity and vision for surgeons, in addition to a facilitated learning curve, compared to conventional laparoscopic surgery, leads to lower rate of complications and higher patient satisfaction. Until recently, these benefits were inaccessible for patients in need of solid organ transplants. Conventional laparoscopy has been seen as non-applicable for such a technically demanding procedure. The introduction of the da Vinci Robotic Surgical System has expanded the ability to complete solid organ transplantation in a minimally invasive fashion. Robotic applications in kidney, pancreas, and liver transplantation have been reported. There are several groups which report their experiences and initial results showing the viability of this technique in the field. The biggest experience has been described in kidney transplantation. One of the main advantages of the robotic surgery is the significantly lower rate of surgical site infections, which in immunosuppressed patients is reflected in superior outcomes. Another proven advantage is that the robotic kidney approach permits transplantation in extreme BMI categories without additional technical complications. The first results in pancreas transplantation and living donor hepatectomies are very promising; however, larger series are needed in order to address the value of the robotic surgery in other areas of solid organ transplantation.

5.1 History of Surgical Robotics

Abdominal surgery has been exposed to many changes in the last two decades and laparoscopic surgery has become the greatest innovation. Minimally invasive surgical technologies have shown benefits that include reduced recovery period, fewer wound complications, and reduced surgical scars [1].

I. Tzvetanov • S.G. Aroz • M. Spaggiari • E. Benedetti (✉)
Division of Transplantation, Department of Surgery, University of Illinois at Chicago, 840 South Wood Street, Chicago, IL 60612, USA
e-mail: Enrico@uic.edu

S.N. Nadig, J.A. Wertheim (eds.), *Technological Advances in Organ Transplantation*, DOI 10.1007/978-3-319-62142-5_5

The world's first surgical robot, the Arthrobot, was developed in Canada. It was used for the first time in Vancouver, in 1983. The Arthrobot was a bone mountable hip arthroplasty surgery robot. At the same time, other projects including medical robots were developed. These projects included a robotic arm involved in eye surgeries and another robotic arm designed to assist the surgeon handling the instruments. The next experience was in 1985, when the PUMA 560 robotic arm was used to perform a neurosurgical biopsy, placing a needle for a brain biopsy using CT guidance. This procedure led to the first laparoscopic surgery involving a robotic system; it was a cholecystectomy performed in 1987. Both PUMA system and the PROBOT, developed in London, were two systems used in 1988 to perform the first urological robotic-assisted surgeries [2].

The Automated Endoscopic System for Optimal Positioning 3000 (AESOP, Computer Motion, Inc., Goleta, California) was the first robotically assisted surgical device approved by the FDA in 1994. The AESOP employs a single mechanical arm with six degrees of freedom to position an endoscope. The AESOP allows the surgeon to position the endoscope through either voice-activated or manual controls.

The Zeus Robotic Surgical System was introduced in 1998. It employs a remote "master" workstation where the surgeon views a flat panel two-dimensional (2-D) or three-dimensional (3-D) display and maneuvers two handheld manipulators to direct movement of two robotic arms that are attached to the operating table. The operator's hand movements are digitized and translated into a robot motion in real time. A camera is mounted on a third arm that is either controlled by foot pedal or by voice activation (AESOP).

5.1.1 Da Vinci Robotic Surgical System

The da Vinci Robotic Surgical System, also introduced in 1998, uses a surgeon-operated console located outside the sterile field. The da Vinci's Insite Vision System affords true stereoscopic 3-D vision of the surgical field. The da Vinci's EndoWrist has seven degrees of freedom and adjustable grip strength. Rotation of the master controls clockwise results in "intuitive" clockwise rotation of the instrument [3, 4]. Himpens and Cadière performed the first robotic-assisted cholecystectomy on a 72-year-old patient on March 3, 1997 [5]. Robotic-assisted surgery is currently used in all surgical disciplines. Practically all gastrointestinal surgical procedures have been performed with a da Vinci Robotic Surgical System, from renal grafts to liver harvesting from a live donor. In 2009, the latest generation arose with a high-definition (HD) vision system. It also allows a second console to the surgeon console, an innovation that allows a novice surgeon to be coached by a mentor during the procedure and thereby increases safety during the learning curve [6]. However, the greatest limitations of the current robotic system are high cost and lack of haptic feedback [7].

In the field of complex procedures such as transplantation, open surgery is still the standard procedure. These operations have been considered too technically demanding by conventional laparoscopy, [8] but the introduction of precise surgical robotic systems such as the da Vinci Robotic Surgical System has expanded our ability to complete complex surgical procedures in a minimally invasive fashion and increase the interest for a robotic application in the field of organ transplantation. Robotic surgery has been successfully used in kidney transplantation and, in a lesser proportion, in pancreas transplantation and donor hepatectomies for living donor liver transplantation.

5.2 Robotic Hand-Assisted Living Donor Nephrectomy

The opportunity to perform a transplant when the recipient is in optimal condition and attain a better quality of kidney allografts makes the living donor transplant an excellent treatment for patients with end-stage renal disease. In addition, time on dialysis can be minimized or even avoided. Given all these advantages, better outcomes such as higher graft and patient survival have been shown as compared with deceased donor transplantation. The main obstacle of living donation remains the exposure of a healthy individual to the inherent risk of a surgical intervention without a direct health benefit. The availability of minimally invasive surgery for donor nephrectomies implies reducing postoperative pain, achieving faster recovery, and minimizing the surgical incisions. These conditions have become significant factors to greatly enhancing living donation rates [9]. The da Vinci Robotic Surgical System has been used in living donor nephrectomies as a logical extension of the widely adapted minimal invasive approach [10]. After acquiring experience with the technique in general surgery, the first worldwide transabdominal hand-assisted robotic donor nephrectomy was performed successfully at the University of Illinois at Chicago in 2000 [11]. Since then, this institution has performed over 950 robotic donor nephrectomies with excellent outcomes [12, 13].

5.2.1 Preoperative Evaluation and Donor Selection

The evaluation for robotic surgery of the potential kidney donors does not differ from the widely accepted approach. The physiological assessment, immunologic compatibility, current medical status, and function and anatomy of the kidneys complete the screening. This evaluation is performed by the transplant team.

The decision regarding which kidney will be harvested is based on the function and anatomy. Usually the left kidney is procured due to its anatomy (longer left renal vein) and the lower complexity of the left nephrectomy. The higher precision of the surgical dissection feasible by the robotic system allows harvesting of kidneys with more complex anatomy, such as multiple arteries.

5.2.2 Surgical Procedure

The donor is usually admitted the day of the procedure. For a left nephrectomy, the patient is rolled into right lateral decubitus position with cushioned beanbag and axillary roll. An adequate fixation of the patient to the operative table is important, because any instability after docking the robotic system could jeopardize the safety of the procedure.

Robotic-assisted donor nephrectomy is a transabdominal procedure. Four laparoscopic ports and one infraumbilical incision are done to perform this procedure. The most common incision is a 7-cm transverse, suprapubic, Pfannenstiel incision, but if the patient prefers another incision, a longitudinal or transverse abdominal incision can be offered.

A 12-mm laparoscopic port is placed above the umbilicus, close to the midline, and at the level of renal hilum. This port is required for the 30° robotic camera system. The left midclavicular line is the appropriate position to place the two 8-mm robotic working ports to achieve good triangulation. These ports are located proximal and distal, 10–12 cm apart from the camera port. The left lower quadrant is the location preferred to place the last 12-mm port which is used to assist with suction, clipping, stapling, and cutting. The robotic system is docked and integrated to the ports and pneumoperitoneum is achieved.

The descending colon is retracted medially and freed from lateral peritoneal attachments allowing the exposure of the left paracolic gutter. The descending and the sigmoid colon are then fully mobilized. The splenocolic ligament is also transected and additionally cauterized with bipolar pickups. Following the exact plan between the mesentery of the left colon and the Gerota's fascia allows bloodless exposure of the anterior surface of the left kidney even in cases with significant intra-abdominal adiposity. Occasionally, if the lower pole of the spleen overlays the upper pole of the kidney, the posterior splenic attachments are transected and the spleen is partially mobilized. The operating surgeon has to be very cautious to avoid injury of the body and tail of the pancreas.

The assistant surgeon retracts to a medial position in the left colon, and this maneuver allows exposure of the retroperitoneal space. The ureteral dissection is performed circumferentially. A correct blood supply of the ureter is assured through a generous amount of adipose tissue conserved around the ureter. During this time, if present, the finding of a lower polar artery originating from the distal abdominal aorta should be identified. Unintentional injury of this vessel would deprive the ureter from an adequate blood supply.

The gonadal vein is followed proximally to its junction with the renal vein. The tissue in front of the renal vein is transected and vein exposed medially to its junction with the inferior vena cava (IVC). The left adrenal vein is identified along the upper border of the renal vein. In most of the cases, at least one lumbar vein will be joining the left renal vein. The precision during the isolation and transection of these veins cannot be overemphasized. In these cases, the articulating skills of the robotic system and the 3-D vision give significant advantage over conventional laparoscopic instruments. All these veins contributing to the left renal vein are clipped and transected.

Proximal to the left renal vein, the adrenal gland is identified and it should be left intact. If a sizable upper polar artery is present, extra care should be taken to preserve this vessel, since it could supply 20–30% of the kidney mass. The upper pole of the kidney is then fully mobilized.

The previously mobilized ureter is clipped with two robotic clips where it crosses the iliac vessels and is sharply transected proximal to the clips. The assisting surgeon's hand helps to divide the posterior attachments of the kidney.

The circumferential dissection of the left renal artery should be carried to its origin from the aorta. If multiple arteries are present, every vessel has to be dissected free as described. When this step is finished, the renal artery and vein are the only connection of the kidney.

The assistant surgeon advances the Endo TA stapler through the 12-mm left lower quadrant port with vascular load. The utilization of Endo TA stapler allows additional length of the artery that facilitates the implantation of the graft. The renal artery is stapled with a TA vascular stapler on its origin from the aorta. The latest da Vinci Xi version, with integrated stapler, allows this step to be completed by the operating surgeon from the console. The artery is sharply divided with robotic scissors at least 3–4 mm distal from the stapler line. The renal vein is divided with an Endo GIA vascular stapler, inserted by the assisting surgeon through the left lower quadrant assisting port.

The cavity must be inspected for bleeding once the kidney graft is removed. The arterial and venous stumps are visualized and the condition of the staple line verified. Last, the presence of chylous and lymphatic leak should be interrogated. The robotic system facilitates significantly if suturing is necessary.

If the right kidney is to be procured, the procedure is efficiently performed with the robotic system in contralateral fashion from the one described.

5.3 Living Donor Totally Robotic Nephrectomy

Giacomoni et al. have recently reported a study describing a modified robotic technique [14]. This technique presents some modifications that allow a completely robotic approach. Their study compares the robotic hand-assisted approach to a totally robotic approach for living donor nephrectomies. The study describes 33 nephrectomies performed with the modified approach. These modifications are described below.

The patient's decubitus position is modified from right 60° lateral decubitus to completely right lateral 90° decubitus. The first surgical step is to perform a Kustner preparatory incision of around 8–10 cm, by opening the fascia and the peritoneum. A camera trocar (12 mm) is moved on the hemiclavicular line and placed using the intra-abdominal hand control through the Kustner incision. Two 8-mm robotic trocars are placed under direct camera vision, and they are moved from the left hemiclavicular line to the left anterior axillary line in this modified technique.

Concerning the surgical console time, leading modifications are the following: introduction of the Ultracision during the kidney dissection and increased application of the metallic clips. At the time of vessel stapling, the kidney is placed into the Endobag previously positioned. After the renal vessels are ligated and divided, specimen extraction occurs by closing the Endobag and cutting the running suture of the fascia, completing the extraction.

In our literature review, the University of Illinois group is the only center with the hand-assisted approach, while the other series that are mentioned such as Hubert et al. in 2007, with 38 cases, describe a totally robotic approach [15].

In all studies, and even if the left kidney has multiple arteries, it is preferred over the right kidney. The main indication for procuring the right kidney is the presence of anatomical defects incompatible with transplantation in the left kidney or significant difference in function between both kidneys.

5.3.1 Discussion

The first living donor nephrectomy was performed in 1954 through a flank incision [16]. Since that initial case, the number of living donor cases have steadily increased and currently represent approximately 40% of the total kidney transplant activity in the United States [17]. The application of minimally invasive surgery has contributed to an increase in the donation rates [9]. Presently, the laparoscopic procedure is available in 90% of the US centers compared to 49% in 2000 [18]. Minimally invasive surgery reduces the surgical trauma to the willing healthy donor, shortens the hospital stay by significantly decreasing the postoperative pain, and allows for early mobilization. Furthermore, the cosmetic results are usually better than open technique. Consequently, minimally invasive techniques are more acceptable for the donors.

The initial transperitoneal laparoscopic donor nephrectomy was described by Ratner in 1995 [19], through a 9-cm infraumbilical incision; the patient recovered uneventfully and was discharged home the following day. Since that time, the procedure has evolved into a wide range of techniques. From the early experiences, laparoscopic technique has shown to offer improved donor recovery compared to open donor nephrectomy, with similar morbidity and mortality and comparable graft function and survival rates in the recipients [5, 20]. The convalescence and return to work are faster after laparoscopic techniques [21].

The da Vinci Robotic Surgical System offers several potential advantages over the laparoscopic technique, including facilitated dissection, easier suturing and knotting, smoother learning curve, and better comfort for the surgeon. These advantages are due to the high-definition three-dimensional view, the EndoWrist technology (improving the freedom of movements and filtration of tremors), and the console conformation and work station. The first worldwide robotic donor nephrectomy was performed in September 2000 at the University of Illinois. The early experience of this procedure was published by Horgan et al. and showed the feasibility of robotic nephrectomy.

Since that time, different reports have described the same procedure with many modifications to the surgical technique [11, 12, 15, 22–24]. Regardless of the different techniques adopted for this procedure, all of the articles published on the topic confirm the safety and feasibility of robotic donor nephrectomy.

5.4 Robotic-Assisted Kidney Transplantation

The first totally robotic transabdominal kidney transplant (RKT) was performed in 2009 by the group from the University of Illinois at Chicago, published by Giulianotti et al. Total operative time was 223 min, with warm ischemia time of 50 min. Boggi et al. published the first successful European RKT in 2011, describing a slightly different technique. Giulianotti's group used a periumbilical incision, while Boggi et al. utilized a 7-cm Pfannenstiel incision. Total operative time was 154 min, with 51 min of warm ischemia. Doumerc et al. described another novel approach to RKT in 2015, using a transvaginal approach to introduce the renal graft into the peritoneum, through a vaginal incision. Mean operative time was 200 min and anastomotic time was 55 min. At that point, the technique was described only for the kidney transplant recipient; however, later in 2015 the same group described the first pure robotic-assisted approach to living donor kidney transplantation utilizing the transvaginal technique for both the donor and the recipient surgeries [25].

Over the last few years, a new technique allowing for intracorporeal regional hypothermia of the graft has also been described [26, 27]. The authors use a periumbilical incision as well but describe the use of a novel gauze jacket filled with ice slush. They report a reduced warm ischemia time and also allowed for atraumatic handling of the graft.

The majority of transplant centers consider BMI $\geq$40 kg/m^2 a contraindication for transplant. Oberholzer et al. presented a study on a cohort of morbid obese robotic kidney recipients compared to the open approach. Initial results showed the advantages and feasibility of the robotic-assisted procedure [28]. This strategy opened the opportunity for transplantation to candidates previously rejected due to obesity. At present, our group has performed 170 robotic kidney transplants for obese recipients. The robotic system allows to minimize the incision by changing the location to the upper abdomen, away from a much more contaminated groin area. With this technical modification, we were able to almost completely eradicate the surgical site infection in this extremely high-risk patient group. This type of complication appears to be very significant factor for the graft survival.

5.4.1 Surgical Technique of RKT

The patient is secured to the operative room table by using shoulder block and tape. Legs are flexed at the knees and kept close together. The hand device is placed through a 7-cm midline incision, approximately 5 cm in length (Fig. 5.1 trocar placement in robotic kidney transplantation) below the xyphoid process. Laparoscopic ports (Fig. 5.1) are positioned in the following manner: (1) one 12-mm port for the 30° robotic scope is inserted to the right of the umbilicus; (2) two 7-mm robotic ports are inserted – one is placed in the right flank and the other one is placed in the left lower quadrant; and (3) a 12-mm assistant port is placed on the left side of the umbilicus between the camera and the left lower quadrant robotic port. The patient is placed in 30° Trendelenburg position with the right side elevated (for implantation to the right external iliac vessels). The robot system is docked into position from the patient's right leg site parallel and slightly diagonal to the body.

The right external iliac artery and vein are exposed and dissected after the peritoneum is incised. A vessel loop is used to retract the artery upward facilitating the exposure and the dissection around the external iliac vein. Similarly, the dissection on the posterior surface of the vein is achieved with another vessel loop, which is placed around the iliac vein.

Two robotic bulldog clamps are used to clamp the external iliac vein. The venotomy (about 15 mm in length) is performed using robotic scissors. A 12-cm, double-needle, 5-0 Gore-Tex suture with a knot in the middle is placed at the corner of the venotomy. Then, the assistant surgeon inserts the kidney graft in the abdominal cavity and orients it properly. The venovenous anastomosis is performed first (Fig. 5.2). This anastomosis is completed in an end-to-side fashion with running suture. Next, the external iliac artery is then clamped between robotic bulldogs,

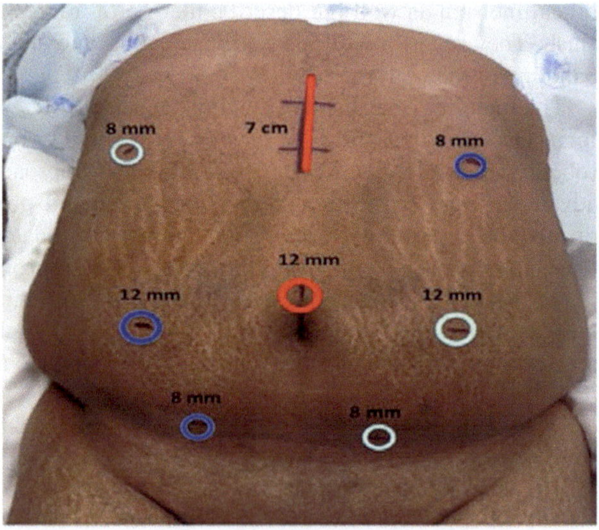

Fig. 5.1 Trocar placement

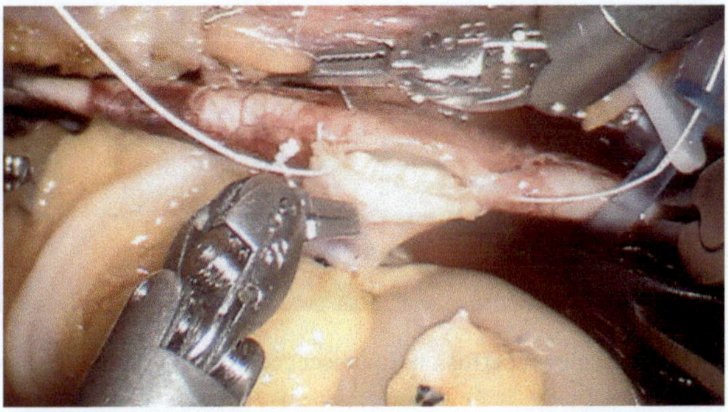

Fig. 5.2 Anastomosis of the renal vein

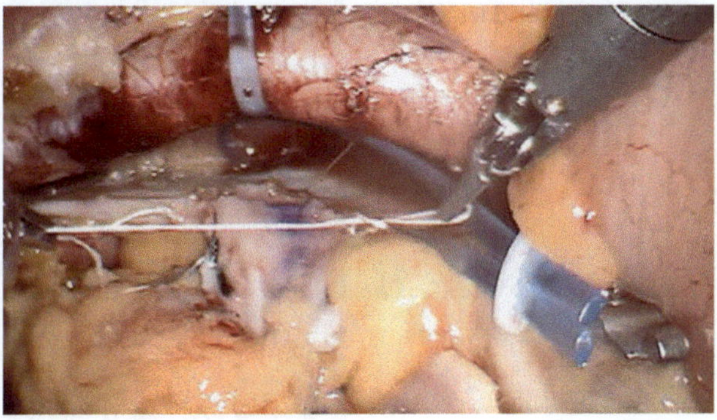

Fig. 5.3 Anastomosis of the renal artery

and an oval-shaped window is made in the anterior wall of the artery using robotic scissors. The arterial anastomosis (Fig. 5.3) is completed in an end-to-side fashion with 12-cm double-needle 6-0 Gore-Tex suture with a knot in the middle. The ease of the fine vascular suturing allowed by the high-definition 3D vision and wrist-like instrument motion of the robotic system are the most important advantages for this procedure.

Once the reconstruction is completed, venous clamps are removed first, followed by immediate removal of the arterial clamps. The reperfusion of the organ (Fig. 5.4) and hemostasis are additionally verified and bleeding points secured with 6-0 Prolene suture. At this point, the pressure of the pneumoperitoneum is decreased to minimize possible negative effect of high intra-abdominal pressure on the graft perfusion.

Diluted methylene blue solution is introduced in to the bladder in order to facilitate its identification. The ureter is anastomosed to the bladder (Fig. 5.5) with running 5-0 Monocryl suture. The typical anti-reflux technique, suturing full thickness of the ureteral wall with the mucosal layer of the bladder, is used. Utilization of ureteral stent is optional.

At the end of the procedure, the minilaparotomy is closed with running 0 PDS, and the two 12-mm port sites are closed with an endoclosure device and 0 Vicryl suture.

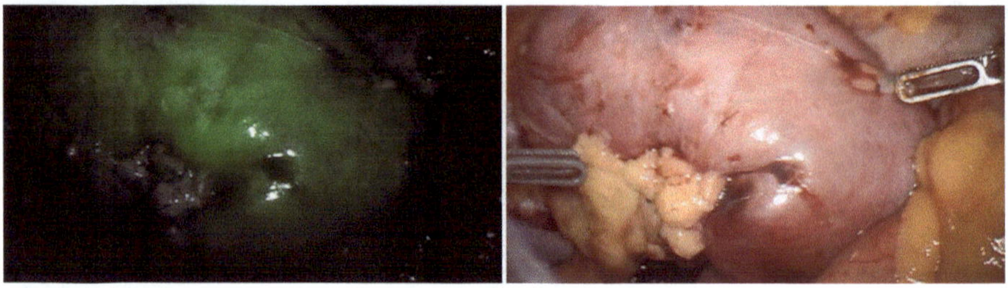

Fig. 5.4 Kidney reperfusion; fluorescence imaging mode (*left*), normal 3DHD illumination mode (*right*)

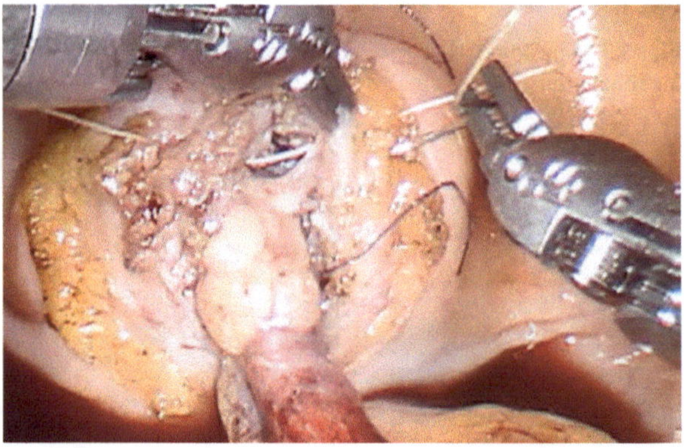

Fig. 5.5 Anastomosis of the ureter to the bladder

5.4.2 Discussion

In line with existing comparative literature evaluating laparoscopic and open surgery, initial experiences at various institutions around the world have now demonstrated lower complication rates for RKT in comparison to similar open renal transplant cohorts [29, 30].

This technique has been applied within the last 7 years at the University of Illinois. During this time, more than 170 robotic-assisted kidney transplants in obese recipients have been performed. We used a BMI >30 kg/m^2 as selection criteria, without an upper limit. The mean BMI of the group was 40 kg/m^2. We reported our early experience in a case–control study [28], where the first 28 robotic-assisted kidney transplants were compared to a frequency-matched retrospective cohort of obese recipients who underwent kidney transplantation by open technique. At 48 months of follow-up, the GFR was 51.5 ± 30.7 ml/min/1.73m^2 in the robotic group and 51.9 ± 21.8 ml/min/1.73m^2 in the control group ($p = 0.83$). The rate of SSI was significantly higher in the control group when compared with the robotic group (28.6% vs. 0%, $p = 0.004$). At 4-year posttransplant, eight patients in the control group (28.6%) experienced graft loss when compared with five patients in the robotic group (17.9%). Three (37.5%) out of the eight patients who lost the graft in the control group had concomitant SSI. The patient survival at 48 months was 92.5% in the robotic group and 92.4% in the control group ($p = 0.97$). Besides the advantages of minimally invasive surgery as early mobilization and high patient satisfaction, excellent graft function was observed.

In another study that was recently performed by our group (Benedetti et al.), the UNOS registry was reviewed for adult living donor kidney transplant recipients with BMI≥40 kg/m^2 performed from September 2009 to December 2014. We compared outcomes in robotic kidney transplantation (RKT) versus open technique kidney transplantation (OKT) at all US centers. Similar patient and graft survival were reported. Renal function, determined by creatinine levels and GFR, was also similar in both groups.

Based on these experiences, we can state that robotic-assisted kidney transplantation is a safe and effective procedure. By achieving excellent kidney graft function and minimizing surgical complications, this surgical technique gives the opportunity to the disadvantaged group of obese patients with ESRD to have more realistic access to transplantation.

The robotic approach has been used in different centers for autotransplantation as a salvage of kidneys with a variety of injuries. The technique published is similar to the one described [31–33].

5.5 Application in Pancreas Transplantation

Pancreas transplantation is a widely accepted procedure that restores the euglycemia in diabetic patients and can prevent progression of complications. Simultaneous pancreas–kidney transplantation (SPK) is successfully used in diabetic patients with end-stage renal disease.

Despite the advantages in surgical technique, pancreas transplantation has the highest rate of surgical complications among solid organ transplantation [34]. Therefore, the introduction of innovating surgical techniques to achieve a reduction of posttransplant morbidity in these patients is needed.

Boggi et al. reported the first three pancreas transplants performed by the assistance of the da Vinci Robotic Surgical System in 2012 [35]. The three grafts were one pancreas alone, one pancreas after the kidney, and one simultaneous kidney–pancreas.

5.5.1 Surgical Technique of Robotic Pancreas Transplantation

The surgery is performed with the patient in supine position with the right flank slightly elevated. The table is then inclined 25° to the left, further elevating the right flank. A 7-cm midline incision is made just above the navel, and a GelPort (Applied Medical, Rancho Santa Margarita, CA) device is inserted. Four ports are positioned in the following manner: (1) one 11-mm port is placed within the GelPort to create the pneumoperitoneum, (2) one 11-mm robotic port is placed slightly to the left of the midline a few centimeters below the navel for insertion and use of the laparoscope, and (3) two 8-mm robotic ports are placed along the right pararectal line 5 cm below the costal margin and 3 cm above the pubis, respectively. The da Vinci Robotic Surgical System is placed to the patient's right side.

The right colon is mobilized until the common iliac artery and the proximal segment of the inferior vena cava are exposed. To perform the vascular anastomosis, the iliac artery is cross clamped and the inferior vena cava is partially occluded using bulldog clamps, manually applied through the GelPort. The graft is then introduced through the hand port access and placed over the psoas muscle. The donor portal vein is anastomosed end-to-side to recipient inferior vena cava using two half running sutures of 7-0 ePTFE. The donor Y graft and recipient common iliac artery are anastomosed, using 6-0 ePTFE. The pneumoperitoneum is interrupted after graft revascularization and exocrine drainage is achieved by means of Roux-en-Y duodenojejunostomy.

In the simultaneous pancreas–kidney transplant, the kidney is transplanted in the ipsilateral iliac fossa, according to the technique previously described for robotic kidney transplantation. The suprapubic robotic port access is converted into a mini-incision, approximately 3 cm long and the ureterovesical anastomosis (Gregoir–Lich extravesical anastomosis) is performed manually.

The experience described proves the feasibility of robotic-assisted laparoscopic surgery in pancreas transplantation. However, it is necessary that further studies and larger series confirm this as an alternative approach to the conventional open technique.

Our group performed two fully robotic pancreas transplants. The first one was pancreas alone and the second simultaneous kidney–pancreas. In both cases, the pancreas was transplanted to the left external iliac vessels, which allowed perfect alignment of the vascular anastomosis. In the first case, the exocrine secretions were drained into the urinary bladder through duedenocystoanastomosis. We used EEA stapler inserted through the jejunal stump of the graft. In the second case, we used enteric drainage (Fig. 5.6). Duodenojejunal anastomosis was completed similarly with EEA stapler. We cannot overemphasize the importance of diligent procurement of the pancreatic graft and meticulous back bench preparation to avoid disturbing bleeding after graft reperfusion.

In our view, the application of robotic technique for minimally invasive pancreas transplantation could significantly decrease complications, which are quite frequent after pancreas transplantation.

5.6 Application in Liver Transplantation

Minimal invasive laparoscopic liver resection has evolved greatly during the past few decades. The experience in minimally invasive liver surgery has been steadily increasing [36]. As described before, the robotic system is believed to be superior to conventional laparoscopic systems because of better visualization of the operative field, wristed instruments with better ergonomics, and greater potential for

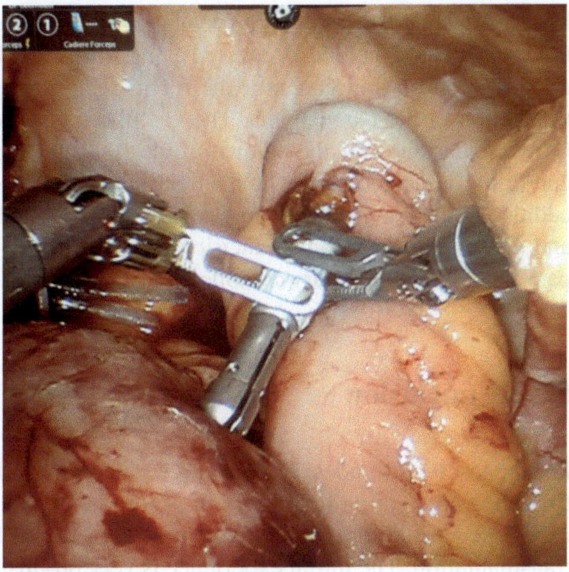

Fig. 5.6 Robotic simultaneous kidney–pancreas transplantation; enteric drainage

intracorporeal suturing techniques. Over 200 robotic-assisted liver resection cases have been published to the date, including living donor right hepatectomy [37–39].

5.6.1 Living Donor Hepatectomy

Soubrane et al. reported the first clinical series of pure laparoscopic left lateral sectionectomy for living liver donors [40]. For living donor right hepatectomies, the introduction of laparoscopic surgery can be traced back to 2006, when Koffron et al. first reported using the laparoscopy-assisted method [41].

Following the advances in robotic surgical technology combined with the extensive institutional experience with major robotic-assisted liver resections [37], this technology was applied to right living lobe donor hepatectomy. The first robotic living donor hepatectomy was performed in 2012 by Giulianotti et al. at the University of Illinois [38]. The liver graft was safely extracted through a lower abdominal incision and the donor recovered without acute complications.

5.6.2 Surgical Technique of Robotic Living Donor Hepatectomy

The entire surgery is performed using the da Vinci Robotic Surgical System. The procedure is started like any other robotic surgery, placing the laparoscopic trocars and installing the robotic system. The first step is the gallbladder removal. After this, the hepatic artery and the right portal vein were dissected free and hepatic duct isolated and transected. The right lobe is retracted along an upward direction to start the retrohepatic caval dissection. Consequently, the robotic Harmonic scalpel is used to perform the transection of the parenchyma. The vascular transections are performed using Endo GIA vascular stapler in the following order: (a) right hepatic artery, (b) right portal vein, and (c) right hepatic vein. The main advantages observed with the robotic system, in addition to the previously mentioned, are the facilitated vascular and biliary dissection and the smaller subumbilical incision for graft extraction. This incision decreases the pain and risk of pulmonary complications associated with an upper midline incision.

5.7 Conclusion

For the last two decades, robotic surgery has come a long way from initial attempts to routine application for variety of procedures in all surgical specialties. The great experience acquired with highly complex operations has allowed its introduction in transplantation. The initial experience is very promising. Candidates for solid organ transplantation, with their long-standing, complex medical illness and immunosuppressed state could benefit the most from this minimally invasive approach as a standard of care in the near future.

References

1. Flowers, J. L., Jacobs, S., Cho, E., Morton, A., Rosenberger, W. F., Evans, D., et al. (1997). Comparison of open and laparoscopic live donor nephrectomy. *Annals of Surgery, 226*(4), 483–489; discussion 9–90.
2. Harris, S. J., Arambula-Cosio, F., Mei, Q., Hibberd, R. D., Davies, B. L., Wickham, J. E., et al. (1997). The Probot – An active robot for prostate resection. *Proceedings of the Institution of Mechanical Engineers. Part H, 211*(4), 317–325.
3. Sung, G. T., & Gill, I. S. (2001). Robotic laparoscopic surgery: A comparison of the DA Vinci and Zeus systems. *Urology, 58*(6), 893–898.
4. Finley, D. S., & Nguyen, N. T. (2005). Surgical robotics. *Current Surgery, 62*(2), 262–272. doi:10.1016/j.cursur.2004.11.005.
5. Matas, A. J., Payne, W. D., Sutherland, D. E., Humar, A., Gruessner, R. W., Kandaswamy, R., et al. (2001). 2,500 living donor kidney transplants: a single-center experience. *Annals of Surgery, 234*(2), 149–164.
6. Pugin, F., Bucher, P., & Morel, P. (2011). History of robotic surgery: From AESOP(R) and ZEUS(R) to da Vinci(R). *Journal of Visceral Surgery, 148*(5 Suppl), e3–e8. doi:10.1016/j.jviscsurg.2011.04.007.
7. Herron, D. M., Marohn, M., & Group S-MRSC. (2008). A consensus document on robotic surgery. *Surgical Endoscopy, 22*(2), 313–325.; discussion 1–2. doi:10.1007/s00464-007-9727-5.
8. Rosales, A., Salvador, J. T., Urdaneta, G., Patino, D., Montlleo, M., Esquena, S., et al. (2010). Laparoscopic kidney transplantation. *European Urology, 57*(1), 164–167. doi:10.1016/j.eururo.2009.06.035.
9. Schweitzer, E. J., Wilson, J., Jacobs, S., Machan, C. H., Philosophe, B., Farney, A., et al. (2000). Increased rates of donation with laparoscopic donor nephrectomy. *Annals of Surgery, 232*(3), 392–400.

10. Horgan, S., Benedetti, E., & Moser, F. (2004). Robotically assisted donor nephrectomy for kidney transplantation. *American Journal of Surgery, 188*(4A Suppl), 45S–51S. doi:10.1016/j.amjsurg.2004.08.028.
11. Horgan, S., Vanuno, D., Sileri, P., Cicalese, L., & Benedetti, E. (2002). Robotic-assisted laparoscopic donor nephrectomy for kidney transplantation. *Transplantation, 73*(9), 1474–1479.
12. Gorodner, V., Horgan, S., Galvani, C., Manzelli, A., Oberholzer, J., Sankary, H., et al. (2006). Routine left robotic-assisted laparoscopic donor nephrectomy is safe and effective regardless of the presence of vascular anomalies. *Transplant International, 19*(8), 636–640. doi:10.1111/j.1432-2277.2006.00315.x.
13. Horgan S, Galvani C, Gorodner MV, Jacobsen GR, Moser F, Manzelli A, et al. (2007). Effect of robotic assistance on the "learning curve" for laparoscopic hand-assisted donor nephrectomy. *Surgical Endoscopy and Other Interventional Techniques, 21*(9), 1512–1517. doi:10.1007/s00464-006-9140-5.
14. Giacomoni, A., Di Sandro, S., Lauterio, A., Concone, G., Buscemi, V., Rossetti, O., et al. (2015). Robotic nephrectomy for living donation: surgical technique and literature systematic review. *American Journal of Surgery*. doi:10.1016/j.amjsurg.2015.08.019.
15. Hubert, J., Renoult, E., Mourey, E., Frimat, L., Cormier, L., & Kessler, M. (2007). Complete robotic-assistance during laparoscopic living donor nephrectomies: an evaluation of 38 procedures at a single site. *International Journal of Urology, 14*(11), 986–989. doi:10.1111/j.1442-2042.2007.01876.x.
16. Toledo-Pereyra, L. H., & Toledo, A. H. (2005). 1954. *Journal of Investigative Surgery, 18*(6), 285–290. doi:10.1080/08941930500433860.
17. OPTN.org. (2015). *Donor, Kidney, by donor type*. http://optn.transplant.hrsa.gov/converge/latestData/rptData.asp. Accessed July 2015.
18. Axelrod, D. A., McCullough, K. P., Brewer, E. D., Becker, B. N., Segev, D. L., & Rao, P. S. (2010). Kidney and pancreas transplantation in the United States, 1999-2008: The changing face of living donation. *American Journal of Transplantation, 10*(4 Pt 2), 987–1002. doi:10.1111/j.1600-6143.2010.03022.x.
19. Ratner, L. E., Ciseck, L. J., Moore, R. G., Cigarroa, F. G., Kaufman, H. S., & Kavoussi, L. R. (1995). Laparoscopic live donor nephrectomy. *Transplantation, 60*(9), 1047–1049.
20. Matas, A. J., Bartlett, S. T., Leichtman, A. B., & Delmonico, F. L. (2003). Morbidity and mortality after living kidney donation, 1999-2001: Survey of United States transplant centers. *American Journal of Transplantation, 3*(7), 830–834.
21. Andersen, M. H., Mathisen, L., Oyen, O., Edwin, B., Digernes, R., Kvarstein, G., et al. (2006). Postoperative pain and convalescence in living kidney donors-laparoscopic versus open donor nephrectomy: A randomized study. *American Journal of Transplantation, 6*(6), 1438–1443. doi:10.1111/j.1600-6143.2006.01301.x.
22. Galvani, C. A., Garza, U., Leeds, M., Kaul, A., Echeverria, A., Desai, C. S., et al. (2012). Single-incision robotic-assisted living donor nephrectomy: Case report and description of surgical technique. *Transplant International, 25*(8), e89–e92. doi:10.1111/j.1432-2277.2012.01493.x.
23. Oberholzer, J., Tzvetanov, I., Mele, A., & Benedetti, E. (2010). Laparoscopic and robotic donor pancreatectomy for living donor pancreas and pancreas-kidney transplantation. *Journal of Hepato-Biliary-Pancreatic Sciences, 17*(2), 97–100. doi:10.1007/s00534-009-0146-y.
24. Liu, X. S., Narins, H. W., Maley, W. R., Frank, A. M., & Lallas, C. D. (2012). Robotic-assistance does not enhance standard laparoscopic technique for right-sided donor nephrectomy. *Journal of the Society of Laparoendscopic Surgeons., 16*(2), 202–207.
25. Doumerc, N., Beauval, J. B., Rostaing, L., & Sallusto, F. (2016). A new surgical area opened in renal transplantation: A pure robot-assisted approach for both living donor nephrectomy and kidney transplantation using transvaginal route. *Transplant International, 29*(1), 122–123. doi:10.1111/tri.12678.
26. Menon, M., Sood, A., Bhandari, M., Kher, V., Ghosh, P., Abaza, R., et al. (2014). Robotic kidney transplantation with regional hypothermia: A step-by-step description of the Vattikuti Urology Institute-Medanta technique (IDEAL phase 2a). *European Urology, 65*(5), 991–1000. doi:10.1016/j.eururo.2013.12.006.

27. Abaza, R., Ghani, K. R., Sood, A., Ahlawat, R., Kumar, R. K., Jeong, W., et al. (2014). Robotic kidney transplantation with intraoperative regional hypothermia. *BJU International, 113*(4), 679–681. doi:10.1111/bju.12572.
28. Oberholzer, J., Giulianotti, P., Danielson, K. K., Spaggiari, M., Bejarano-Pineda, L., Bianco, F., et al. (2013). Minimally invasive robotic kidney transplantation for obese patients previously denied access to transplantation. *American Journal of Transplantation, 13*(3), 721–728. doi:10.1111/ajt.12078.
29. Menon, M., Abaza, R., Sood, A., Ahlawat, R., Ghani, K. R., Jeong, W., et al. (2014). Robotic kidney transplantation with regional hypothermia: Evolution of a novel procedure utilizing the IDEAL guidelines (IDEAL phase 0 and 1). *European Urology, 65*(5), 1001–1009. doi:10.1016/j.eururo.2013.11.011.
30. Tsai, M. K., Lee, C. Y., Yang, C. Y., Yeh, C. C., RH, H., & Lai, H. S. (2014). Robot-assisted renal transplantation in the retroperitoneum. *Transplant International, 27*(5), 452–457. doi:10.1111/tri.12279.
31. Gordon, Z. N., Angell, J., & Abaza, R. (2014). Completely intracorporeal robotic renal autotransplantation. *The Journal of Urology, 192*(5), 1516–1522. doi:10.1016/j.juro.2014.02.2589.
32. Lee, J., & Ordon, M. (2016). Innovative applications of robotic surgery: Renal allograft and autologous transplantation. *F1000Res, 5*. doi:10.12688/f1000research.7343.1.
33. Lee, J. Y., Alzahrani, T., & Ordon, M. (2015). Intra-corporeal robotic renal auto-transplantation. *Canadian Urological Association Journal, 9*(9–10), E748–E749. doi:10.5489/cuaj.3015.
34. Troppmann, C., Gruessner, A. C., Dunn, D. L., Sutherland, D. E., & Gruessner, R. W. (1998). Surgical complications requiring early relaparotomy after pancreas transplantation: A multivariate risk factor and economic impact analysis of the cyclosporine era. *Annals of Surgery, 227*(2), 255–268.
35. Boggi, U., Signori, S., Vistoli, F., D'Imporzano, S., Amorese, G., Consani, G., et al. (2012). Laparoscopic robot-assisted pancreas transplantation: First world experience. *Transplantation, 93*(2), 201–206. doi:10.1097/TP.0b013e318238daec.
36. Buell, J. F., Cherqui, D., Geller, D. A., O'Rourke, N., Iannitti, D., Dagher, I., et al. (2009). The international position on laparoscopic liver surgery: The Louisville Statement, 2008. *Annals of Surgery, 250*(5), 825–830.
37. Giulianotti, P. C., Coratti, A., Sbrana, F., Addeo, P., Bianco, F. M., Buchs, N. C., et al. (2011). Robotic liver surgery: Results for 70 resections. *Surgery, 149*(1), 29–39. doi:10.1016/j.surg.2010.04.002.
38. Giulianotti, P. C., Tzvetanov, I., Jeon, H., Bianco, F., Spaggiari, M., Oberholzer, J., et al. (2012). Robot-assisted right lobe donor hepatectomy. *Transplant International, 25*(1), e5–e9. doi:10.1111/j.1432-2277.2011.01373.x.
39. Lai, E. C., Tang, C. N., & Li, M. K. (2012). Robot-assisted laparoscopic hemi-hepatectomy: Technique and surgical outcomes. *International Journal of Surgery, 10*(1), 11–15. doi:10.1016/j.ijsu.2011.10.005.
40. Soubrane, O., Cherqui, D., Scatton, O., Stenard, F., Bernard, D., Branchereau, S., et al. (2006). Laparoscopic left lateral sectionectomy in living donors: Safety and reproducibility of the technique in a single center. *Annals of Surgery, 244*(5), 815–820. doi:10.1097/01.sla.0000218059.31231.b6.
41. Koffron, A. J., Kung, R., Baker, T., Fryer, J., Clark, L., & Abecassis, M. (2006). Laparoscopic-assisted right lobe donor hepatectomy. *American Journal of Transplantation, 6*(10), 2522–2525. doi:10.1111/j.1600-6143.2006.01498.x.

Chapter 6
Cellular Therapy in Transplantation and Tolerance

Gavin M. Mason, Jayna Patel, Leena Halim, Niloufar Safinia and Giovanna Lombardi

Abstract The adoptive transfer of human regulatory T cells (Tregs) in transplantation offers an attractive therapeutic alternative in the current struggle to improve long-term outcomes.

$CD4^+CD25^+FOXP3^+$ (Tregs) play an important role in immunoregulation and have been shown in animal models to promote transplantation tolerance. Phase I trials in bone marrow transplantation and type I diabetes have already shown that ex vivo expanded Tregs have an excellent safety profile, which is encouraging for their current use as novel therapeutic strategies in solid organ transplantation.

As such, the practicality of Treg adoptive cell therapy is now widely accepted, provided that tailor-made clinical grade procedures for the isolation and ex vivo cell handling are available. Here we present a review on the concept of Treg biology and heterogeneity, the desire to isolate and expand a functionally superior Treg population and report on the effect of differing culture conditions.

We will summarise some of the protocols used for their ex vivo expansion, outline the clinical trials to date and discuss the future directions of Treg cell therapy.

6.1 Introduction

Solid organ transplantation has become an established therapeutic approach in the treatment of end-stage failure of multiple organs. This is in part due to the synchronous improvements in surgical techniques used to transplant the graft and additionally the immunosuppressive therapies that specifically target T cell-mediated responses to avoid immunological rejection of the transplanted organ. However, treatment is still associated with toxicity due to the immunosuppressive regimens required post-transplantation. Consequently, current approaches in the field of transplantation aim to improve the long-term health of both patient and graft by

G.M. Mason • J. Patel • L. Halim • N. Safinia • G. Lombardi (✉)
MRC Centre for Transplantation, Division of Transplantation Immunology and Mucosal Biology, King's College London, Guy's Hospital, London SE1 9RT, UK
e-mail: giovanna.lombardi@kcl.ac.uk

S.N. Nadig, J.A. Wertheim (eds.), *Technological Advances in Organ Transplantation*, DOI 10.1007/978-3-319-62142-5_6

establishing or promoting immunological tolerance, whereby the recipients' immune system, in the absence of immunosuppressive therapy, is conditioned to accept the graft organ without engaging the deleterious allospecific immune response, graft injury and finally organ rejection. These approaches aim to rebalance the immune response post-transplant from immuno-activation to immuno-regulation by adoptive transfer of in vitro expanded Tregs to increase Treg numbers and the co-administration of therapeutics to further expand the number of stable and functional regulatory T cells (Tregs).

6.2 Regulatory T Cells

A multitude of T cell subtypes have been demonstrated to elicit regulatory function including $CD8^+$ T cells [1], $CD8^+CD28^-$ T cells [2], $CD4^-CD8^-$ T cells [3], natural killer (NK) T cells [4] and γδ T cells [5]. However, these all remain less well studied compared to $CD4^+$ Tregs, which have been shown to be essential for immune-homeostasis and are critical mediators of immunological tolerance [6]. Tregs are functionally determined by their regulatory properties and suppressive function, but routinely are phenotypically characterised as $CD4^+$ T cells that constitutively express high levels of CD25 (IL2 receptor α chain) [7] and the canonical transcription factor forkhead box protein 3 (FoxP3), which is known to be essential for the development and maintenance of Tregs [8, 9]. Mutations in the *FoxP3* gene, such as those seen in patients with immune dysregulation polyendocrinopathy enteropathy and X-linked syndrome (IPEX), are known to result in defective generation of Tregs with reduced functional capacity, loss of peripheral tolerance and the development of a range of severe autoimmune conditions [8, 10–12]. $CD4^+CD25^{hi}FoxP3^+$ Tregs are derived in the thymus and represent stable and committed cells referred to as thymic Tregs (tTregs), in contrast to peripheral Tregs (pTreg) that are generated in the periphery by T cell receptor engagement (reviewed in [13]). It has been shown that tTregs are more stable and suppressive as compared to pTreg in both the prevention of autoimmunity and maintaining tolerance [14]. Furthermore, data from murine models and clinical studies have clearly demonstrated that reduced Treg numbers or function may contribute to autoimmune disease development and pathology [15]. Consequently, the therapeutic modulation of Tregs represents a novel approach in the treatment of a range of immune-mediated pathologies including autoimmunity, transplantation and graft versus host disease (GVHD).

In the context of transplantation, therapeutic strategies that modulate Tregs are currently being investigated as an attractive approach to promote tolerance and suppress alloreactive T cell responses. To implement such strategies, it is essential to consider the antigen presentation pathways. Recipient T cells can respond to alloantigens presented by two main pathways [16]. In the direct pathway of allorecognition, recipient T cells respond directly to intact major histocompatibility (MHC) on antigen-presenting cells (APCs), derived from the graft. In the indirect pathway of allorecognition, recipient T cells respond to processed alloantigens derived from donor MHC presented on recipient APC in the context of self-MHC.

Effector T cells (Teff) with direct alloreactivity are present at high frequency, and their activation post-transplant is rapid, resulting in vigorous T cell-mediated immune-pathology [17–19]. The direct pathway of allorecognition was thought to be associated with early rejection after transplantation, whilst the indirect pathway was thought to be more associated with chronic rejection. However, recently our group has described a third pathway of allorecognition suggesting that recipient APC by trafficking trough the graft can acquire intact MHC molecules, leading to presentation of both intact donor MHC molecules and graft-derived peptide in the context of recipient MHC. This pathway was termed semi-direct allorecognition (reviewed in [20]). The consequence of this discovery is that the recognition of intact allogeneic MHC molecules by the recipient T cells with direct allorecognition can last for the entire duration of the graft. As such, to counterbalance this and promote regulation instead of alloreactivity, sufficient numbers of functional and stable Tregs would be required to modulate both direct and indirect alloimmune responses and promote tolerance [21, 22]. Approaches to increase the available pool of Tregs and tip the immunological balance towards regulation and tolerance have involved therapeutic modulation of Tregs and Teff in vivo but also ex vivo purification of Tregs, followed by in vitro expansion and infusion of expanded Tregs by adoptive cell therapy approaches.

6.2.1 *Phenotypic Diversity of Human Tregs*

Tregs are most correctly defined by their regulatory function but are currently best phenotypically defined as $CD4^{+}CD25^{+}FOXP3^{+}$ cells [23]. Unlike in the murine system, where constitutive expression of these markers defines pure bona fide Tregs, in humans, this does not result in sufficiently pure populations of Tregs, as activated Teff can also temporarily express both FOXP3 and CD25 and constitute non-suppressive Teff [24]. Furthermore, the intranuclear localisation of FOXP3 complicates matters, as it cannot be used to isolate viable cells due to the required fixation and permeabilisation needed to detect its presence. However, FOXP3-expressing Treg has been shown to express low levels of CD127 (IL-7 receptor α chain), which in combination with the constitutive high levels of CD25 allows identification and isolation of highly pure populations of human Tregs by flow cytometry [25]. However, even using current flow cytometry sorting strategies, it is still not possible to isolate absolutely pure populations of functionally suppressive Tregs. Despite decades of work, we are yet to identify markers that define the lineage of Tregs and as such a phenotype that would allow isolation of pure Tregs without contaminating non-regulatory T cells.

It is becoming clear that the Treg cell compartment in humans is highly heterogeneous. Seminal work from Miyara *et al.*, was able to split the human $FOXP3^{+}$

Treg compartment into three phenotypically and functionally distinct groups of cells based on their differential expression of CD45RA and FOXP3, describing naïve/resting Treg (CD45RA^{+}FOXP3^{low}) and effector Treg (CD45RA^{+}FOXP3^{high}), both of which displayed suppressive functionality and stable expression of FOXP3 as measured by epigenetic modification of the *FOXP3* promoter [6, 26]. The third population, non-Tregs (CD45RA^{-} FOXP3^{low}), was poorly suppressive, displayed unstable FOXP3 expression and were found to largely express pro-inflammatory cytokines. More recently, we have identified a subpopulation of suppressive Tregs within this 'non-Treg' population (CD45RA^{-} FOXP3^{low}) that additionally express CD161 and produce pro-inflammatory cytokines such as IL-17 [27, 28]. This demonstrated that the Treg compartment in humans was composed of phenotypically and functionally distinct populations of Tregs. Interestingly, populations of FOXP3^{+} Tregs have since been described that resemble the classic Teff subsets including Th1, Th2 and Th17 cells, co-expressing their respective canonical transcription factors (Tbet, Gata3 and RORγT), associated chemokine receptors (CXCR3, CCR4, CCR6) and secrete their characteristic cytokines (IFNγ, IL4 and IL17), whilst still retaining their regulatory capacity [29]. In conclusion, human Tregs have been described to express a broad range of markers, many of which can be co-expressed by Teff and so limit their use in isolating human Tregs (summarised in Table 6.1) (reviewed in [30]).

Work from our laboratory has further dissected the phenotypic complexity of the human peripheral blood Treg compartment using novel deep immunophenotypic approaches [59]. Using an extensive panel of Treg markers and mass cytometry, we described 22 distinct populations of human Treg, demonstrating a dynamic, complex and diverse structure of subpopulations within the human Treg compartment [59].

Table 6.1 Established phenotypic markers of regulatory T cells (Treg)

Marker	Reference	Marker	Reference	Marker	Reference
CD39	[31]	CCR4	[29, 32]	KLRG1	[33]
CD73	[34]	CD28	[35]	CD15S	[36]
CTLA-4	[37]	CD62L	[38]	CD5	[39, 40]
LAP	[41]	CXCR3	[29]	IL-10	[42, 43]
GITR	[44]	CD161	[28]	IL-35	[45]
GARP	[46]	ICOS	[47]	TGF-β	[48]
HLA-DR	[49]	OX40	[50]	FoxP3	[9]
CCR6	[29]	CCR10	[29]	Helios	[51]
CD45RA	[26]	CD137	[52]	Eos	[28]
CD103	[53]	CD25	[7]	Perforin	[54, 55]
CD31	[56]	CD38	[57]	Granzyme	[58]

6.2.2 Mechanisms of Treg-Mediated Suppression

Central to the definition and essential role Tregs play is their regulatory function. Currently, there are five general mechanisms of suppression that have been identified, utilised by Tregs (summarised in Fig. 6.1). Firstly, Tregs have been shown to secrete IL10, TGFβ and IL35, each of which can have immunomodulatory/

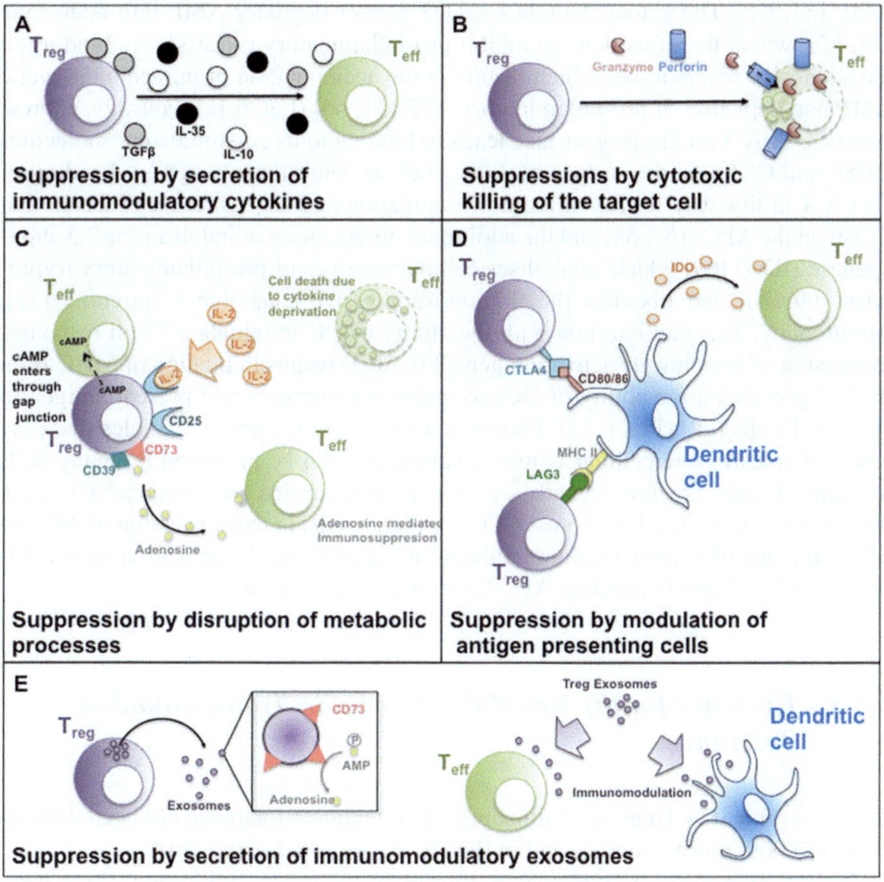

Fig. 6.1 Mechanisms of regulatory T cell (Treg) suppression. There are five main potential mechanisms of suppression used by Treg to modulate target cell function: the release of immunomodulatory cytokines including IL10, TGFβ and IL35 (**a**); cytotoxic targeting and killing of the target cells by the secretion of cytotoxic mediators including perforin and granzyme (**b**); metabolic disruption of the target cell by IL2 deprivation, cyclic AMP production or CD39-/CD73-mediated extracellular generation of immune-suppressive adenosine (**c**); modulation of APCs by interfering with the maturation by LAG3/MHC II inhibition of maturation, CTLA4/CD80/CD86 induction of indoleamine 2,3-deoxygenase (IDO) production (**d**); and release of immunomodulatory exosomes bearing CD73 which mediates reduction of AMP to adenosine resulting in immunosuppression of T cells and dendritic cells (**e**). (Figures generated by Gavin)

suppressive effects on the target cell [42, 45, 60] (Fig. 6.1a). Secondly, Tregs can express and release cytotoxic mediators including perforin and granzymes, which results in the cell contact dependent cytolysis of the target cell [55, 58, 61] (Fig. 6.1b). Thirdly, Tregs constitutively express high levels of CD25 and as such compete for IL2 with other T cells, resulting in deprivation of IL2 and subsequent apoptosis of the deprived Teff cells [62, 63]. Alternatively, the expression of the ectoenzyme CD39 (nucleoside triphosphate diphosphohydrolase-1) is expressed on a subset of memory Tregs that exhibit superior suppressive function and degrades extracellular ATP into AMP [31, 64]. The ectonucleotidase CD73 further degrades AMP into adenosine [34, 65], which then functions to inhibit pro-inflammatory processes by binding to the adenosine receptor 2A, which results in the accumulation of intracellular cyclic AMP and depletion of pro-inflammatory ATP [31, 66] (Fig. 6.1c). Fourthly, expression of CTLA-4 on the Treg surface leads to binding to its co-stimulatory molecules CD80 and CD86 on the surface of APC, such as dendritic cells (DCs). Binding of CTLA-4 in this way results in the downregulation and sequestration of CD80 and CD86 on the APCs [67, 68] and the additional upregulation of indoleamine 2,3-dioxygenase (IDO) [69] which catabolises cellular reserves of pro-inflammatory tryptophan [69, 70] and liberates the immunosuppressive catabolite kynurenine [71]. Additionally, Treg can interfere with the ability of DC to initiate a T cell response; expression of lymphocyte activation gene 3 (LAG3) results in ligation of MHC II on the DC preventing the ability of DCs to undergo maturation and present antigen to effector T cells [72] (Fig. 6.1d). Finally, exosomes are membrane vesicles that play essential roles in intercellular communication and can be produced by many cells including T cells. We have recently shown that when murine Tregs become activated, they secrete exosomes that express CD73, which results in the conversion of AMP to adenosine and ultimately immunomodulation, suppressing the activation and proliferation of T cells and inhibiting APC function [65] (Fig. 6.1e).

6.2.3 Evidence for the Key Role of Tregs in Transplantation Tolerance

The involvement of Tregs in establishing transplantation tolerance has been demonstrated in both murine models and in the clinic (reviewed in [73, 74]).

In murine models of spontaneous transplant tolerance, infiltrating Tregs were found to increase in number in the transplanted grafts. Subsequent depletion of these Tregs led to disruption of the regulatory balance, prompting graft rejection [75]. Importantly, in-depth characterisation of these Tregs, which were central to determining the balance between tolerance and rejection, revealed they had indirect anti-donor specificity [76–78].

In humans, studies from a proportion of liver transplant recipients deemed ‘operationally tolerant’, a phenomenon where patients spontaneously accept their graft independent of immunosuppressive regimens, revealed increased proportions of

$FOXP3^{+}$ cells in liver biopsies, a finding which was mirrored in the peripheral blood [79]. Correspondingly, reduced numbers of circulating Tregs have been reported in liver transplant patients undergoing acute rejection [80]. Taken together, these studies establish the role of Tregs in tolerance post-transplantation and provide evidence towards the clinical potential of Tregs in the establishment of transplant tolerance.

6.3 Adoptive Treg Therapy from Murine Models to the Clinic

The concept of 'adoptive immunity', describing the passive transfer of immune cells in order to influence immune responses, has been at the forefront of biological medicine. On account of the overwhelming evidence supporting the role of Tregs in the development of tolerance, research has very much been directed to investigate the feasibility and efficacy of adoptive transfer of these cells in the setting of organ transplantation. A number of animal models have demonstrated the potential of adoptive Treg transfer in achieving transplant tolerance (reviewed in [74]). In murine models, we and others have shown that murine Treg lines with self-specificity expanded in vitro can prolong allograft survival; however, Tregs generated to be specific for alloantigen were more effective at promoting survival after MHC-mismatched heart allograft transplants [81, 82]. Similarly, in murine models of skin transplantation, we also report that Tregs with indirect allospecificity prolong graft survival [81, 83]. Initially, in a murine model of islet transplant, it was demonstrated that Tregs with specificity for the graft are ten times more efficient than self-specific Tregs in prolonging allograft survival [84]. More recent refinement has suggested that donor antigen-specific Tregs are five times more potent than self-specific, polyclonal Tregs in inducing indefinite islet graft survival [85].

Humanised mouse models have been developed for both tissue transplantation and GVHD and have been used to additionally demonstrate the importance of Tregs in the promotion of transplant tolerance [86, 87]. Adoptive transfer of cell-sorted and cell-expanded Tregs has been shown to ameliorate vasculopathy associated with graft rejection in a human vessel transplant model [88, 89]. We and others have demonstrated in humanised mouse model of human skin transplantation that polyclonal Tregs are capable of protecting the skin from graft damage although using the same number of Tregs, they are less efficacious in doing so as compared to Tregs with direct allospecificity [90, 91].

Altogether, data from animal models in which Tregs have been adoptively transferred suggest that adoptive transfer of Tregs in the context of transplantation represents an efficacious and viable approach to establish transplant tolerance and prevent graft rejection. In addition, it is clear that Tregs with specificity for the graft offer an advantage compared to polyclonal Tregs in protecting from graft rejection [77, 81, 82, 92, 93]. It is also clear that to effectively suppress immune responses and potentially promote tolerance, high numbers of Tregs are required, to alter the Teff/Treg ratio [94, 95], although as discussed above, less Tregs are needed if they are specific for the alloantigen [87]. As presented in a very elegant review by Tang *et al.*, it has been estimated that the endogenous pool of Tregs needs to be increased by 30% to tip the balance in favour of regulation and to achieve tolerance (reviewed in [96]). The frequency of human Tregs in the peripheral blood compartment is estimated between 4 and 10% of $CD4^+$ T cells. As such, protocols to isolate and then expand suppressive and stable Tregs with efficient regulatory function are central to adoptive Treg therapy approaches and protocols. Clinical protocols to generate Tregs with direct allospecificity have already emerged [91], whilst the generation of Tregs with indirect allospecificity has as of yet proved to be more challenging. Reasons for this discrepancy in the ease of generation may lie with the fact that Tregs with direct allospecificity tend to exist at higher frequencies in the circulation as compared to those with indirect allospecificity [85, 97, 98].

6.3.1 Isolation and Expansion of Human Treg for Clinical Use

Isolation strategies for the purification of human Tregs have been under intense scrutiny; however, the marker or combination of markers that delineates a phenotypically and functionally pure population of Tregs currently remains elusive. As mentioned previously in this chapter, the canonical transcription factor FOXP3 is currently the best delineator of Tregs in human $CD4^+$ T cells; however, its intranuclear localisations preclude its use in the isolation of viable cells. Current approaches to characterise and isolate human Tregs for clinical use are now focused on $CD4^+CD25^+$ Tregs. Their selection is centred on the use of the CliniMACS Plus system, with a strategy composed of a two-step isolation process of $CD8^+$ cell depletion and $CD25^+$ cell enrichment. However, the purity of Tregs obtained in this way stands at only around 60–80% [99]. A more technical approach of obtaining phenotypically pure Tregs involves using the cell sorter. The primary advantage of this technique is the ability to isolate Tregs with a purity of over 95%.

Based on the discovery that low expression of CD127 molecules characterise Tregs with potent suppressor function, flow-based cell sorting of $CD4^+CD25^{hi}CD127^{low}$ Tregs has proven to yield a highly pure and functionally superior population of Tregs [25]. However, this strategy requires GMP compatible cell-sorting facilities, which have only recently been made avaibale in a few centres around the world. We are currently validating the new MACSQuant® Tyto™ cell sorter for use in our GMP facility at Guy's Hospital.

Following isolation, purified Tregs need to then be expanded to increase the total number. We have reported the successful isolation and expansion of Tregs for clinical use from both healthy donors and patients on the waiting list for kidney and liver transplants using magnetic beads separation and polyclonal (anti-CD3/CD28 beads) expansion [100–102]. In fact, published clinical trials have relied on the polyclonal activation induced expansion to generate high doses of Tregs [87, 103–106]. These large-scale expansions of human Tregs for clinical application were achievable in line with a protocol detailing the use of anti-CD3/CD28 beads, high concentrations of IL2 and rapamycin, an immunosuppressant shown to favour Treg survival.

Altogether, these methods to identify, analyse and isolate human Tregs are now well established [25, 59, 107] alongside a number of protocols to expand these cells ex vivo [94, 95, 100, 101, 108].

6.4 Clinical Trials of Adoptive Treg Therapy in Solid Organ Transplantation

The initial phase I clinical trials examining adoptive transfer of in vitro expanded Tregs into patients have been performed in the context of graft versus host disease, transplantation and autoimmunity (summarised in Table 6.2) with all studies establishing Treg cell therapy as a safe and well-tolerated approach.

In the context of GVHD, adoptive Treg therapy has been used to both treat patients with active GVHD [104] and as prophylaxis in patients undergoing haematopoietic stem cell transplantation [87, 105]. Whilst these studies were too small to establish clinical efficacy, they did show safety of adoptive Treg therapy in humans. Whilst not within the field of transplantation, adoptive Treg therapy has also been established as safe, with some limited efficacy in the context of paediatric [103] and adult [106] type 1 diabetes. More recently, a clinical trial in liver transplant patients has been published. However, the adoptive transferred T cells were different from previous studies as they were generated using a 2-week coculture of recipient lymphocytes with irradiated donor cells in the presence of anti-CD80/CD86 monoclonal antibodies ('anergic' T cells). More than seven patients out of ten have completed successful weaning and cessation of immunosuppressive agents [109]. Other stud-

Table 6.2 Summary of adoptive Treg studies in humans

Pathology	Patient numbers	Tregs infused	Determined effect/result	Reference
GVHD	2	1×10^5 – 3×10^6/kg	(i) Immunosuppression discontinued in chronic GVHD (ii) Slowed clinical deterioration in acute GVHD	[104]
GVHD	23	1–30 $\times 10^5$/kg	Reduction in acute GVHD compared to historical controls	[87]
GVHD	28	2–4 $\times 10^6$/kg	Onset of acute GVHD in only 2/28 patients, no cases of chronic GVHD	[105]
T1D	10	10–20 $\times 10^6$/kg	Reported reduced insulin requirement and C-peptide levels in treated patients	[103]
T1D	14	0.05–26 $\times 10^8$cells	Transient increases in peripheral blood Treg numbers and C-peptide levels persisted for >2 years post-treatment	[106]

Table 6.3 Summary of current adoptive Treg clinical trials in humans

Study ID	Phase	Product	Indication	Centre
NCT02129881	I/II	Expanded polyclonal T_{regs}	Living donor kidney transplant	London, Oxford
NCT02371434	I/II	Expanded polyclonal T_{regs}	Living donor kidney transplant	Berlin
NCT02244801	I/II	Donor alloantigen-reactive T_{regs}	Living donor kidney transplant	San Francisco
NCT02091232	I/II	Belatacept conditioned	Living donor kidney transplant	Boston
Planned	I/II	Antigen-specific Tr1	Living donor kidney transplant	Milan
NCT02166177	I	Expanded polyclonal T_{regs}	Liver transplant	London
NCT02188719	I	Donor alloantigen-reactive T_{regs}	Liver transplant	San Francisco
NCT02088931	I	Expanded polyclonal T_{regs}	Subclinical rejection in kidney transplantation	San Francisco
NCT02474199	I	Donor alloantigen-reactive T_{regs}	CNI reduction in liver transplantation	San Francisco

ies in the context of solid organ transplantation, using more 'conventional' Tregs, are underway, including the ThRIL study (discussed below) (current clinical trials summarised in Table 6.3).

We have explored the use of adoptive Treg therapy in the context of solid organ transplantation in both kidney (the ONE study) and liver (ThRIL) transplant recipients [101, 102, 110]. The ONE study was a European union, FP7 programme-funded multicentre phase I/II study investigating the safety of adoptive cell therapy in kidney transplant recipients. The ThRIL study was a UK Medical Research Council-funded, open-label, randomised, controlled, parallel-group, single ascending dose study, investigating the safety and tolerability of adoptive Treg therapy in liver transplant recipients. These two phase I trials using polyclonal in vitro expanded recipient Treg represent the first-in-man trials for adoptive Treg therapy in solid organ transplantation.

6.5 Combined Therapy

Although Treg therapy has been demonstrated safe and has shown some signs of efficacy, it is more and more clear that there is a need to start to think about combined therapies in which Tregs are used in conjunction with other strategies. Here we discuss two of these possible combined therapies.

6.5.1 Rapamycin

Rapamycin is an immune-suppressive drug that targets the mammalian target of rapamycin (mTOR) kinase signalling pathway [111]. mTOR is a serine-threonine protein kinase involved in cell growth, proliferation and differentiation. Rapamycin inactivates this signalling cascade in Teff and not Treg, preferentially inhibiting Teff resulting in the selective inhibition of non-Tregs both in vitro and in vivo. Additionally, metabolic pathways can set the responsiveness of Tregs and have impact upon the establishment of tolerance. Specifically, leptin induces increased activity in the mTOR pathway and subsequent anergy in Tregs. However, inhibition of mTOR with rapamycin prior to TCR stimulation resulted in highly proliferative Tregs even in the absence of IL2 [112].

Administration of rapamycin in a humanised mouse model firstly resulted in the apoptosis of $CD4^+$ Teff cells and, secondly, in conjunction with the adoptive transfer of subtherapeutic doses of Tregs, was shown to specifically inhibit T cell proliferation and suppressed transplant arteriosclerosis [113]. In agreement with what we know about the effect of rapamycin on Tregs, in the context of liver and kidney transplantation, maintenance therapy using rapamycin over several years resulted in increased Treg numbers compared to those treated with calcineurin inhibitors [114, 115]. Furthermore, Tregs isolated from rapamycin-treated patients demonstrated increased expression of CTLA-4, highly suggestive of Tregs with a more potent suppressive capacity [114].

It has been suggested that rapamycin may have a dual effect in terms of Treg biology. The mechanism of rapamycin involves inhibition of the mammalian target of rapamycin (mTOR) pathway. This pathway plays a predominant role in the activation of the majority of inflammatory T cells; thus, upon treatment with rapamycin, T cell proliferation and survival are curtailed. However, in Tregs, this pathway is underactive rendering Tregs 'immune' to the effects of rapamycin, conveying the impression of Treg preservation.

Furthermore, in addition to its unique relationship in promoting Treg survival, rapamycin has also been shown to play a significant role in safeguarding Treg stability through promoting the expansion of nonplastic Treg populations and preventing the release of inflammatory cytokines in inflammatory microenvironments [100]. Additionally, treatment with rapamycin stabilises FOXP3 gene expression through epigenetic mechanisms, maintaining the demethylation status in the Treg-specific demethylated region (TSDR) in the *FOXP3* promoter region supporting the stable and active transcription of FOXP3 in Tregs [116]. In the ThRIL trial, in vitro expanded Tregs are injected 1 month after the start of rapamycin treatment.

6.5.2 Interleukin 2 (IL2)

The therapeutic use of low-dose IL2 has been explored with the aim of promoting tolerance and regulation by supporting the endogenous Treg pool in vivo. The cytokine IL2 is essential for Treg proliferation, maintenance and survival, and the administration of IL2 has been shown to induce the in vivo expansion of Tregs in patients with metastatic cancer and chronic myelogenous leukaemia after receiving allogeneic haematopoietic stem cell transplantation [117, 118]. Similarly, administration of low-dose IL2 in patients with active GVHD induced sustained Treg expansion and reduction in disease severity and manifestation of GVHD symptoms [119]. Furthermore, the same group, in a phase II clinical trial, showed that IL2 therapy over the course of 2 years induced persistent Treg responses, demonstrating normalisation of Treg/Teff ratios and clinical improvement [119]. In the context of a murine model of type 1 diabetes, the administration of low-dose IL2 induced the in vivo activation and expansion of Tregs, which restricted inflammation in the pancreas with partial amelioration of disease [120]. Dual treatment with rapamycin and IL2, in the context of an experimental skin graft model, resulted in significantly delayed rejection when compared with IL2 monotherapy, which appeared to be due to increased numbers of activated Treg and concomitant reduction in Teff numbers [121]. Based on the evidence that IL2 is beneficial for Tregs, many centres including ours have started to propose new clinical trials in which Treg therapy will be combined with low-dose IL2.

6.6 Conclusions

Significant advances have been made in the field of Tregs and their biology over the past decade. The advent of methods to routinely identify, isolate, expand and characterise Tregs in humans has accelerated our understanding of their importance and function during immune-homeostasis and crucially in the development and maintenance of immunological tolerance. Consequently, the potential modulation of Tregs clinically represents a novel and attractive approach in the context of transplantation and establishment of transplant tolerance. Adoptive Treg therapy has the potential to revolutionise the approach to treatment after solid organ transplantation. The results from the first adoptive Treg therapy clinical trials in GVHD and T1D are encouraging in terms of both being safe and also showing some efficacy in the stem cell transplantation and treatment of autoimmunity, respectively. Furthermore, the world's first clinical trials of adoptive Treg therapy in solid organ transplantation have been completed, and recently the first clinical trial with 'anergic' T cells in liver transplant patients has been published [109]. The results from the ONE study and ThRIL will be available in the next few months. Altogether we await the data from these and other studies to determine if adoptive Treg therapy will make it possible to reduce or replace standard immune-suppressive regimens and whether Treg therapy alone or in conjunction with other therapeutic avenues may facilitate the establishment of transplantation tolerance in the future.

References

1. Reibke, R., et al. (2006). CD8+ regulatory T cells generated by neonatal recognition of peripheral self-antigen. *Proceedings of the National Academy of Sciences of the United States of America, 103*(41), 15142–15147.
2. Haribhai, D., et al. (2007). Regulatory T cells dynamically control the primary immune response to foreign antigen. *Journal of Immunology, 178*(5), 2961–2972.
3. Zhang, Z. X., et al. (2000). Identification of a previously unknown antigen-specific regulatory T cell and its mechanism of suppression. *Nature Medicine, 6*(7), 782–789.
4. Monteiro, M., et al. (2010). Identification of regulatory Foxp3+ invariant NKT cells induced by TGF-beta. *Journal of Immunology, 185*(4), 2157–2163.
5. Hayday, A., & Tigelaar, R. (2003). Immunoregulation in the tissues by gammadelta T cells. *Nature Reviews. Immunology, 3*(3), 233–242.
6. Miyara, M., & Sakaguchi, S. (2011). Human FoxP3(+)CD4(+) regulatory T cells: Their knowns and unknowns. *Immunology and Cell Biology, 89*(3), 346–351.
7. Sakaguchi, S., et al. (1995). Immunologic self-tolerance maintained by activated T cells expressing IL-2 receptor alpha-chains (CD25). Breakdown of a single mechanism of self-tolerance causes various autoimmune diseases. *Journal of Immunology, 155*(3), 1151–1164.
8. Bennett, C. L., et al. (2001). The immune dysregulation, polyendocrinopathy, enteropathy, X-linked syndrome (IPEX) is caused by mutations of FOXP3. *Nature Genetics, 27*(1), 20–21.
9. Hori, S., Nomura, T., & Sakaguchi, S. (2003). Control of regulatory T cell development by the transcription factor Foxp3. *Science, 299*(5609), 1057–1061.
10. Kobayashi, I., et al. (2001). Novel mutations of FOXP3 in two Japanese patients with immune dysregulation, polyendocrinopathy, enteropathy, X linked syndrome (IPEX). *Journal of Medical Genetics, 38*(12), 874–876.
11. Le Bras, S., & Geha, R. S. (2006). IPEX and the role of Foxp3 in the development and function of human Tregs. *The Journal of Clinical Investigation, 116*(6), 1473–1475.
12. van der Vliet, H. J., & Nieuwenhuis, E. E. (2007). IPEX as a result of mutations in FOXP3. *Clinical & Developmental Immunology, 2007*, 89017.
13. Abbas, A. K., et al. (2013). Regulatory T cells: Recommendations to simplify the nomenclature. *Nature Immunology, 14*(4), 307–308.
14. Gupta, S., Shang, W., & Sun, Z. (2008). Mechanisms regulating the development and function of natural regulatory T cells. *Archivum Immunologiae et Therapiae Experimentalis (Warsz), 56*(2), 85–102.
15. Miyara, M., et al. (2011). Human FoxP3+ regulatory T cells in systemic autoimmune diseases. *Autoimmunity Reviews, 10*(12), 744–755.
16. Afzali, B., Lechler, R. I., & Hernandez-Fuentes, M. P. (2007). Allorecognition and the alloresponse: Clinical implications. *Tissue Antigens, 69*(6), 545–556.
17. Lindahl, K. F., & Wilson, D. B. (1977). Histocompatibility antigen-activated cytotoxic T lymphocytes. II. Estimates of the frequency and specificity of precursors. *The Journal of Experimental Medicine, 145*(3), 508–522.
18. Lindahl, K. F., & Wilson, D. B. (1977). Histocompatibility antigen-activated cytotoxic T lymphocytes. I. Estimates of the absolute frequency of killer cells generated in vitro. *The Journal of Experimental Medicine, 145*(3), 500–507.
19. Suchin, E. J., et al. (2001). Quantifying the frequency of alloreactive T cells in vivo: New answers to an old question. *Journal of Immunology, 166*(2), 973–981.
20. Smyth, L. A., et al. (2006). A novel pathway of antigen presentation by dendritic and endothelial cells: Implications for allorecognition and infectious diseases. *Transplantation, 82*(1 Suppl), S15–S18.
21. Lechler, R. I., Garden, O. A., & Turka, L. A. (2003). The complementary roles of deletion and regulation in transplantation tolerance. *Nature Reviews. Immunology, 3*(2), 147–158.
22. Walsh, P. T., Taylor, D. K., & Turka, L. A. (2004). Tregs and transplantation tolerance. *The Journal of Clinical Investigation, 114*(10), 1398–1403.

23. Sakaguchi, S., et al. (2013). The plasticity and stability of regulatory T cells. *Nature Reviews. Immunology, 13*(6), 461–467.
24. Allan, S. E., et al. (2007). Activation-induced FOXP3 in human T effector cells does not suppress proliferation or cytokine production. *International Immunology, 19*(4), 345–354.
25. Liu, W., et al. (2006). CD127 expression inversely correlates with FoxP3 and suppressive function of human CD4+ T reg cells. *The Journal of Experimental Medicine, 203*(7), 1701–1711.
26. Miyara, M., et al. (2009). Functional delineation and differentiation dynamics of human CD4+ T cells expressing the FoxP3 transcription factor. *Immunity, 30*(6), 899–911.
27. Pesenacker, A. M., et al. (2013). CD161 defines the subset of FoxP3+ T cells capable of producing proinflammatory cytokines. *Blood, 121*(14), 2647–2658.
28. Afzali, B., et al. (2013). CD161 expression characterizes a subpopulation of human regulatory T cells that produces IL-17 in a STAT3-dependent manner. *European Journal of Immunology, 43*(8), 2043–2054.
29. Duhen, T., et al. (2012). Functionally distinct subsets of human FOXP3+ Treg cells that phenotypically mirror effector Th cells. *Blood, 119*(19), 4430–4440.
30. Povoleri, G. A., et al. (2013). Thymic versus induced regulatory T cells – Who regulates the regulators? *Frontiers in Immunology, 4*, 169.
31. Borsellino, G., et al. (2007). Expression of ectonucleotidase CD39 by Foxp3+ Treg cells: Hydrolysis of extracellular ATP and immune suppression. *Blood, 110*(4), 1225–1232.
32. Sugiyama, D., et al. (2013). Anti-CCR4 mAb selectively depletes effector-type FoxP3+CD4+ regulatory T cells, evoking antitumor immune responses in humans. *Proceedings of the National Academy of Sciences of the United States of America, 110*(44), 17945–17950.
33. Tauro, S., et al. (2013). Diversification and senescence of Foxp3+ regulatory T cells during experimental autoimmune encephalomyelitis. *European Journal of Immunology, 43*(5), 1195–1207.
34. Deaglio, S., et al. (2007). Adenosine generation catalyzed by CD39 and CD73 expressed on regulatory T cells mediates immune suppression. *The Journal of Experimental Medicine, 204*(6), 1257–1265.
35. Guo, F., et al. (2008). CD28 controls differentiation of regulatory T cells from naive CD4 T cells. *Journal of Immunology, 181*(4), 2285–2291.
36. Miyara, M., et al. (2015). Sialyl Lewis x (CD15s) identifies highly differentiated and most suppressive FOXP3high regulatory T cells in humans. *Proceedings of the National Academy of Sciences of the United States of America, 112*(23), 7225–7230.
37. Takahashi, T., et al. (2000). Immunologic self-tolerance maintained by CD25(+)CD4(+) regulatory T cells constitutively expressing cytotoxic T lymphocyte-associated antigen 4. *The Journal of Experimental Medicine, 192*(2), 303–310.
38. Fu, S., et al. (2004). CD4+ CD25+ CD62+ T-regulatory cell subset has optimal suppressive and proliferative potential. *American Journal of Transplantation, 4*(1), 65–78.
39. Henderson, J. G., & Hawiger, D. (2015). Regulation of extrathymic Treg cell conversion by CD5. *Oncotarget, 6*(29), 26554–26555.
40. Henderson, J. G., et al. (2015). CD5 instructs extrathymic regulatory T cell development in response to self and tolerizing antigens. *Immunity, 42*(3), 471–483.
41. Nakamura, K., et al. (2004). TGF-beta 1 plays an important role in the mechanism of CD4+CD25+ regulatory T cell activity in both humans and mice. *Journal of Immunology, 172*(2), 834–842.
42. Hara, M., et al. (2001). IL-10 is required for regulatory T cells to mediate tolerance to alloantigens in vivo. *Journal of Immunology, 166*(6), 3789–3796.
43. Rubtsov, Y. P., et al. (2008). Regulatory T cell-derived interleukin-10 limits inflammation at environmental interfaces. *Immunity, 28*(4), 546–558.
44. McHugh, R. S., et al. (2002). CD4(+)CD25(+) immunoregulatory T cells: Gene expression analysis reveals a functional role for the glucocorticoid-induced TNF receptor. *Immunity, 16*(2), 311–323.

45. Collison, L. W., et al. (2007). The inhibitory cytokine IL-35 contributes to regulatory T-cell function. *Nature, 450*(7169), 566–569.
46. Stockis, J., et al. (2009). Membrane protein GARP is a receptor for latent TGF-beta on the surface of activated human Treg. *European Journal of Immunology, 39*(12), 3315–3322.
47. Ito, T., et al. (2008). Two functional subsets of FOXP3+ regulatory T cells in human thymus and periphery. *Immunity, 28*(6), 870–880.
48. Fu, S., et al. (2004). TGF-beta induces Foxp3 + T-regulatory cells from CD4 + CD25 - precursors. *American Journal of Transplantation, 4*(10), 1614–1627.
49. Baecher-Allan, C., Wolf, E., & Hafler, D. A. (2006). MHC class II expression identifies functionally distinct human regulatory T cells. *Journal of Immunology, 176*(8), 4622–4631.
50. So, T., & Croft, M. (2007). Cutting edge: OX40 inhibits TGF-beta- and antigen-driven conversion of naive CD4 T cells into CD25+Foxp3+ T cells. *Journal of Immunology, 179*(3), 1427–1430.
51. Thornton, A. M., et al. (2010). Expression of Helios, an Ikaros transcription factor family member, differentiates thymic-derived from peripherally induced Foxp3+ T regulatory cells. *Journal of Immunology, 184*(7), 3433–3441.
52. Till, B. G., & Press, O. W. (2012). Depletion of Tregs for adoptive T-cell therapy using CD44 and CD137 as selection markers. *Immunotherapy, 4*(5), 483–485.
53. Annacker, O., et al. (2005). Essential role for CD103 in the T cell-mediated regulation of experimental colitis. *The Journal of Experimental Medicine, 202*(8), 1051–1061.
54. Grossman, W. J., et al. (2004). Human T regulatory cells can use the perforin pathway to cause autologous target cell death. *Immunity, 21*(4), 589–601.
55. Grossman, W. J., et al. (2004). Differential expression of granzymes A and B in human cytotoxic lymphocyte subsets and T regulatory cells. *Blood, 104*(9), 2840–2848.
56. Haas, J., et al. (2007). Prevalence of newly generated naive regulatory T cells (Treg) is critical for Treg suppressive function and determines Treg dysfunction in multiple sclerosis. *Journal of Immunology, 179*(2), 1322–1330.
57. Patton, D. T., et al. (2011). The PI3K p110delta regulates expression of CD38 on regulatory T cells. *PLoS ONE, 6*(3), e17359.
58. Gondek, D. C., et al. (2005). Cutting edge: Contact-mediated suppression by CD4+CD25+ regulatory cells involves a granzyme B-dependent, perforin-independent mechanism. *Journal of Immunology, 174*(4), 1783–1786.
59. Mason, G. M., et al. (2015). Phenotypic complexity of the human regulatory T cell compartment revealed by mass cytometry. *Journal of Immunology, 195*(5), 2030–2037.
60. Sakaguchi, S., et al. (2009). Regulatory T cells: How do they suppress immune responses? *International Immunology, 21*(10), 1105–1111.
61. Cao, X., et al. (2007). Granzyme B and perforin are important for regulatory T cell-mediated suppression of tumor clearance. *Immunity, 27*(4), 635–646.
62. de la Rosa, M., et al. (2004). Interleukin-2 is essential for CD4+CD25+ regulatory T cell function. *European Journal of Immunology, 34*(9), 2480–2488.
63. Pandiyan, P., et al. (2007). CD4+CD25+Foxp3+ regulatory T cells induce cytokine deprivation-mediated apoptosis of effector CD4+ T cells. *Nature Immunology, 8*(12), 1353–1362.
64. Dwyer, K. M., et al. (2010). Expression of CD39 by human peripheral blood CD4+ CD25+ T cells denotes a regulatory memory phenotype. *American Journal of Transplantation, 10*(11), 2410–2420.
65. Smyth, L. A., et al. (2013). CD73 expression on extracellular vesicles derived from CD4+ CD25+ Foxp3+ T cells contributes to their regulatory function. *European Journal of Immunology, 43*(9), 2430–2440.
66. Sitkovsky, M. V., & Ohta, A. (2005). The 'danger' sensors that STOP the immune response: The A2 adenosine receptors? *Trends in Immunology, 26*(6), 299–304.
67. Cederbom, L., Hall, H., & Ivars, F. (2000). CD4+CD25+ regulatory T cells down-regulate co-stimulatory molecules on antigen-presenting cells. *European Journal of Immunology, 30*(6), 1538–1543.

68. Sakaguchi, S., Wing, K., & Yamaguchi, T. (2009). Dynamics of peripheral tolerance and immune regulation mediated by Treg. *European Journal of Immunology, 39*(9), 2331–2336.
69. Grohmann, U., et al. (2002). CTLA-4-Ig regulates tryptophan catabolism in vivo. *Nature Immunology, 3*(11), 1097–1101.
70. Munn, D. H., et al. (1999). Inhibition of T cell proliferation by macrophage tryptophan catabolism. *The Journal of Experimental Medicine, 189*(9), 1363–1372.
71. Belladonna, M. L., et al. (2007). Immunosuppression via tryptophan catabolism: The role of kynurenine pathway enzymes. *Transplantation, 84*(1 Suppl), S17–S20.
72. Liang, B., et al. (2008). Regulatory T cells inhibit dendritic cells by lymphocyte activation gene-3 engagement of MHC class II. *Journal of Immunology, 180*(9), 5916–5926.
73. Shalev, I., et al. (2012). Role of regulatory T cells in the promotion of transplant tolerance. *Liver Transplantation, 18*(7), 761–770.
74. Wood, K. J., & Sakaguchi, S. (2003). Regulatory T cells in transplantation tolerance. *Nature Reviews. Immunology, 3*(3), 199–210.
75. Li, W., et al. (2006). The role of Foxp3+ regulatory T cells in liver transplant tolerance. *Transplantation Proceedings, 38*(10), 3205–3206.
76. Graca, L., Cobbold, S. P., & Waldmann, H. (2002). Identification of regulatory T cells in tolerated allografts. *The Journal of Experimental Medicine, 195*(12), 1641–1646.
77. Wise, M. P., et al. (1998). Linked suppression of skin graft rejection can operate through indirect recognition. *Journal of Immunology, 161*(11), 5813–5816.
78. Sanchez-Fueyo, A., et al. (2002). Tracking the immunoregulatory mechanisms active during allograft tolerance. *Journal of Immunology, 168*(5), 2274–2281.
79. Li, Y., et al. (2004). Analyses of peripheral blood mononuclear cells in operational tolerance after pediatric living donor liver transplantation. *American Journal of Transplantation, 4*(12), 2118–2125.
80. Demirkiran, A., et al. (2006). Low circulating regulatory T-cell levels after acute rejection in liver transplantation. *Liver Transplantation, 12*(2), 277–284.
81. Tsang, J. Y., et al. (2009). Indefinite mouse heart allograft survival in recipient treated with CD4(+)CD25(+) regulatory T cells with indirect allospecificity and short term immunosuppression. *Transplant Immunology, 21*(4), 203–209.
82. Joffre, O., et al. (2008). Prevention of acute and chronic allograft rejection with CD4+CD25+Foxp3+ regulatory T lymphocytes. *Nature Medicine, 14*(1), 88–92.
83. Golshayan, D., et al. (2007). In vitro-expanded donor alloantigen-specific CD4+CD25+ regulatory T cells promote experimental transplantation tolerance. *Blood, 109*(2), 827–835.
84. Tang, Q., et al. (2004). In vitro-expanded antigen-specific regulatory T cells suppress autoimmune diabetes. *The Journal of Experimental Medicine, 199*(11), 1455–1465.
85. Lee, K., et al. (2014). Attenuation of donor-reactive T cells allows effective control of allograft rejection using regulatory T cell therapy. *American Journal of Transplantation, 14*(1), 27–38.
86. Shultz, L. D., Ishikawa, F., & Greiner, D. L. (2007). Humanized mice in translational biomedical research. *Nature Reviews. Immunology, 7*(2), 118–130.
87. Brunstein, C. G., et al. (2011). Infusion of ex vivo expanded T regulatory cells in adults transplanted with umbilical cord blood: Safety profile and detection kinetics. *Blood, 117*(3), 1061–1070.
88. Issa, F., et al. (2010). Ex vivo-expanded human regulatory T cells prevent the rejection of skin allografts in a humanized mouse model. *Transplantation, 90*(12), 1321–1327.
89. Nadig, S. N., et al. (2010). In vivo prevention of transplant arteriosclerosis by ex vivo-expanded human regulatory T cells. *Nature Medicine, 16*(7), 809–813.
90. Sagoo, P., et al. (2012). Alloantigen-specific regulatory T cells prevent experimental chronic graft-versus-host disease by simultaneous control of allo- and autoreactivity. *European Journal of Immunology, 42*(12), 3322–3333.
91. Putnam, A. L., et al. (2013). Clinical grade manufacturing of human alloantigen-reactive regulatory T cells for use in transplantation. *American Journal of Transplantation, 13*(11), 3010–3020.

92. Spadafora-Ferreira, M., et al. (2007). CD4+CD25+Foxp3+ indirect alloreactive T cells from renal transplant patients suppress both the direct and indirect pathways of allorecognition. *Scandinavian Journal of Immunology, 66*(2–3), 352–361.
93. Yamada, A., et al. (2001). Recipient MHC class II expression is required to achieve long-term survival of murine cardiac allografts after costimulatory blockade. *Journal of Immunology, 167*(10), 5522–5526.
94. Taylor, P. A., Lees, C. J., & Blazar, B. R. (2002). The infusion of ex vivo activated and expanded CD4(+)CD25(+) immune regulatory cells inhibits graft-versus-host disease lethality. *Blood, 99*(10), 3493–3499.
95. Hoffmann, P., et al. (2006). Only the CD45RA+ subpopulation of CD4+CD25high T cells gives rise to homogeneous regulatory T-cell lines upon in vitro expansion. *Blood, 108*(13), 4260–4267.
96. Tang, Q., & Bluestone, J. A. (2013). Regulatory T-cell therapy in transplantation: Moving to the clinic. *Cold Spring Harb Perspect Med, 3*(11), 1–15.
97. Veerapathran, A., et al. (2013). Human regulatory T cells against minor histocompatibility antigens: Ex vivo expansion for prevention of graft-versus-host disease. *Blood, 122*(13), 2251–2261.
98. Noyan, F., et al. (2014). Isolation of human antigen-specific regulatory T cells with high suppressive function. *European Journal of Immunology, 44*(9), 2592–2602.
99. Canavan, J. B., et al. (2016). Developing in vitro expanded CD45RA+ regulatory T cells as an adoptive cell therapy for Crohn's disease. *Gut, 65*(4), 584–594.
100. Scotta, C., et al. (2013). Differential effects of rapamycin and retinoic acid on expansion, stability and suppressive qualities of human CD4(+)CD25(+)FOXP3(+) T regulatory cell subpopulations. *Haematologica, 98*(8), 1291–1299.
101. Safinia, N., Vaikunthanathan, T., Fraser, H., Thirkell, S., Lowe, K., Blackmore, L., Whitehouse, G., Martinez-Llordella, M., Jassem, W., Sanchez-Fueyo, A., Lechler, R. I., & Lombardi, G. (2016). Successful expansion of functional and stable regulatory T cells for immunotherapy in liver transplantation. *Oncotarget, 7*(7), 7563–7577.
102. Afzali, B., et al. (2013). Comparison of regulatory T cells in hemodialysis patients and healthy controls: Implications for cell therapy in transplantation. *Clinical Journal of the American Society of Nephrology, 8*(8), 1396–1405.
103. Marek-Trzonkowska, N., et al. (2012). Administration of CD4+CD25highCD127- regulatory T cells preserves beta-cell function in type 1 diabetes in children. *Diabetes Care, 35*(9), 1817–1820.
104. Trzonkowski, P., et al. (2009). First-in-man clinical results of the treatment of patients with graft versus host disease with human ex vivo expanded CD4+CD25+CD127- T regulatory cells. *Clinical Immunology, 133*(1), 22–26.
105. Di Ianni, M., et al. (2011). Tregs prevent GVHD and promote immune reconstitution in HLA-haploidentical transplantation. *Blood, 117*(14), 3921–3928.
106. Bluestone, J. A., et al. (2015). Type 1 diabetes immunotherapy using polyclonal regulatory T cells. *Science Translational Medicine, 7*(315), 315ra189.
107. Jonuleit, H., et al. (2001). Identification and functional characterization of human CD4(+) CD25(+) T cells with regulatory properties isolated from peripheral blood. *The Journal of Experimental Medicine, 193*(11), 1285–1294.
108. Levings, M. K., Sangregorio, R., & Roncarolo, M. G. (2001). Human cd25(+)cd4(+) t regulatory cells suppress naive and memory T cell proliferation and can be expanded in vitro without loss of function. *The Journal of Experimental Medicine, 193*(11), 1295–1302.
109. Todo, S., et al. (2016). A pilot study of operational tolerance with a regulatory T cell-based cell therapy in living donor liver transplantation. *Hepatology, 64*(2), 632–643.
110. Safinia, N., et al. (2015). Regulatory T cells: Serious contenders in the promise for immunological tolerance in transplantation. *Frontiers in Immunology, 6*, 438.
111. Ballou, L. M., & Lin, R. Z. (2008). Rapamycin and mTOR kinase inhibitors. *Journal of Chemical Biology, 1*(1–4), 27–36.

112. Procaccini, C., et al. (2010). An oscillatory switch in mTOR kinase activity sets regulatory T cell responsiveness. *Immunity, 33*(6), 929–941.
113. Hester, J., et al. (2012). Low-dose rapamycin treatment increases the ability of human regulatory T cells to inhibit transplant arteriosclerosis in vivo. *American Journal of Transplantation, 12*(8), 2008–2016.
114. Akimova, T., et al. (2012). Differing effects of rapamycin or calcineurin inhibitor on T-regulatory cells in pediatric liver and kidney transplant recipients. *American Journal of Transplantation, 12*(12), 3449–3461.
115. Zhao, T., et al. (2013). Comparison of regulatory T cells and FoxP3-positive T-cell subsets in the peripheral blood of renal transplant recipients with sirolimus versus cyclosporine: A preliminary study. *Transplantation Proceedings, 45*(1), 148–152.
116. Rossetti, M., et al. (2015). Ex vivo-expanded but not in vitro-induced human regulatory T cells are candidates for cell therapy in autoimmune diseases thanks to stable demethylation of the FOXP3 regulatory T cell-specific demethylated region. *Journal of Immunology, 194*(1), 113–124.
117. Liao, W., Lin, J. X., & Leonard, W. J. (2013). Interleukin-2 at the crossroads of effector responses, tolerance, and immunotherapy. *Immunity, 38*(1), 13–25.
118. Zorn, E., et al. (2006). IL-2 regulates FOXP3 expression in human CD4+CD25+ regulatory T cells through a STAT-dependent mechanism and induces the expansion of these cells in vivo. *Blood, 108*(5), 1571–1579.
119. Koreth, J., et al. (2016). Efficacy, durability, and response predictors of low-dose interleukin-2 therapy for chronic graft vs. host disease. *Blood, 128*(1), 130–137.
120. Grinberg-Bleyer, Y., et al. (2010). IL-2 reverses established type 1 diabetes in NOD mice by a local effect on pancreatic regulatory T cells. *The Journal of Experimental Medicine, 207*(9), 1871–1878.
121. Pilon, C. B., et al. (2014). Administration of low doses of IL-2 combined to rapamycin promotes allogeneic skin graft survival in mice. *American Journal of Transplantation, 14*(12), 2874–2882.

Chapter 7
Organ Preservation, Ischemia Reperfusion Injury, and Nanotherapeutics in Transplantation

Kunal J. Patel, Carl Atkinson, Ann-Marie Broome, and Satish N. Nadig

Abstract Over the past 30 years, solid organ transplantation has advanced tremendously in many facets, with great strides made in organ selection and allocation to the monitoring and treatment of rejection. Despite these breakthroughs, organ preservation and chronic rejection rates have remained largely unchanged. Recent trends in translational research have begun to address this issue, and future directions for organ preservation will have far-reaching implications for the field of transplantation as a whole. In this chapter, we will highlight the current trends and innovations in preservation techniques, with a special focus on the role that nanotechnology is playing in this frontier.

7.1 Introduction

The standard of care for organ preservation has long been cold storage in one of a various number of perfusion solutions. While the intuitive benefits of hypothermic organ storage had long been corroborated with scientific data demonstrating decreased metabolic demand (and subsequent ischemic tissue damage), there were also significant deleterious effects caused by induction of hypothermia itself in donor organs that needed to be addressed [1]. The answer to these issues came largely through the advent of preservation solutions – first with the widespread adoption of Collins and eventually Euro-Collins (EC) and then with the introduction of University of Wisconsin (UW) solution, which has become the standard of care for many solid organ transplants. Cold storage perfusion solutions were able to reshape the field of transplantation by allowing longer preservation times and minimizing allograft damage [2–4].

K.J. Patel • C. Atkinson • A.-M. Broome • S.N. Nadig (✉)
Medical University of South Carolina, Charleston, SA, USA
e-mail: nadigsn@musc.edu

S.N. Nadig, J.A. Wertheim (eds.), *Technological Advances in Organ Transplantation*, DOI 10.1007/978-3-319-62142-5_7

7.2 Ionic Composition

In the design of perfusion solutions, several obstacles presented themselves. Specifically, the design of the solutions needed to encompass multiple objectives: ionic composition (including a buffer system for acid-base perturbations), free radical scavenging, and the prevention or reduction of organ edema [1]. Table 7.1 summarizes the compositions of the most commonly encountered and popular perfusion solutions available in transplantation today. Initial considerations of composition surrounded the ionic makeup of the solutions, with the debate revolving around mimicking an intracellular environment versus an extracellular environment. At least part of the rationale for simulating the intracellular environment stems from the desire to prevent activity of the membrane-associated Na+/K+ ATPase, thereby minimizing depletion of cellular ATP and cellular edema [5, 9, 10]. As a result, early preservation solutions such as UW and EC contain a relatively low concentration of sodium (10–27 mmol/L) and a high concentration of potassium (115–125 mmol/L) [5, 6, 10–12].

While UW solution remains the standard for organ preservation, some have challenged whether optimal outcomes could be achieved with formulations consisting of potassium concentrations approaching 10–25 mmol/L. Extrapolating from the documented adverse effects of hyperkalemic cardioplegia solutions on vascular endothelium, concerns arose surrounding intracellular perfusion solutions [10]. These concerns ultimately bore out only in the lung transplantation literature, where multiple studies have been able to demonstrate beneficial outcomes of extracellular potassium concentrations. Multiple studies demonstrated improved structural integrity and function, particularly in oxygenating blood, in low-potassium dextran (LPD) perfusion solutions in comparison with standard EC solution in animal models of lung transplantation [5, 13, 14]. As further clinical research corroborated these early findings, the clinical lung transplant arena began shifting toward Perfadex solution (a type of LPD) and away from EC (Table 7.1). Although the clinical data were not as dramatic, they did suggest that low-potassium Perfadex provided a degree of protection to the allograft not conferred by either EC or Papworth solutions (another extracellular perfusion solution which has largely fallen out of favor) and prevented moderate to severe primary graft dysfunction (16% versus 46% versus 42%, $p < 0.01$) [8, 15–18]. As such, Perfadex remains the standard perfusion solution utilized in lung transplantation today.

7.3 Buffering Capacity

In close concert with the ionic environment of a perfusion solution is the pH and, subsequently, the buffering capacity. As the ischemic time increases, there is a natural acidification of the cellular environment in the allograft secondary to anaerobic metabolism. As such, a robust buffering system is required to ameliorate the propagation of acidosis-induced graft injury. Hypothermia tends to offset some of this acidification, but cellular metabolism is not fully arrested during cold storage.

Table 7.1 Compositions of popular currently available perfusion solutions (mmol/L)

		UW [1]	EC [5]	HTK [6]	IGL-1 [7]	Celsior [1]	Perfadex [8]
	pH	**7.40**	**7.25–7.52**	**7.40**	**7.40**	**7.30**	**7.40**
Electrolyte composition	Na^+	25	10	15	120	100	138
	K^+	125	115	10	25	15	6
	Cl^-	–	15	50	–	–	142
	Ca^{2+}	–	–	0.015	0.5	0.25	–
	Mg^{2+}	5	–	4	5	13	0.8
Buffers	HCO_3^-	–	10	–	–	–	–
	SO_4^{2-}	5	–	–	5	–	0.8
	PO_4^{3-}	25	57.5	–	25	–	0.8
	Histidine	–	–	198	–	30	–
	mOsm/L	**320**	**355**	**310**	**290**	**320**	**292**
Impermeants	Glucose	–	180	–	–	–	5
	Lactobionate	100	–	–	100	80	–
	Mannitol	–	–	30	–	60	–
	Raffinose	30	–	–	30	–	–
Colloids	Pentafraction (HES) (g/L)	50	–	–	–	–	–
	PEG-35	–	–	–	0.03	–	–
Antioxidant	Allopurinol	1	–	–	1	–	–
	Glutathione	3	–	–	3	3	–
	Adenosine	5	–	–	5	–	–
	Additional Additives	Dexamethasone, penicillin G, insulin		Ketoglutarate, tryptophan		Glutamic acid	Dextran 40 (50 g)

While no number in the physiologic range has been identified as definitively superior pH for ex vivo hypothermic preservation, it is well documented that extreme perturbations in the pH are to be avoided [19]. Additionally, under hypothermic conditions, standard buffers such as phosphates and sulfates possess different efficacies from normothermic conditions. The overall buffering capacity of preservation solutions is largely dependent on the main buffer in the solution, but ultimately, the collective buffering of all the constituents in the solution plays a significant role [19, 20]. As such, the choice of buffers and additives is very deliberate for each perfusion solution. The most common base buffers still remain bicarbonate, phosphate, and sulfate derivatives. However, recent data have shown that zwitterions such as 4-(2-hydroxyethyl)-1-piperazineethanesulfonic acid (HEPES buffer), containing both positive and negative charges, confer the largest buffering capacities and efficacies and, importantly, that solutions supplemented with HEPES better preserve cellular integrity [20, 21].

Solutions such as histidine-tryptophan-ketoglutarate (HTK) and Celsior utilize histidine as the primary buffer. While preliminary small animal studies show a high buffering capacity and suggest improved outcomes, subsequent head-to-head clinical comparisons with standard UW solution have failed to demonstrate superiority in heart, pancreas, liver, and kidney transplantations [22–26]. Conversely, a recent retrospective analysis of lung transplantation at a single institution in Germany showed that Celsior-preserved lungs decreased the rate of bronchiolitis obliterans when compared with Perfadex preservation despite longer ischemic times in the Celsior group ($p = 0.03$) [27]. More comprehensive trials need to be performed before larger conclusions can be drawn.

7.4 Impermeants and Colloids

As mentioned previously, allograft edema is a prime concern during hypothermic storage, and while partially addressed by the ionic composition of the preservation solution (mainly sodium and potassium concentrations), the task of preventing fluid from leaving the extracellular compartment has fallen largely to additive impermeants. The necessity for impermeants should not be underestimated, as cellular edema represents a manifestation of ischemic injury that has far-reaching implications for allograft function and longevity, which will be covered later in this chapter. For the purposes of this discussion, though, it should be recognized that depletion of ATP and the subsequent ionic and electrical derangements that occur across the cell membrane results in cellular swelling. Downstream of this process, the consequences of edema have been shown to initiate a proapoptotic signaling cascade via the disruption of the mitochondrial membrane [28]. In addition, post-reperfusion oxygen delivery may also be altered with the compression of capillary beds by parenchymal cell expansion, further propagating allograft injury [29, 30]. Indeed, on a macroscopic scale, significant graft injury from cellular edema has been demonstrated in rat livers, which manifested decreased bile flow and transaminitis in the post-reperfusion period [31].

Impermeants work by inducing an osmotic gradient against the cell membrane, thereby maintaining fluid balance in the extracellular spaces. Using a standard plasma osmolarity equation, $2[Na^+] + [Glucose]/18 + [BUN]/2.8$, it can be seen that much of the osmotic pressure in the human body is naturally attributable to the ionic gradient and the monosaccharide glucose. As discussed previously and seen in Table 7.1, other than HTK, many perfusion solutions contain large concentrations of cations, and earlier perfusion solutions such as EC solution utilized glucose as an impermeant. However, small saccharides, such as glucose and mannitol, quickly fell out of favor secondary to the realization that glucose was still small enough to penetrate the cell membrane, reverse the osmotic effect, and degrade to lactate (even under hypothermic conditions), ultimately contributing to lactic acidosis [32, 33]. Current popular perfusion solutions utilize larger saccharides and polysaccharides, such as raffinose, in UW solution or dextran in Perfadex. The larger size prevents entry into the cell and the subsequent lactic acid cascade.

In addition to saccharide-based impermeants, many solutions use an additional ionically charged molecule. Lactobionate (4-O-β-D-galactopyranosyl-D-gluconic acid) is a 358 dalton anion most commonly known for its use in UW solution and has been shown to have a significant allograft protective effect, mainly by abrogating edematous changes. Interestingly, it also appears to decrease free radical production by chelating iron ions [34–36]. As an additional bonus, lactobionate has also been shown to have an inhibitory effect on matrix metalloproteinases (specifically MMP-2 and MMP-9), which may further serve to protect against ischemia-reperfusion injury (IRI) [37].

While these additives aimed to shift the osmotic gradient, the inclusion of colloids into preservation solutions functions by inducing significant oncotic pressure. It should be noted that while colloids improve allograft function post-reperfusion, they are more responsible for reducing interstitial edema rather than cellular edema [38]. The use of colloids, such as hydroxyethyl starch (HES) in UW, is the topic of some debate, in small part, not only because evidence has raised questions regarding the necessity and efficacy of HES in organ preservation but also, and perhaps more importantly, because HES has been associated with significant renal injury in the resuscitative setting that may have implications for its use as a perfusate [39–41]. Additional concerns have been voiced based on the observation of the hypercoagulable effects of HES on blood, particularly with high molecular weight HES and under hypothermic conditions when its viscosity is significantly greater than normal [42, 43]. Other than the documented increase in blood viscosity, the exact mechanism by which these molecules increase aggregation remains to be elucidated as the molecules have no apparent effect on either platelets or clotting factors [43, 44].

As a proposed alternative to HES, some newer storage solutions, such as Institute Georges Lopez solution (IGL-1), have substituted with polyethylene glycol (PEG) with a fair amount of compelling data to support the switch. In addition to its known colloidal properties, PEG appears to have no increased risk of hyper-aggregation of red blood cells (especially at lower molecular weights [20 kDa versus 35 kDa]), and multiple other allograft protective properties have been noted [43]. Perhaps the most notable, particularly in the transplant setting, is its ability to protect against IRI. This attribute has been demonstrated in several models, ranging from rat cardiac ischemia and steatotic livers to porcine renal and pancreas transplants [7, 45–50]. Other compelling findings have included an ability to decrease T-cell-antigen-presenting cell (APC) interactions by direct interference to blunt effector T-cell proliferation [51]. Thus far, clinical trials comparing IGL-1 to UW have only been able to demonstrate similar efficacy between the two. Despite the potential of PEG use as a colloid, it is clear that the sum is greater than any of the individual parts [7, 52, 53].

7.5 Free Radical Scavengers

The addition of free radical scavengers to allograft preservation solutions is covered in detail in the *ischemia-reperfusion injury (IRI)* portion of this chapter (below).

7.6 Ancillary Additives

A number of additional additives have been suggested and tried with varying degrees of evidential support to justify their inclusion. UW solution often includes dexamethasone, insulin, and penicillin, but mixed evidence exists to corroborate their effect on outcomes [54]. The other main additive class, for which only marginal beneficial data exists, is the energy precursor class, that is, compounds that will contribute directly to ATP production. In theory, these substances make intuitive sense in the transplant context considering the prolonged ischemic period devoid of nutrients. However, the inclusion of adenosine in UW and IGL-1 or glutamate/glutamic acid in HTK and Celsior, for the purposes of increasing ATP production, has gone largely unsubstantiated [55–57].

7.7 Allograft Preservation and Ischemia-Reperfusion Injury (IRI)

IRI is a certainty of current organ preservation tactics. While the long-term effects of IRI from cold storage are still being elucidated and appear to be dependent on a number of factors, the goal of all current preservation solutions is to minimize the damage caused by this phenomenon. Herein, we attempt to summarize what is currently known regarding the pathogenesis and consequences of IRI, including a discussion of the future trends and directions of research in this field as it applies to solid organ transplantation.

7.8 Ischemia-Reperfusion-Associated Pathology

IRI, as an entity, manifests itself in a number of different clinical scenarios including, but not limited to, myocardial infarction, stroke, peripheral vascular occlusive disease, and transplantation. Indeed, much of what is known comes from research in these other fields but remains just as relevant to the field of transplantation. Increasingly, recent basic science research has demonstrated greater correlations between the initial insults of IRI and both short- and long-term allograft outcomes, making the pathogenesis of IRI key to the understanding and prevention of rejection. Utilizing the knowledge gained from many disparate research disciplines involved in IRI investigation is vital to addressing the ill effects of this deleterious process.

The period of ischemia that occurs during organ preservation facilitates further graft damage post-reperfusion. The exact mechanisms whereby this process is initiated are currently being defined in the literature but is known to originate from the production of reactive oxygen species (ROS) from the mitochondria [58–60]. The cascade of events ending in IRI appears to begin with complexes I and III in the mitochondrial electron transport chain (ETC), which are the main sources of ROS (specifically superoxide and peroxide, respectively). This finding is somewhat variable depending on the type of tissue tested and, compellingly, on the age of the tissue tested, but still largely holds true [61–65]. The logical correlation that exists is between tissues with high metabolic activity, and a subsequently high oxidative capacity (e.g., cardiac tissue), and with high free radical production. Accordingly, in mouse studies of various tissue oxidative capacities, complex I levels are approximately tenfold higher in cardiac muscle than the liver and threefold higher than the kidney with similar trends noted for complex III [62, 65].

The majority of the previously cited data was studied under physiologic conditions of mitochondrial respiration, and while valuable as an indicator of potential mediators of ischemic injury, further validation via a recapitulation of the clinical scenario is required. Studies have shown that the majority of mitochondrial ROS under physiologic conditions tend to either remain in the mitochondrial matrix or are catalyzed by superoxide dismutase (SOD). Under ischemic and IRI conditions, several studies have demonstrated that biochemical changes of ETC dynamics, and complex I in particular, result in the generation of pathologic levels of cytosolic ROS rather than mitochondrial ROS [62, 66, 67]. ROS generation occurs in a positive feedback loop, whereby mitochondrial membrane-associated cardiolipin becomes damaged and results in further disruption of ETC function, causing further production of ROS [66, 68]. For a full, detailed overview of mitochondrial generation of ROS in response to IRI, Chouchani et al. have recently published a very compelling review hypothesizing a unifying mechanism [58].

Once this cycle of ROS-induced damage has begun, the cellular and molecular effects are wide ranging. Within the mitochondria themselves, at the source of ROS production, free radical damage to the mitochondrial membrane not only disrupts components of the ETC but also the organelle's ability to maintain transmembrane Ca^{2+}-gradients [67]. This damage has been mechanistically implicated in secondary energy failure of the cell (i.e., inability to produce ATP), resulting from an inability to generate a proton-motive force. Indeed, this entire process of ROS-driven mitochondrial disruption plays a role in necrotic cell death via a signaling cascade heavily dependent on intramitochondrial cyclophilin D, a mitochondrial calcium regulator [69–71]. Alternative apoptotic pathways initiated by ROS-induced DNA damage, protein misfolding, and lipid peroxidation also play a role in IRI mediated through the release of molecules, such as cytochrome *c*, and caspase activation. The immunologic and pathogenic implications from necrotic pathways are likely more significant for reasons discussed below. (As an aside, cyclosporine A is a known cyclophilin inhibitor and has been shown to interrupt this Ca^{2+}-dependent pathway [72–75].)

It should be noted that while the above information represents a brief overview of available information regarding IRI in multiple different tissues, the very real possibility remains that alternate pathways and mechanisms are responsible for similar pathologies in a tissue-dependent manner. As ever, the data should be approached with a critical eye and accepted for what they are rather than for what they might imply.

Damage-associated molecular patterns (DAMPs) are a direct consequence of cellular necrosis. These by-products of cell death mimic pathogen-associated molecular patterns (PAMPs) which are recognized by highly conserved receptors, such as toll-like receptors (TLRs), on immune cells and lead to the initiation and propagation of the inflammatory cascade [76]. A number of different intracellular molecules have been identified as DAMPs, the DNA-binding protein high-mobility group box 1 (HMGB-1), the cytoplasmic Ca^{2+}-regulating S100 family, ATP (and its metabolites), uric acid, DNA, and heat shock proteins, and recent studies have begun to show the role of mitochondrial fragments as DAMPs, including in the transplant-specific setting [77–81]. Via the activation of their respective receptors, DAMPs then upregulate inflammatory cytokines such as interferons, tumor necrosis factor (TNF), interleukins (IL)-1 and (IL)-6, as well as endothelial and leukocytic adhesion molecules [76, 79]. The consequence of all this is significant tissue damage from vascular leakage and edema, intravascular thrombosis and injury, and ultimately the activation of the innate and adaptive immune responses.

7.9 Clinical Impact of IRI in Transplantation

Donor organs are exposed to a series of injurious events prior to and during the transplant operative period: brain death, cold and warm ischemia, and eventually reperfusion injury as described above. The cumulative effect is to damage and immunologically prime the donor organ for alloimmune recognition.

The first manifestations of these insults are seen clinically in the immediate postoperative period, and clinical signs and symptoms vary by organ (Fig. 7.1). Different terminology is utilized for this type of injury depending on the organ being discussed.

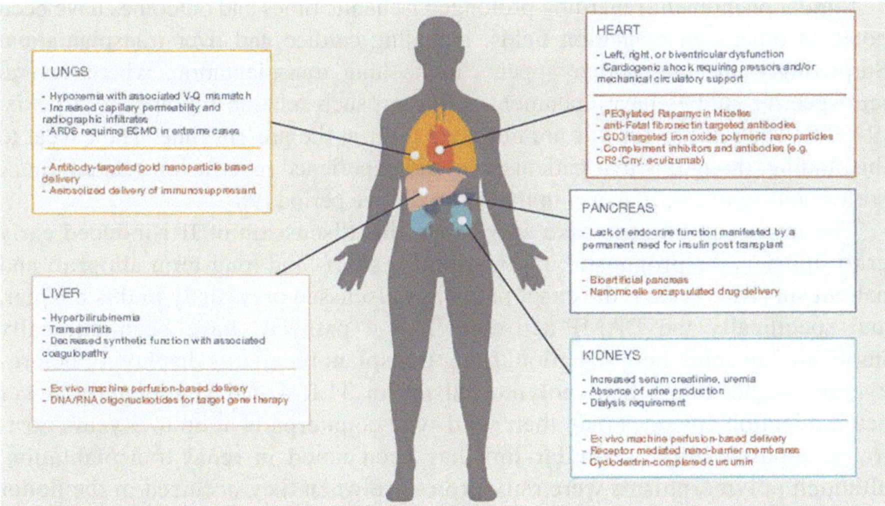

Fig. 7.1 Acute manifestations of ischemia-reperfusion injury and emerging nanotherapeutics currently being investigated, by organ system

Here we broadly term this postoperative pathology as "delayed graft function" (DGF). Despite the organ-dependent variations of DGF, a clear relationship has been observed between IRI (with ischemic time as an indicator for more severe IRI) and severity of DGF in most settings. The consequences of DGF are well known and include longer hospital stays, poorer overall allograft function, increased frequency of acute rejection episodes, and ultimately, decreased allograft and patient survival [82–85]. As such, it is paramount to delineate the correlation between IRI and DGF in an attempt to address this as a root cause.

In renal transplantation, where clinical data are readily available for both living and deceased organ donors, ischemic time correlates with DGF (defined as the need for dialysis within 1 week of transplant) and poor outcomes in both settings [86]. Cadaveric kidney donors, which have a reported DGF rate ranging from 20 to 30%, tend to have poorer outcomes with ischemic times longer than 20–24 h; however, every hour an organ is in cold storage compounds the risk of DGF and graft failure [87–91]. Interestingly, living donor studies have demonstrated similar trends, but the cutoff time for increased DGF and graft failure were significantly lower at 4–8 h of cold ischemic time. This occurs despite the fact that live-donor kidneys do not experience the additional insults associated with brain death (e.g., complement activation, dysautonomia) [84, 92]. These are clearly very complex scenarios with multiple confounding factors. For each additional hour of ischemic time, the hazard ratios for live-donor kidneys were 1.14 versus 1.013 for cadaveric kidneys [84, 91]. It should be noted that recent advances in pulsatile perfusion technology have been able to offset some but not all of the deleterious effects of cold ischemia on graft outcomes [93].

Similar phenomena regarding prolonged ischemic times and outcomes have been noted in other transplantation fields, including cardiac and liver transplantation. Surprisingly, the one outlier appears to be lung transplantation, where several retrospective studies have documented that no such relationship appears to exist [94–97]. The reasons for this are not quite clear at the present time. The caveat to this finding, though, is that patients who did experience reperfusion-related injury performed significantly worse in the perioperative period [98].

The important point to take away from this discussion of IRI-induced early graft injury is the prognostic implication for short- and long-term allograft and patient survival. In fact, the exact pathways discussed previously in this chapter, and specifically the DAMP-activated TLR-4 pathway, have been clinically implicated in acute lung rejection. Lung transplant recipients displaying heterozygous single amino acid polymorphisms in TLR-4 had significantly fewer acute rejection episodes than their wild-type counterparts at up to 3 years posttransplantation [99]. A similar link has been noted in renal transplantation, although polymorphisms were only protective when they occurred in the donor graft [100].

The ultimate goal of transplantation research is to prolong allograft survival, which to a large extent means preventing chronic rejection. As intimated before, chronic rejection has a close association with the frequency and severity of acute rejection episodes, and especially early acute rejection. In cardiac transplantation, where chronic rejection manifests as cardiac allograft vasculopathy (CAV), acute cellular rejection occurring within the first year posttransplantation is associated with an increased incidence of CAV in the long term (as high as 100% in some series), and outcomes were strongly correlated with the grade and frequency of the acute rejection episodes [101–103]. The same trends exist in both chronic renal nephropathy and lung transplantation, where chronic rejection manifests as bronchiolitis obliterans [104, 105]. The exception here is liver transplantation.

Given the cascade of events that begins with the generation and propagation of free radicals in the mitochondria, antioxidant additives to perfusion solutions logically make sense. Unfortunately, very little data exist indicating that free radical scavengers have any impact on outcomes. Allopurinol, which is the predominant antioxidant in UW solution, has recently been shown to ameliorate renal IRI when used as a preconditioning agent. However, rather than its established mechanism of antioxidant defense via xanthine oxidase inhibition, allopurinol's protective effects seem to originate from interference with HMGB-1, mentioned previously [106]. Regardless, the effect may not necessarily apply to perfusates as the study was performed in rats who were administered approximately 10–15 mg of allopurinol intraperitoneal daily for 2 weeks prior to IRI, whereas the concentration of allopurinol in UW solution is around 136 mg total per liter.

While a paucity of compelling data is also present for other common antioxidants such as glutathione, several additives used for other purposes have demonstrated additional capabilities as free radical scavengers. For instance, mannitol, tryptophan, and histidine all have documented antioxidant properties [107]. Molecules such as PEG-35 in solutions similar to IGL-1 have demonstrated significant IRI-protective effects and an ability to protect rat livers undergoing 24 h of cold storage, even when only used as a rinse solution immediately pre-reperfusion [108]. However, as with any complex system, it is extremely difficult to isolate these effects to a single property of these additives, and nothing can truly be said of their antioxidant abilities except that they certainly do not appear to be detrimental.

7.10 Emergence of Nanotherapy in Transplantation

Due to the harmful sequelae of long-term systemic immunosuppression as well as increased risks for malignancies and concomitant toxicities, therapies utilizing nano- and microparticles to deliver immunosuppressive medications directly to the allograft are currently being researched. The impact of such drug delivery vehicles includes the ability to locally treat an organ while maintaining systemic immunosurveillance as well as potentially obviating the side effects that are so common with many of these medications. Herein, we describe in further detail how nanotherapeutics and emerging technologies such as these are being applied to a wide range of solid organ transplant models.

7.11 Construction and Modification of Nanoparticles

Poor biopharmaceutical properties, such as low bioavailability, minimal solubility, low permeability, high molecular weight, and high pre-systemic metabolism, of drug formulations contribute to difficulties in delivering systemic dosing of immunosuppressive drugs [109–111]. A limited number of direct drug modifications have been utilized to formulate improved drugs. Therefore, the development of drug nanocarriers, such as emulsions, liposomes, and polymeric micelles, are considered alternative methods to deliver water-insoluble drugs.

The attraction of nanocarriers is, in large part, attributed to their unique physiochemical properties, such as their small size, stability, and the ability for tailoring with various functionalities. Nanocarriers can prolong drug circulation in the blood and offer new methods for targeted delivery after intravenous administration. For example, cyclosporine has been emulsified to form micelles and further modified into microemulsion to enhance bioavailability [112]. The liposomal formulation of tacrolimus also has been demonstrated as an effective strategy to increase efficacy and decrease toxicity [113, 114]. The formulation of amphiphilic block copolymer micelles has been used to prepare injectable formulations of cyclosporine and the highly lipophilic sirolimus (rapamycin) [115–117].

Liposomes represent one of the most promising systemic delivery strategies. Since their initial development around 1980, the related technology has continued to improve, and several formulations are commercially available for clinical applications. Liposomes can be classified according to their lamellarity, size, or preparation method [118]. Liposomes can carry either hydrophilic or hydrophobic drugs, partitioned either in the aqueous or organic lamellar layers to protect them from unfavorable conditions [119, 120]. The physicochemical properties of liposomes, such as net surface charge, hydrophobicity, size, fluidity, and packing of the lipid bilayers, influence their stability [121]. Instability of the liposomes in plasma is primarily due to their interaction with high (HDL) and low density (LDL) lipoproteins [122]. Modifications of the liposome surface can improve stability. Particularly, liposomes conjugated with PEG can evade recognition by opsonins and the subsequent clearance by the reticuloendothelial system (RES), and hence they stay longer in circulation with sustained drug release [123–127]. Size-wise liposomes between 100 and 300 nm tend to accumulate within the interstitium and promote drug delivery efficiency, owing to an organ's vascular features, referred to as enhanced penetration and retention (EPR) effect [128–130]. Further, liposomes can be constructed with special moieties for targeted delivery, achieving higher delivery efficiency and minimizing side effects [131–135].

Micelles, even smaller than liposomes and ranging in size from 10 to 50 nm, can also be altered on their exterior surface with functional moieties, such as ligands or peptides, to provide targeting capability. The inner micelle core can be used as a container for many hydrophobic drugs. Environmental-sensitive lipids that take advantage of pH (or temperature) can be used to formulate the micelle shell to provide responsive drug release. Novel, pH-sensitive, targeted micelles encapsulating rapamycin have been developed to attenuate responses of human EC inflammation and allopresentation when subjected to oxidative stress (described below) [136]. These micelles were decorated with cyclic arginine-glycine-aspartate (cRGD) moieties to facilitate targeting to integrin alpha v beta 3 ($\alpha V\beta 3$) on the endothelium and loaded with the immunosuppressive rapamycin, a lipophilic microcyclic lactone (sirolimus). Rapamycin, a potent mToR inhibitor, was selected due to its ability to not only inhibit T-cell effector cell functions but also protect the endothelium and promote tolerance [136].

Polymeric nanocarriers, based on polylactide (PLA) or polylactide-co-glycolide (PLGA), have been developed to deliver drugs like cyclosporine or sirolimus [137–139]. The nanocarrier formulations developed to date have improved the overall bioavailability of the drug and protected the milieu from the toxicity of the drug. Many of the published studies have focused on controlling the formulation by maintaining therapeutic blood levels. The release of drug from the polymeric nanocarriers is relatively slow (only <20% drug release over a 2-week period). As a result, maintenance of drug concentrations within the therapeutic window after intravenous administration is difficult.

7.12 Recent Advances and the Current State of Nanotherapeutics

7.12.1 Cardiac Transplantation

According to the International Society for Heart and Lung Transplantation registry data, approximately 4000 cardiac transplants are performed worldwide each year with a median survival time of 11 years [140]. CAV, a manifestation of chronic rejection on vascular endothelial and smooth muscle cells, remains a major cause of long-term graft loss and mortality, with up to 50% of allografts demonstrating some amount of CAV by 10 years posttransplantation [141, 142]. While improvements in immunosuppressive agents have greatly improved short-term survival, long-term survival has remained essentially unchanged, largely due to CAV [140–142]. Other potentially modifiable causes of mortality include renal failure, malignancy, and opportunistic infections; all of these could potentially be addressed with graft-specific drug delivery and nanotherapeutics [140, 143].

Graft endothelial cell activation (as occurs with IRI) is a key component to the pathogenesis and progression of CAV [142, 144]. As such, the protection of allograft vasculature and prevention of vasculopathy, and ultimately chronic rejection, intuitively hinges on the protection of the endothelial cell lining. Our group created a novel drug delivery molecule that preferentially targets endothelial cells in the ex vivo setting and promotes the intracellular uptake of rapamycin, a highly effective immunosuppressant with a multitude of systemic side effects (Fig. 7.2) [136]. By packaging rapamycin in a protective polyethylene glycol (PEG)-modified micelle, the unwanted molecular attributes and systemic side effects may largely be obviated [136, 145–147]. This study utilized a previously described endothelial cell targeting moiety (Arg-Gly-Asp or RGD) to demonstrate the drug system's ability to depress endothelial cell inflammation [148]. In the ex vivo model utilized in their study, the relatively nonspecific RGD was able to target only the allograft vascular endothelial cells of interest, and this may go a long way toward preventing the initial insults that pave the way toward CAV and chronic rejection. However, for longer-term immunosuppression, a more applicable and graft-specific in vivo target may be required, such as platelet-endothelial cell adhesion molecule 1 (PECAM-1) and intercellular adhesion molecule 1 (ICAM-1), which are upregulated on pathologically activated endothelial cells [149].

Beyond the endothelial cell's role in the initiation and propagation of chronic rejection, associated changes to the allograft vasculature in CAV include a marked expansion and remodeling of the extracellular matrix during the process of neointimal hyperplasia. The unlikely positive outcome from this process has been the identification of disease-specific molecules, namely, fetal variants of fibronectin and tenascin-C, which have become targets for immunotherapeutics in cardiac chronic rejection models [142, 150, 151]. Indeed, the re-expression of fetal fibronectin

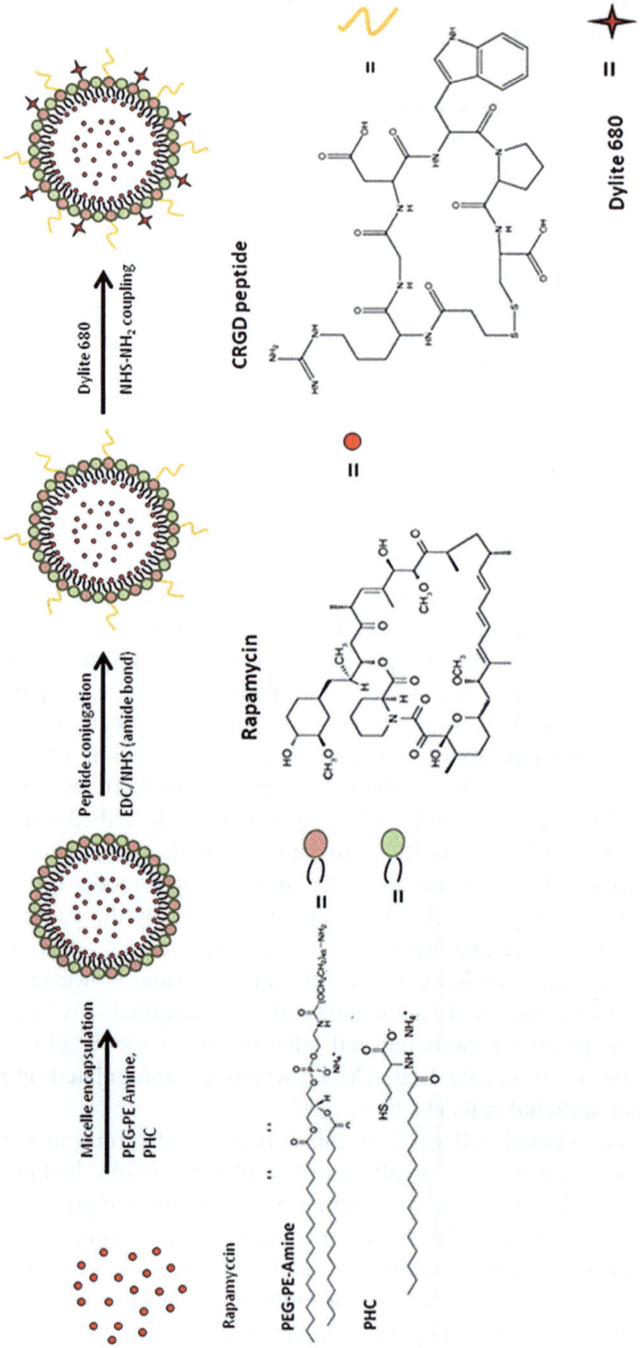

Fig. 7.2 Bioengineered nanotherapeutic for targeted drug delivery

(ED-A+ Fn) and tenascin-C following heart transplantation was directly correlated to rejection and inflammation and was markedly absent in healthy adult organs [150, 151]. Franz et al. were, accordingly, able to successfully design and validate a targeted ED-A+ Fn antibody to deliver IL-10 in a rat heterotopic heart transplant model. While the therapeutic effects of IL-10 delivery were found wanting in this study, there was a clear demonstration of the targeting ability and specificity of the delivery system on fluorescence imaging and microscopy [150, 152]. Nonetheless, this represents a very important molecular maker for chronic cardiac rejection and drug targeting, and the creation of more effective therapy will only be a matter of time. It is worth noting here that clinical studies have documented significant antiproliferative effects of sirolimus (rapamycin) and a reduction of intimal hyperplasia on ultrasound imaging after being started in patients several years posttransplantation, and this may pose as a highly beneficial therapeutic if properly developed [153].

Not to be forgotten for their function in transplant immunopathology, effector T-cells play a prominent role in both acute and chronic rejection. While many current therapies exist to suppress T-cell proliferation and function, they are largely hampered by the pitfalls of systemic immunosuppression, poor specificity (i.e., B-cell suppression), and associated toxicities. Monoclonal and polyclonal antibodies have been developed with this purpose in mind; however, their use beyond induction therapy is still not well delineated. To date, no current therapies are able to selectively inhibit T-cells sensitized to allograft-specific antigen. Despite this, some new advancements in T-cell targeting and suppression have recently been reported in a rat cardiac transplant model. Guo et al. created a CD3-targeting nanoparticle, with high MRI signal intensity, that presents a lot of interesting potential [154]. While the molecule effectively immunosuppressed, there remained issues of nonspecificity; however, the truly exciting prospect is accurately and noninvasively identifying acute rejection episodes (which are significant risk factors for subsequent chronic rejection) while simultaneously treating the patient.

7.12.2 Lung Transplantation

Lung transplantation, both single and double, offers a unique opportunity for targeted drug delivery beyond the other mechanisms previously discussed. Specifically, the ability to deliver aerosolized medications to the lung has the potential to appropriately immunosuppress the lung allograft while almost entirely avoiding entering the systemic circulation. Inhaled steroids are already a staple for the treatment of numerous pulmonary disorders; however, the developments of aerosolized calcineurin inhibitors are at the forefront of more tailored pulmonary immunosuppression [155].

While aerosolized versions of cyclosporine A have been in development since 1997, applications for lung transplantation have very recently come close to clinical use, with current Phase II trials ongoing for cyclosporine inhalation solution (CIS) use in transplant recipients suffering from bronchiolitis obliterans. CIS is a propylene glycol suspension that had been demonstrated preclinically to be safe with favorable pharmacokinetics in dogs and rats [156, 157]. Building on this and clinical data first published in the New England Journal of Medicine, CIS had previously made its way to Phase III trials but failed to show efficacy in disease-free survival or overall mortality [158, 159]. However, that trial did produce valuable information, including the finding that this aerosolized mechanism of delivery can produce highly selective pulmonary delivery and substantially lower systemic levels of cyclosporine than oral administration and that minimal coaching could assist in avoiding alimentary ingestion of the drug. Equally as important, the patients tolerated the therapy fairly well [160]. Should further studies of CIS not prove fruitful, several other aerosolized variations including liposomal and glycerol monooleate-based dry-emulsion powder formulations have already been developed with promising preliminary data [161, 162].

A little farther behind on the pharmaceutical development trail lie the inhalational formulations for tacrolimus and rapamycin. Recently reported, a rapamycin powder, compounded with lactose for stabilization, was demonstrated to successfully absorb and distribute throughout the lung [163]. The study, performed in a non-transplant rat model, examined both the crystallized compound and a preparation done by thin film freezing (TFF), a technique for increasing solubility and bioavailability. Ultimately, both formulations targeted the lung for comparable amounts of time; however, systemic levels of TFF were higher [163]. Similarly early in its development, inhaled tacrolimus tested in a rat lung transplant model demonstrated efficacy in attenuating IRI, despite the fact the study was somewhat hampered by poor delivery of the therapeutic. This phenomenon was attributed mostly to upper airway interference in a spontaneously breathing animal, a situation circumvented in humans by virtue of endotracheal intubation [164].

Creating an aerosolized medication that will distribute deep into the lungs requires an appropriately sized nanomolecule (~1–5 μm effective diameter) that can be effectively incorporated into a liquid or powder form, while simultaneously accounting for differential levels of humidity at different points in the airway affecting hydrophilic properties [165, 166]. This is clearly complicated; however, utilizing a nanoparticle delivery system to encapsulate medications could confer pharmacokinetic predictability to a wide range of drugs, thereby simplifying this process. In the vast majority of cases, nanoparticles could also be modified more easily than native drugs to produce sustained-release formulations of medications, as well as more specific targeting [166].

To this end, Cova et al. have begun utilizing everolimus-loaded gold nanoparticles to treat damaged lung allografts [167]. Utilizing antibody targeting, they were able to demonstrate mesenchymal cell inhibition and induction of apoptosis in in vitro models. In unpublished preliminary data, they have also been able to demonstrate safety and efficacy of nebulized administration of their nanoparticles in mice, with no detectable levels of drug in off-site organs. While this is still very early days, the work represents yet another example of the exciting direction of nanoparticle therapy in transplantation.

7.12.3 Kidney Transplantation

IRI is a significant contributor to allograft rejection, both acute and chronic. Renal transplantation is no exception to this rule and, in fact, may be particularly important to the pathogenesis of subsequent graft failure when compounded with the inherent nephrotoxicities of current standard immunosuppressive medications (e.g., tacrolimus), as well as the increasing use of extended criteria donors. Consequently, many of the emerging developments in renal transplant medicine have revolved around these two accepted phenomena with targeted therapies representing the ideal.

Intuitively, the kidney's sensitivity to ischemia/reperfusion injury could be inferred from the high rate with which acute kidney injury occurs in the general patient population. Indeed, the kidney appears to be highly reactive to alterations in perfusion and blood flow. In the setting of deceased donor transplantation, the ischemic insult begins even before the organ has been harvested, continues through cold and warm ischemic periods, and ultimately manifests itself with delayed graft function and rejection episodes [168–170]. To combat this inevitability, minimizing the ischemic insult and related damage has been studied in depth. Great strides have been made in ex vivo machine perfusion of renal transplants to address IRI. Hypothermic machine perfusion has been utilized for many years, and, increasingly, normothermic perfusion has been demonstrated to benefit organs. These modalities have been able to decrease the rates of delayed graft function, expand the donor pools, and even rehabilitate previously damaged kidneys [171–174]. The relevance of machine perfusion to our discussion largely lies in its ability to deliver novel drug therapies in the unique ex vivo setting available in transplantation [168, 171, 175].

Among the many types of perfusion solutions currently in use and in development for renal transplantation, there have been a handful of molecular additives designed as targeted therapies for pretransplantation immunomodulation. In 2010, Brasile et al. utilized a canine renal transplant model to demonstrate the efficacy of a novel receptor-mediated bioengineered nano-barrier membrane (NB-LVF4) in conferring immunoprotection to vascular endothelial cells [175]. They utilized an ex vivo warm perfusion model in which the nanomolecules were administered over a 3 h period at 32 °C prior to re-implantation. In the posttransplantation period, the animals were monitored in the absence of systemic immunosuppression and the treatment group manifested rejection later (30 days versus 6 days) while maintaining normal renal function [175]. While it appears their nano-barrier does not function in preventing IRI, there was more than adequate amelioration of the post-reperfusion sequelae.

In an effort to address IRI, a number of additives have been studied showing great promise. Most recently, the utilization of hydrogen sulfide in University of Wisconsin solution was demonstrated to improve early graft survival in rat renal transplants when incubated for 6 h at 4 °C [176]. The presumed mechanism of protection is related to the mitigation of oxidative stress and subsequent inflammatory cascades.

Similarly, curcumin, an inhibitory molecule for the NF-kB pathway, has been regarded as a promising therapeutic secondary to its immunosuppressive properties and the protection provided from oxidative stress [170, 177]. Multiple preclinical models exist expounding on curcumin's potential, and several delivery systems have been designed to improve hydrophilicity of the molecule. In a non-transplant IRI model, Rogers et al. delivered liposomal-packaged curcumin to mice prior to a renal ischemic insult and were able to successfully reduce renal injury as demonstrated by improved serum creatinine and serum urea, reduced pro-inflammatory markers, and better preserved histology [170]. Thuillier et al. formed a cyclodextrin-complexed curcumin compound to improve solubility and subsequently demonstrated improved in vitro and in vivo function, survival, and markers of cellular and parenchymal damage in a porcine renal autotransplant model [177].

7.12.4 Liver and Pancreas Transplantation

Currently, a paucity of research and data exists on the immense potential of nanotechnology in liver and pancreas transplantation. A multitude of factors may be responsible for this. For instance, while large animal models of liver transplantation exist (e.g., porcine models), more manageable (both logistically and monetarily) small animal models do not. The current murine orthotopic liver transplant model is extremely difficult from a technical standpoint, and few labs have undertaken the endeavor, though recently, Pan et al. described a potentially easier method for reanastomosis which may turn the tide [178]. This does not appear to be the case in regard to pancreatic transplantation; there are a plethora of animal models including recent descriptions of nonhuman primate models of transplantation. However, there still remains a dearth of research into nanotherapy [179].

Despite the absence of transplant-related developments, both the hepatic and pancreatic literature do contain an extensive amount of nanotherapeutic innovation in malignant and nonmalignant diseases afflicting these organs. From the use of siRNA delivering nanosomes to combat hepatitis C to the creation of nanomicelles for the treatment of pancreatic cancer, there exists a wide range of research in these areas [180, 181]. There have even been breakthroughs in nanotherapy toward the creation of bioartificial pancreases as a potential cure for diabetes [182]. These types of ingenuity clearly demonstrate the potential for where the fields of liver and pancreas transplantation could potentially go. The current voids will eventually be filled, and in the meantime, the continuing emergence of more sophisticated nanoparticle therapy and drug delivery systems in the surrounding fields will only serve as a higher platform from which to launch.

7.13 Conclusion

The utilization of nanotechnology represents the next frontier of medical breakthroughs with the fields of oncology and transplantation slated to appreciate the largest gains. Solid organ transplantation is uniquely positioned by its very nature to harness this technology to selectively reshape pharmaceutical therapies. The ability to manipulate three disparate environments (donor, ex vivo, and recipient) to the patient's and the allograft's advantage is unparalleled. As scientific research becomes more sophisticated and our knowledge of the molecular dynamics of the transplantation process becomes more complete, the role nanotechnology plays will become ever greater. The efficacy and specificity nanotechnology offers will only be enhanced as more is learned about IRI and its impact on allo-immunogenicity. The discussion in this chapter is only the tip of the iceberg with regard to the potential in this field, and the future prospects truly are reasons for genuine excitement and optimism.

References

1. Guibert, E. E., et al. (2011). Organ preservation: Current concepts and new strategies for the next decade. *Transfusion Medicine and Hemotherapy : Offizielles Organ der Deutschen Gesellschaft fur Transfusionsmedizin und Immunhamatologie, 38*, 125–142. doi:10.1159/000327033.
2. Collins, G. M., Hartley, L. C., & Clunie, G. J. (1972). Kidney preservation for transportation. Experimental analysis of optimal perfusate composition. *The British Journal of Surgery, 59*, 187–189.
3. Andrews, P. M., & Bates, S. B. (1985). Improving Euro-Collins flushing solution's ability to protect kidneys from normothermic ischemia. *Mineral and Electrolyte Metabolism, 11*, 309–313.
4. Kalayoglu, M., et al. (1988). Extended preservation of the liver for clinical transplantation. *Lancet (London, England), 1*, 617–619.
5. Fukuse, T., et al. (1996). Comparison of low potassium Euro-Collins solution and standard Euro-Collins solution in an extracorporeal rat heart-lung model. *European Journal of Cardio-Thoracic Surgery : Official Journal of the European Association for Cardio-thoracic Surgery, 10*, 621–627.
6. Adam, R., et al. (2015). Compared efficacy of preservation solutions in liver transplantation: A long-term graft outcome study from the European Liver Transplant Registry. *American Journal of Transplantation : Official Journal of the American Society of Transplantation and the American Society of Transplant Surgeons, 15*, 395–406. doi:10.1111/ajt.13060.
7. Garcia-Gil, F. A., et al. (2014). Evaluation of Institut Georges Lopez-1 preservation solution in pig pancreas transplantation: A pilot study. *Transplantation, 97*, 901–907. doi:10.1097/tp.0000000000000050.
8. Aziz, T. M., et al. (2003). Perfadex for clinical lung procurement: Is it an advance? *The Annals of Thoracic Surgery, 75*, 990–995.
9. Belzer, F. O., & Southard, J. H. (1988). Principles of solid-organ preservation by cold storage. *Transplantation, 45*, 673–676.
10. Drinkwater, D. C., Jr., et al. (1995). Extracellular and standard University of Wisconsin solutions provide equivalent preservation of myocardial function. *The Journal of Thoracic and Cardiovascular Surgery, 110*, 738–745. doi:10.1016/s0022-5223(95)70106-0.
11. Roskott, A. M., et al. (2011). Small bowel preservation for intestinal transplantation: A review. *Transplant International : Official Journal of the European Society for Organ Transplantation, 24*, 107–131. doi:10.1111/j.1432-2277.2010.01187.x.
12. Latchana, N., et al. (2015). Preservation solutions used during abdominal transplantation: Current status and outcomes. *World Journal of Transplantation, 5*, 154–164. doi:10.5500/wjt.v5.i4.154.

13. Binns, O. A., et al. (1996). Both blood and crystalloid-based extracellular solutions are superior to intracellular solutions for lung preservation. *The Journal of Thoracic and Cardiovascular Surgery, 112*, 1515–1521. doi:10.1016/s0022-5223(96)70010-7.
14. Keshavjee, S. H., et al. (1992). The role of dextran 40 and potassium in extended hypothermic lung preservation for transplantation. *The Journal of Thoracic and Cardiovascular Surgery, 103*, 314–325.
15. Oto, T., et al. (2006). Early outcomes comparing Perfadex, Euro-Collins, and Papworth solutions in lung transplantation. *The Annals of Thoracic Surgery, 82*, 1842–1848. doi:10.1016/j.athoracsur.2006.05.088.
16. Marasco, S. F., et al. (2011). Effect of donor preservation solution and survival in lung transplantation. *The Journal of Heart and Lung Transplantation : The Official Publication of the International Society for Heart Transplantation, 30*, 414–419. doi:10.1016/j.healun.2010.10.002.
17. Divisi, D., et al. (2001). A comparative study of Euro-Collins, low potassium University of Wisconsin and cold modified blood solutions in lung preservation in acute autotransplantations in the pig. *European Journal of Cardio-Thoracic Surgery : Official Journal of the European Association for Cardio-thoracic Surgery, 19*, 333–338.
18. Nath, D. S., et al. (2005). Does Perfadex affect outcomes in clinical lung transplantation? *The Journal of Heart and Lung Transplantation : The Official Publication of the International Society for Heart Transplantation, 24*, 2243–2248. doi:10.1016/j.healun.2005.06.019.
19. Baicu, S. C., Taylor, M. J., & Brockbank, K. G. (2006). The role of preservation solution on acid-base regulation during machine perfusion of kidneys. *Clinical Transplantation, 20*, 113–121. doi:10.1111/j.1399-0012.2005.00451.x.
20. Baicu, S. C., & Taylor, M. J. (2002). Acid-base buffering in organ preservation solutions as a function of temperature: New parameters for comparing buffer capacity and efficiency. *Cryobiology, 45*, 33–48.
21. Peltz, M., Milchgrub, S., Jessen, M. E., & Meyer, D. M. (2010). Effect of pyruvate and HEPES on rat lung allograft acidosis and cell death after long-term hypothermic storage. *Transplantation Proceedings, 42*, 2771–2776. doi:10.1016/j.transproceed.2010.06.004.
22. Sumimoto, R., et al. (1992). A comparison of histidine lactobionate solution with University of Wisconsin solution for rat liver and heart preservation. *Transplant International : Official Journal of the European Society for Organ Transplantation, 5*(Suppl 1), S408–S410.
23. Becker, T., et al. (2007). Pancreas transplantation with histidine-tryptophan-ketoglutarate (HTK) solution and University of Wisconsin (UW) solution: Is there a difference? *JOP : Journal of the Pancreas, 8*, 304–311.
24. Feng, L., et al. (2007). Histidine-tryptophan-ketoglutarate solution vs. University of Wisconsin solution for liver transplantation: A systematic review. *Liver Transplantation : Official Publication of the American Association for the Study of Liver Diseases and the International Liver Transplantation Society, 13*, 1125–1136. doi:10.1002/lt.21208.
25. O'Callaghan, J. M., Knight, S. R., Morgan, R. D., & Morris, P. J. (2012). Preservation solutions for static cold storage of kidney allografts: A systematic review and meta-analysis. *American Journal of Transplantation : Official Journal of the American Society of Transplantation and the American Society of Transplant Surgeons, 12*, 896–906. doi:10.1111/j.1600-6143.2011.03908.x.
26. Li, Y., et al. (2015). Three preservation solutions for cold storage of heart allografts: A systematic review and meta-analysis. *Artificial Organs*. doi:10.1111/aor.12585.
27. Gohrbandt, B., et al. (2015). Lung preservation with perfadex or celsior in clinical transplantation: A retrospective single-center analysis of outcomes. *Transplantation, 99*, 1933–1939. doi:10.1097/tp.0000000000000578.
28. Petit, P. X., et al. (1998). Disruption of the outer mitochondrial membrane as a result of large amplitude swelling: The impact of irreversible permeability transition. *FEBS Letters, 426*, 111–116.
29. Parrish, D., et al. (2015). New low-volume resuscitation solutions containing PEG-20k. *The Journal of Trauma and Acute Care Surgery, 79*, 22–29. doi:10.1097/ta.0000000000000682.

30. Parrish, D., Lindell, S. L., Reichstetter, H., Aboutanos, M., & Mangino, M. J. (2016). Cell Impermeant-based low-volume resuscitation in hemorrhagic shock: A biological basis for injury involving cell swelling. *Annals of Surgery, 263*, 565–572. doi:10.1097/sla.0000000000001049.
31. Sundberg, R., Ar'rajab, A., Ahren, B., & Bengmark, S. (1991). The functional effects of suppression of hypothermia-induced cell swelling in liver preservation by cold storage. *Cryobiology, 28*, 150–158.
32. Muhlbacher, F., Langer, F., & Mittermayer, C. (1999). Preservation solutions for transplantation. *Transplantation Proceedings, 31*, 2069–2070.
33. Jamart, J., & Lambotte, L. (1983). Efficiency and limitation of Euro-Collins solution in kidney preservation. *The Journal of Surgical Research, 34*, 195–204.
34. Fukuda, C., Kollmar, O., Schafer, T., Tian, Y. H., & Schilling, M. K. (2002). Anionic polysaccharides. A class of substances with hepatoprotective and antiadhesive properties in rat liver preservation. *Transplant International : Official Journal of the European Society for Organ Transplantation, 15*, 17–23. doi:10.1007/s00147-001-0374-9.
35. Ahmed, I., Attia, M. S., Corps, C. L., Lodge, J. P., & Potts, D. J. (2001). Protective effects of lactobionate in modified phosphate-buffered sucrose. *Transplantation Proceedings, 33*, 950–951.
36. Mees, N., Southard, J. H., & Belzer, F. O. (1982). Inhibition of ischemic induced cellular swelling in kidney cortex tissue by lactobionate anions. *The Journal of Trauma, 22*, 118–120.
37. Upadhya, G. A., & Strasberg, S. M. (2000). Glutathione, lactobionate, and histidine: Cryptic inhibitors of matrix metalloproteinases contained in University of Wisconsin and histidine/tryptophan/ketoglutarate liver preservation solutions. *Hepatology (Baltimore, MD), 31*, 1115–1122. doi:10.1053/he.2000.6780.
38. Ar'Rajab, A., Ahren, B., Sundberg, R., & Bengmark, S. (1991). The function of a colloid in liver cold-storage preservation. *Transplantation, 52*, 34–38.
39. Howden, B. O., et al. (1990). Liver preservation with UW solution. I. Evidence that hydroxyethyl starch is not essential. *Transplantation, 49*, 869–872.
40. Biguzas, M., et al. (1990). Evaluation of UW solution in a rat kidney preservation model. I. Effect of hydroxyethyl starch and electrolyte composition. *Transplantation, 49*, 872–875.
41. Mutter, T. C., Ruth, C. A., & Dart, A. B. (2013). Hydroxyethyl starch (HES) versus other fluid therapies: Effects on kidney function. *The Cochrane Database of Systematic Reviews, 7*, CD007594. doi:10.1002/14651858.CD007594.pub3.
42. Morariu, A. M., et al. (2003). Hyperaggregating effect of hydroxyethyl starch components and University of Wisconsin solution on human red blood cells: A risk of impaired graft perfusion in organ procurement? *Transplantation, 76*, 37–43. doi:10.1097/01.tp.0000068044.84652.9f.
43. Mosbah, I. B., et al. (2006). Effects of polyethylene glycol and hydroxyethyl starch in University of Wisconsin preservation solution on human red blood cell aggregation and viscosity. *Transplantation Proceedings, 38*, 1229–1235. doi:10.1016/j.transproceed.2006.02.068.
44. Bakaltcheva, I., Ganong, J. P., Holtz, B. L., Peat, R. A., & Reid, T. (2000). Effects of high-molecular-weight cryoprotectants on platelets and the coagulation system. *Cryobiology, 40*, 283–293. doi:10.1006/cryo.2000.2247.
45. Bejaoui, M., et al. (2015). Protective effect of intravenous high molecular weight polyethylene glycol on fatty liver preservation. *BioMed Research International, 2015*, 794287. doi:10.1155/2015/794287.
46. Bejaoui, M., et al. (2016). Polyethylene glycol preconditioning: An effective strategy to prevent liver ischemia reperfusion injury. *Oxidative Medicine and Cellular Longevity, 2016*, 9096549. doi:10.1155/2016/9096549.
47. Xu, X., et al. (2015). High-molecular-weight polyethylene glycol inhibits myocardial ischemia-reperfusion injury in vivo. *The Journal of Thoracic and Cardiovascular Surgery, 149*, 588–593. doi:10.1016/j.jtcvs.2014.10.074.
48. Malhotra, R., et al. (2011). High-molecular-weight polyethylene glycol protects cardiac myocytes from hypoxia- and reoxygenation-induced cell death and preserves ventricular function. *American Journal of Physiology. Heart and Circulatory Physiology, 300*, H1733–H1742. doi:10.1152/ajpheart.01054.2010.

49. Hauet, T., et al. (2002). Polyethylene glycol reduces the inflammatory injury due to cold ischemia/reperfusion in autotransplanted pig kidneys. *Kidney International, 62*, 654–667. doi:10.1046/j.1523-1755.2002.00473.x.
50. Faure, J. P., et al. (2004). Protective roles of polyethylene glycol and trimetazidine against cold ischemia and reperfusion injuries of pig kidney graft. *American Journal of Transplantation : Official Journal of the American Society of Transplantation and the American Society of Transplant Surgeons, 4*, 495–504. doi:10.1111/j.1600-6143.2004.00365.x.
51. Murad, K. L., Gosselin, E. J., Eaton, J. W., & Scott, M. D. (1999). Stealth cells: Prevention of major histocompatibility complex class II-mediated T-cell activation by cell surface modification. *Blood, 94*, 2135–2141.
52. Dondero, F., et al. (2010). A randomized study comparing IGL-1 to the University of Wisconsin preservation solution in liver transplantation. *Annals of Transplantation, 15*, 7–14.
53. Codas, R., et al. (2009). IGL-1 solution in kidney transplantation: First multi-center study. *Clinical Transplantation, 23*, 337–342. doi:10.1111/j.1399-0012.2009.00959.x.
54. Yu, W. M., Coddington, D., & Bitter-Suermann, H. (1990). Rat liver preservation. I. The components of UW solution that are essential to its success. *Transplantation, 49*, 1060–1066.
55. Petsikas, D., et al. (1990). Enhanced 24-hour in vitro heart preservation with adenosine and adenosine monophosphate. *The Journal of Heart Transplantation, 9*, 114–118.
56. Corps, C. L., et al. (2009). Influence on energy kinetics and histology of different preservation solutions seen during cold ischemia in the liver. *Transplantation Proceedings, 41*, 4088–4093. doi:10.1016/j.transproceed.2009.07.107.
57. Janssen, H., Janssen, P. H., & Broelsch, C. E. (2004). UW is superior to Celsior and HTK in the protection of human liver endothelial cells against preservation injury. *Liver Transplantation : Official Publication of the American Association for the Study of Liver Diseases and the International Liver Transplantation Society, 10*, 1514–1523. doi:10.1002/lt.20309.
58. Chouchani, E. T., et al. (2016). A unifying mechanism for mitochondrial superoxide production during ischemia-reperfusion injury. *Cell Metabolism, 23*, 254–263. doi:10.1016/j.cmet.2015.12.009.
59. Quarrie, R., et al. (2014). Mitochondrial uncoupling does not decrease reactive oxygen species production after ischemia-reperfusion. *American Journal of Physiology. Heart and Circulatory Physiology, 307*, H996–h1004. doi:10.1152/ajpheart.00189.2014.
60. Jaeschke, H., & Woolbright, B. L. (2012). Current strategies to minimize hepatic ischemia-reperfusion injury by targeting reactive oxygen species. *Transplantation Reviews (Orlando, Fla.), 26*, 103–114. doi:10.1016/j.trre.2011.10.006.
61. Murphy, M. P. (2009). How mitochondria produce reactive oxygen species. *The Biochemical Journal, 417*, 1–13. doi:10.1042/bj20081386.
62. St-Pierre, J., Buckingham, J. A., Roebuck, S. J., & Brand, M. D. (2002). Topology of superoxide production from different sites in the mitochondrial electron transport chain. *The Journal of Biological Chemistry, 277*, 44784–44790. doi:10.1074/jbc.M207217200.
63. Barja, G. (1999). Mitochondrial oxygen radical generation and leak: Sites of production in states 4 and 3, organ specificity, and relation to aging and longevity. *Journal of Bioenergetics and Biomembranes, 31*, 347–366.
64. Grivennikova, V. G., & Vinogradov, A. D. (2006). Generation of superoxide by the mitochondrial Complex I. *Biochimica et Biophysica Acta, 1757*, 553–561. doi:10.1016/j.bbabio.2006.03.013.
65. Kwong, L. K., & Sohal, R. S. (2000). Age-related changes in activities of mitochondrial electron transport complexes in various tissues of the mouse. *Archives of Biochemistry and Biophysics, 373*, 16–22. doi:10.1006/abbi.1999.1495.
66. Paradies, G., et al. (2004). Decrease in mitochondrial complex I activity in ischemic/reperfused rat heart: Involvement of reactive oxygen species and cardiolipin. *Circulation Research, 94*, 53–59. doi:10.1161/01.res.0000109416.56608.64.
67. Niatsetskaya, Z. V., et al. (2012). The oxygen free radicals originating from mitochondrial complex I contribute to oxidative brain injury following hypoxia-ischemia in neonatal mice. *The Journal of Neuroscience : The Official Journal of the Society for Neuroscience, 32*, 3235–3244. doi:10.1523/jneurosci.6303-11.2012.

68. Paradies, G., et al. (1999). Lipid peroxidation and alterations to oxidative metabolism in mitochondria isolated from rat heart subjected to ischemia and reperfusion. *Free Radical Biology & Medicine, 27*, 42–50.
69. Nakagawa, T., et al. (2005). Cyclophilin D-dependent mitochondrial permeability transition regulates some necrotic but not apoptotic cell death. *Nature, 434*, 652–658. doi:10.1038/nature03317.
70. Baines, C. P., et al. (2005). Loss of cyclophilin D reveals a critical role for mitochondrial permeability transition in cell death. *Nature, 434*, 658–662. doi:10.1038/nature03434.
71. Schinzel, A. C., et al. (2005). Cyclophilin D is a component of mitochondrial permeability transition and mediates neuronal cell death after focal cerebral ischemia. *Proceedings of the National Academy of Sciences of the United States of America, 102*, 12005–12010. doi:10.1073/pnas.0505294102.
72. Broekemeier, K. M., Dempsey, M. E., & Pfeiffer, D. R. (1989). Cyclosporin A is a potent inhibitor of the inner membrane permeability transition in liver mitochondria. *The Journal of Biological Chemistry, 264*, 7826–7830.
73. Waldmeier, P. C., Zimmermann, K., Qian, T., Tintelnot-Blomley, M., & Lemasters, J. J. (2003). Cyclophilin D as a drug target. *Current Medicinal Chemistry, 10*, 1485–1506.
74. Halestrap, A. P., Connern, C. P., Griffiths, E. J., & Kerr, P. M. (1997). Cyclosporin A binding to mitochondrial cyclophilin inhibits the permeability transition pore and protects hearts from ischaemia/reperfusion injury. *Molecular and Cellular Biochemistry, 174*, 167–172.
75. Warne, J., et al. (2016). Selective inhibition of the mitochondrial permeability transition pore protects against neurodegeneration in experimental multiple sclerosis. *The Journal of Biological Chemistry, 291*, 4356–4373. doi:10.1074/jbc.M115.700385.
76. Takeuchi, O., & Akira, S. (2010). Pattern recognition receptors and inflammation. *Cell, 140*, 805–820. doi:10.1016/j.cell.2010.01.022.
77. Tsung, A., et al. (2005). The nuclear factor HMGB1 mediates hepatic injury after murine liver ischemia-reperfusion. *The Journal of Experimental Medicine, 201*, 1135–1143. doi:10.1084/jem.20042614.
78. Lotze, M. T., et al. (2007). The grateful dead: Damage-associated molecular pattern molecules and reduction/oxidation regulate immunity. *Immunological Reviews, 220*, 60–81. doi:10.1111/j.1600-065X.2007.00579.x.
79. Hu, Q., Wood, C. R., Cimen, S., Venkatachalam, A. B., & Alwayn, I. P. (2015). Mitochondrial Damage-Associated Molecular Patterns (MTDs) are released during hepatic ischemia reperfusion and induce inflammatory responses. *PLoS ONE, 10*, e0140105. doi:10.1371/journal.pone.0140105.
80. Zhang, Q., et al. (2010). Circulating mitochondrial DAMPs cause inflammatory responses to injury. *Nature, 464*, 104–107. doi:10.1038/nature08780.
81. Krysko, D. V., et al. (2011). Emerging role of damage-associated molecular patterns derived from mitochondria in inflammation. *Trends in Immunology, 32*, 157–164. doi:10.1016/j.it.2011.01.005.
82. Ditonno, P., et al. (2013). Effects of ischemia-reperfusion injury in kidney transplantation: Risk factors and early and long-term outcomes in a single center. *Transplantation Proceedings, 45*, 2641–2644. doi:10.1016/j.transproceed.2013.07.025.
83. Yarlagadda, S. G., Coca, S. G., Formica, R. N., Jr., Poggio, E. D., & Parikh, C. R. (2009). Association between delayed graft function and allograft and patient survival: A systematic review and meta-analysis. *Nephrology, Dialysis, Transplantation : Official Publication of the European Dialysis and Transplant Association - European Renal Association, 24*, 1039–1047. doi:10.1093/ndt/gfn667.
84. Krishnan, A. R., et al. (2016). Prolonged ischemic time, delayed graft function, graft and patient outcomes in live-donor kidney transplant recipients. *American Journal of Transplantation : Official Journal of the American Society of Transplantation and the American Society of Transplant Surgeons*. doi:10.1111/ajt.13817.
85. Kayler, L. K., Srinivas, T. R., & Schold, J. D. (2011). Influence of CIT-induced DGF on kidney transplant outcomes. *American Journal of Transplantation : Official Journal of the American Society of Transplantation and the American Society of Transplant Surgeons, 11*, 2657–2664. doi:10.1111/j.1600-6143.2011.03817.x.

86. Wu, W. K., Famure, O., Li, Y., & Kim, S. J. (2015). Delayed graft function and the risk of acute rejection in the modern era of kidney transplantation. *Kidney International, 88*, 851–858. doi:10.1038/ki.2015.190.
87. Mikhalski, D., et al. (2008). Cold ischemia is a major determinant of acute rejection and renal graft survival in the modern era of immunosuppression. *Transplantation, 85*, S3–S9. doi:10.1097/TP.0b013e318169c29e.
88. Perez Valdivia, M. A., et al. (2011). Impact of cold ischemia time on initial graft function and survival rates in renal transplants from deceased donors performed in Andalusia. *Transplantation Proceedings, 43*, 2174–2176. doi:10.1016/j.transproceed.2011.06.047.
89. Salahudeen, A. K., Haider, N., & May, W. (2004). Cold ischemia and the reduced long-term survival of cadaveric renal allografts. *Kidney International, 65*, 713–718. doi:10.1111/j.1523-1755.2004.00416.x.
90. Roodnat, J. I., et al. (2003). Ischemia times and donor serum creatinine in relation to renal graft failure. *Transplantation, 75*, 799–804. doi:10.1097/01.tp.0000056632.00848.8d.
91. Debout, A., et al. (2015). Each additional hour of cold ischemia time significantly increases the risk of graft failure and mortality following renal transplantation. *Kidney International, 87*, 343–349. doi:10.1038/ki.2014.304.
92. Simpkins, C. E., et al. (2007). Cold ischemia time and allograft outcomes in live donor renal transplantation: Is live donor organ transport feasible? *American Journal of Transplantation : Official Journal of the American Society of Transplantation and the American Society of Transplant Surgeons, 7*, 99–107. doi:10.1111/j.1600-6143.2006.01597.x.
93. Gill, J., Dong, J., Eng, M., Landsberg, D., & Gill, J. S. (2014). Pulsatile perfusion reduces the risk of delayed graft function in deceased donor kidney transplants, irrespective of donor type and cold ischemic time. *Transplantation, 97*, 668–674. doi:10.1097/01.tp.0000438637.29214.10.
94. Schechter, M. A., et al. (2016). Elevated cardiac troponin I in preservation solution is associated with primary graft dysfunction. *Journal of Cardiac Failure, 22*, 158–162. doi:10.1016/j.cardfail.2015.08.339.
95. Chen, X. B., & Xu, M. Q. (2014). Primary graft dysfunction after liver transplantation. *Hepatobiliary & Pancreatic Diseases International : HBPD INT, 13*, 125–137.
96. Grimm, J. C., et al. (2015). Association between prolonged graft ischemia and primary graft failure or survival following lung transplantation. *JAMA Surgery, 150*, 547–553. doi:10.1001/jamasurg.2015.12.
97. Fiser, S. M., et al. (2001). Influence of graft ischemic time on outcomes following lung transplantation. *The Journal of Heart and Lung Transplantation : The Official Publication of the International Society for Heart Transplantation, 20*, 1291–1296.
98. King, R. C., et al. (2000). Reperfusion injury significantly impacts clinical outcome after pulmonary transplantation. *The Annals of Thoracic Surgery, 69*, 1681–1685.
99. Palmer, S. M., et al. (2005). Innate immunity influences long-term outcomes after human lung transplant. *American Journal of Respiratory and Critical Care Medicine, 171*, 780–785. doi:10.1164/rccm.200408-1129OC.
100. Palmer, S. M., et al. (2006). Donor polymorphisms in Toll-like receptor-4 influence the development of rejection after renal transplantation. *Clinical Transplantation, 20*, 30–36. doi:10.1111/j.1399-0012.2005.00436.x.
101. Imamura, T., et al. (2015). Late rejection occurred in recipients who experienced acute cellular rejection within the first year after heart transplantation. *International Heart Journal, 56*, 174–179. doi:10.1536/ihj.14-187.
102. Soderlund, C., et al. (2014). Acute cellular rejection the first year after heart transplantation and its impact on survival: A single-centre retrospective study at Skane University Hospital in Lund 1988-2010. *Transplant International : Official Journal of the European Society for Organ Transplantation, 27*, 482–492. doi:10.1111/tri.12284.
103. Kubo, S. H., et al. (1995). Risk factors for late recurrent rejection after heart transplantation: A multiinstitutional, multivariable analysis. Cardiac Transplant Research Database Group. *The Journal of Heart and Lung Transplantation : The Official Publication of the International Society for Heart Transplantation, 14*, 409–418.

104. Sharples, L. D., McNeil, K., Stewart, S., & Wallwork, J. (2002). Risk factors for bronchiolitis obliterans: A systematic review of recent publications. *The Journal of Heart and Lung Transplantation : The Official Publication of the International Society for Heart Transplantation, 21*, 271–281.
105. Almond, P. S., et al. (1993). Risk factors for chronic rejection in renal allograft recipients. *Transplantation, 55*, 752–756.; discussion 756–757.
106. Zhou, J. Q., et al. (2016). Allopurinol preconditioning attenuates renal ischemia/reperfusion injury by inhibiting HMGB1 expression in a rat model. *Acta cirurgica brasileira / Sociedade Brasileira para Desenvolvimento Pesquisa em Cirurgia, 31*, 176–182. doi:10.1590/s0102-865020160030000005.
107. Janssen, H., Janssen, P. H., & Broelsch, C. E. (2003). Celsior solution compared with University of Wisconsin solution (UW) and histidine-tryptophan-ketoglutarate solution (HTK) in the protection of human hepatocytes against ischemia-reperfusion injury. *Transplant International : Official Journal of the European Society for Organ Transplantation, 16*, 515–522. doi:10.1007/s00147-003-0583-5.
108. Zaouali, M. A., et al. (2014). Polyethylene glycol rinse solution: An effective way to prevent ischemia-reperfusion injury. *World Journal of Gastroenterology, 20*, 16203–16214. doi:10.3748/wjg.v20.i43.16203.
109. Faulds, D., Goa, K. L., & Benfield, P. (1993). Cyclosporin. A review of its pharmacodynamic and pharmacokinetic properties, and therapeutic use in immunoregulatory disorders. *Drugs, 45*, 953–1040.
110. Patel, P., Patel, H., Panchal, S., & Mehta, T. (2012). Formulation strategies for drug delivery of tacrolimus: An overview. *International Journal of Pharmaceutical Investigation, 2*, 169–175. doi:10.4103/2230-973X.106981.
111. Morelon, E., Mamzer-Bruneel, M. F., Peraldi, M. N., & Kreis, H. (2001). Sirolimus: A new promising immunosuppressive drug. Towards a rationale for its use in renal transplantation. *Nephrology, Dialysis, Transplantation : Official Publication of the European Dialysis and Transplant Association - European Renal Association, 16*, 18–20.
112. Thomas, K., Koelwel, C., Machei, U., Farber, L., & Gopferich, A. (2005). Three generations of cyclosporine a formulations: An in vitro comparison. *Drug Development and Industrial Pharmacy, 31*, 357–366. doi:10.1081/DDC-54311.
113. McAlister, V. C., Keshavamurthy, M., & Lee, T. D. (1999). Oral delivery of liposomal tacrolimus: Increased efficacy and reduced toxicity. *Transplantation Proceedings, 31*, 1110.
114. Alemdar, A. Y., Sadi, D., McAlister, V. C., & Mendez, I. (2004). Liposomal formulations of tacrolimus and rapamycin increase graft survival and fiber outgrowth of dopaminergic grafts. *Cell Transplantation, 13*, 263–271.
115. Ashok, B., Arleth, L., Hjelm, R. P., Rubinstein, I., & Onyuksel, H. (2004). In vitro characterization of PEGylated phospholipid micelles for improved drug solubilization: Effects of PEG chain length and PC incorporation. *Journal of Pharmaceutical Sciences, 93*, 2476–2487. doi:10.1002/jps.20150.
116. Forrest, M. L., Won, C. Y., Malick, A. W., & Kwon, G. S. (2006). In vitro release of the mTOR inhibitor rapamycin from poly(ethylene glycol)-b-poly(epsilon-caprolactone) micelles. *Journal of Controlled Release: Official Journal of the Controlled Release Society, 110*, 370–377. doi:10.1016/j.jconrel.2005.10.008.
117. Tang, L., et al. (2012). Immunosuppressive activity of size-controlled PEG-PLGA nanoparticles containing encapsulated cyclosporine A. *Journal of Transplantation, 2012*, 896141. doi:10.1155/2012/896141.
118. Immordino, M. L., Dosio, F., & Cattel, L. (2006). Stealth liposomes: Review of the basic science, rationale, and clinical applications, existing and potential. *International Journal of Nanomedicine, 1*, 297–315.
119. Pinto-Alphandary, H., Andremont, A., & Couvreur, P. (2000). Targeted delivery of antibiotics using liposomes and nanoparticles: Research and applications. *International Journal of Antimicrobial Agents, 13*, 155–168.

120. Brasnjevic, I., Steinbusch, H. W., Schmitz, C., Martinez-Martinez, P., & European NanoBioPharmaceutics Research, I. (2009). Delivery of peptide and protein drugs over the blood-brain barrier. *Progress in Neurobiology, 87*, 212–251. doi:10.1016/j.pneurobio.2008.12.002.
121. Chonn, A., Semple, S. C., & Cullis, P. R. (1992). Association of blood proteins with large unilamellar liposomes in vivo. Relation to circulation lifetimes. *The Journal of Biological Chemistry, 267*, 18759–18765.
122. Senior, J., & Gregoriadis, G. (1982). Stability of small unilamellar liposomes in serum and clearance from the circulation: The effect of the phospholipid and cholesterol components. *Life Sciences, 30*, 2123–2136.
123. Bae, Y. H., & Park, K. (2011). Targeted drug delivery to tumors: Myths, reality and possibility. *Journal of Controlled Release: Official Journal of the Controlled Release Society, 153*, 198–205. doi:10.1016/j.jconrel.2011.06.001.
124. Boerman, O. C., et al. (1995). Sterically stabilized liposomes labeled with indium-111 to image focal infection. *Journal of Nuclear Medicine : Official Publication, Society of Nuclear Medicine, 36*, 1639–1644.
125. Li, S. D., & Huang, L. (2008). Pharmacokinetics and biodistribution of nanoparticles. *Molecular Pharmaceutics, 5*, 496–504. doi:10.1021/mp800049w.
126. Price, M. E., Cornelius, R. M., & Brash, J. L. (2001). Protein adsorption to polyethylene glycol modified liposomes from fibrinogen solution and from plasma. *Biochimica et Biophysica Acta, 1512*, 191–205.
127. Woodle, M. C., & Lasic, D. D. (1992). Sterically stabilized liposomes. *Biochimica et Biophysica Acta, 1113*, 171–199.
128. Fukumura, D., & Jain, R. K. (2007). Tumor microvasculature and microenvironment: Targets for anti-angiogenesis and normalization. *Microvascular Research, 74*, 72–84. doi:10.1016/j.mvr.2007.05.003.
129. Saha, R. N., Vasanthakumar, S., Bende, G., & Snehalatha, M. (2010). Nanoparticulate drug delivery systems for cancer chemotherapy. *Molecular Membrane Biology, 27*, 215–231. doi:10.3109/09687688.2010.510804.
130. Juillerat-Jeanneret, L. (2008). The targeted delivery of cancer drugs across the blood-brain barrier: Chemical modifications of drugs or drug-nanoparticles? *Drug Discovery Today, 13*, 1099–1106. doi:10.1016/j.drudis.2008.09.005.
131. Martuza, R. L. (1992). Molecular neurosurgery for glial and neuronal disorders. *Stereotactic and Functional Neurosurgery, 59*, 92–99.
132. Okada, H., Okamoto, S., & Yoshida, J. (1994). [Gene therapy for brain tumors: cytokine gene therapy using DNA/liposome (series 3)]. *No shinkei geka. Neurological Surgery, 22*, 999–1004.
133. Sharma, U. S., Sharma, A., Chau, R. I., & Straubinger, R. M. (1997). Liposome-mediated therapy of intracranial brain tumors in a rat model. *Pharmaceutical Research, 14*, 992–998.
134. Siegal, T., Horowitz, A., & Gabizon, A. (1995). Doxorubicin encapsulated in sterically stabilized liposomes for the treatment of a brain tumor model: Biodistribution and therapeutic efficacy. *Journal of Neurosurgery, 83*, 1029–1037. doi:10.3171/jns.1995.83.6.1029.
135. Kakinuma, K., et al. (1996). Targeting chemotherapy for malignant brain tumor using thermosensitive liposome and localized hyperthermia. *Journal of Neurosurgery, 84*, 180–184. doi:10.3171/jns.1996.84.2.0180.
136. Nadig, S. N., et al. (2015). Immunosuppressive nano-therapeutic micelles downregulate endothelial cell inflammation and immunogenicity. *RSC Advances, 5*, 43552–43562. doi:10.1039/c5ra04057d.
137. Ankola, D. D., Battisti, A., Solaro, R., & Kumar, M. N. (2010). Nanoparticles made of multiblock copolymer of lactic acid and ethylene glycol containing periodic side-chain carboxyl groups for oral delivery of cyclosporine A. *Journal of the Royal Society, Interface / the Royal Society, 7 Suppl 4*, S475–S481. doi:10.1098/rsif.2010.0046.focus.

138. Gref, R., et al. (2001). Development and characterization of CyA-loaded poly(lactic acid)-poly(ethylene glycol)PEG micro- and nanoparticles. Comparison with conventional PLA particulate carriers. *European Journal of Pharmaceutics and Biopharmaceutics : Official Journal of Arbeitsgemeinschaft fur Pharmazeutische Verfahrenstechnik e.V, 51*, 111–118.
139. Italia, J. L., Bhatt, D. K., Bhardwaj, V., Tikoo, K., & Kumar, M. N. (2007). PLGA nanoparticles for oral delivery of cyclosporine: Nephrotoxicity and pharmacokinetic studies in comparison to Sandimmune Neoral. *Journal of Controlled Release: Official Journal of the Controlled Release Society, 119*, 197–206. doi:10.1016/j.jconrel.2007.02.004.
140. Lund, L. H., et al. (2013). The Registry of the International Society for Heart and Lung Transplantation: Thirtieth Official Adult Heart Transplant Report--2013; focus theme: Age. *The Journal of Heart and Lung Transplantation : The Official Publication of the International Society for Heart Transplantation, 32*, 951–964. doi:10.1016/j.healun.2013.08.006.
141. Hollis, I. B., Reed, B. N., & Moranville, M. P. (2015). Medication management of cardiac allograft vasculopathy after heart transplantation. *Pharmacotherapy, 35*, 489–501. doi:10.1002/phar.1580.
142. Pober, J. S., Jane-wit, D., Qin, L., & Tellides, G. (2014). Interacting mechanisms in the pathogenesis of cardiac allograft vasculopathy. *Arteriosclerosis, Thrombosis, and Vascular Biology, 34*, 1609–1614. doi:10.1161/atvbaha.114.302818.
143. Khush, K. K., & Valantine, H. A. (2009). New developments in immunosuppressive therapy for heart transplantation. *Expert Opinion on Emerging Drugs, 14*, 1–21. doi:10.1517/14728210902791605.
144. Jane-Wit, D., et al. (2013). Alloantibody and complement promote T cell-mediated cardiac allograft vasculopathy through noncanonical nuclear factor-kappaB signaling in endothelial cells. *Circulation, 128*, 2504–2516. doi:10.1161/circulationaha.113.002972.
145. Photos, P. J., Bacakova, L., Discher, B., Bates, F. S., & Discher, D. E. (2003). Polymer vesicles in vivo: Correlations with PEG molecular weight. *Journal of Controlled Release: Official Journal of the Controlled Release Society, 90*, 323–334.
146. Singh, B., Garg, T., Goyal, A. K., & Rath, G. (2014). Recent advancements in the cardiovascular drug carriers. *Artificial Cells, Nanomedicine, and Biotechnology*, 1–10. doi:10.3109/21691401.2014.937868.
147. Muralidharan, P., Mallory, E., Malapit, M., Hayes, D., Jr., & Mansour, H. M. (2014). Inhalable PEGylated phospholipid nanocarriers and PEGylated therapeutics for respiratory delivery as aerosolized colloidal dispersions and dry powder inhalers. *Pharmaceutics, 6*, 333–353. doi:10.3390/pharmaceutics6020333.
148. Danhier, F., Le Breton, A., & Preat, V. (2012). RGD-based strategies to target alpha(v) beta(3) integrin in cancer therapy and diagnosis. *Molecular Pharmaceutics, 9*, 2961–2973. doi:10.1021/mp3002733.
149. Muzykantov, V., & Muro, S. (2011). Targeting delivery of drugs in the vascular system. *International Journal of Transport Phenomena, 12*, 41–49.
150. Franz, M., et al. (2015). Targeted delivery of interleukin-10 to chronic cardiac allograft rejection using a human antibody specific to the extra domain A of fibronectin. *International Journal of Cardiology, 195*, 311–322. doi:10.1016/j.ijcard.2015.05.144.
151. Franz, M., et al. (2014). De novo expression of fetal ED-A(+) fibronectin and B (+) tenascin-C splicing variants in human cardiac allografts: Potential impact for targeted therapy of rejection. *Journal of Molecular Histology, 45*, 519–532. doi:10.1007/s10735-014-9573-4.
152. Franz, M., et al. (2013). Selective imaging of chronic cardiac rejection using a human antibody specific to the alternatively spliced EDA domain of fibronectin. *The Journal of Heart and Lung Transplantation : The Official Publication of the International Society for Heart Transplantation, 32*, 641–650. doi:10.1016/j.healun.2013.04.003.
153. Raichlin, E., et al. (2007). Conversion to sirolimus as primary immunosuppression attenuates the progression of allograft vasculopathy after cardiac transplantation. *Circulation, 116*, 2726–2733. doi:10.1161/circulationaha.107.692996.

154. Guo, Y., et al. (2012). Simultaneous diagnosis and gene therapy of immuno-rejection in rat allogeneic heart transplantation model using a T-cell-targeted theranostic nanosystem. *ACS Nano, 6*, 10646–10657. doi:10.1021/nn3037573.
155. Watts, A. B., Williams, R. O., 3rd, & Peters, J. I. (2009). Recent developments in drug delivery to prolong allograft survival in lung transplant patients. *Drug Development and Industrial Pharmacy, 35*, 259–271. doi:10.1080/03639040802282904.
156. Niven, R., et al. (2011). Safety and toxicology of cyclosporine in propylene glycol after 9-month aerosol exposure to beagle dogs. *Journal of Aerosol Medicine and Pulmonary Drug Delivery, 24*, 205–212. doi:10.1089/jamp.2010.0863.
157. Wang, T., et al. (2007). Preclinical safety evaluation of inhaled cyclosporine in propylene glycol. *Journal of Aerosol Medicine : The Official Journal of the International Society for Aerosols in Medicine, 20*, 417–428. doi:10.1089/jam.2007.0626.
158. Iacono, A. T., et al. (2006). A randomized trial of inhaled cyclosporine in lung-transplant recipients. *The New England Journal of Medicine, 354*, 141–150. doi:10.1056/NEJMoa043204.
159. Johnson, B. A., et al. (2012). Cyclosporine inhalation solution does not improve bronchiolitis obliterans syndrome-free survival following lung transplant: Results from the CYCLIST trial. *Journal of Heart and Lung Transplantation, 31*, S66. doi:10.1016/j.healun.2012.01.177.
160. Corcoran, T. E., Niven, R., Verret, W., Dilly, S., & Johnson, B. A. (2014). Lung deposition and pharmacokinetics of nebulized cyclosporine in lung transplant patients. *Journal of Aerosol Medicine and Pulmonary Drug Delivery, 27*, 178–184. doi:10.1089/jamp.2013.1042.
161. Behr, J., et al. (2009). Lung deposition of a liposomal cyclosporine A inhalation solution in patients after lung transplantation. *Journal of Aerosol Medicine and Pulmonary Drug Delivery, 22*, 121–130.
162. Sato, H., et al. (2013). Development of cyclosporine A-loaded dry-emulsion formulation using highly purified glycerol monooleate for safe inhalation therapy. *International Journal of Pharmaceutics, 448*, 282–289. doi:10.1016/j.ijpharm.2013.03.026.
163. Carvalho, S. R., et al. (2014). Characterization and pharmacokinetic analysis of crystalline versus amorphous rapamycin dry powder via pulmonary administration in rats. *European Journal of Pharmaceutics and Biopharmaceutics : Official Journal of Arbeitsgemeinschaft fur Pharmazeutische Verfahrenstechnik e.V, 88*, 136–147. doi:10.1016/j.ejpb.2014.05.008.
164. Bayer, J., et al. (2013). Effect of inhaled tacrolimus on ischemia reperfusion injury in rat lung transplant model. *The Journal of Thoracic and Cardiovascular Surgery, 146*, 1213–1219.; discussion 1219. doi:10.1016/j.jtcvs.2013.07.030.
165. Heyder, J. (2004). Deposition of inhaled particles in the human respiratory tract and consequences for regional targeting in respiratory drug delivery. *Proceedings of the American Thoracic Society, 1*, 315–320. doi:10.1513/pats.200409-046TA.
166. Sung, J. C., Pulliam, B. L., & Edwards, D. A. (2007). Nanoparticles for drug delivery to the lungs. *Trends in Biotechnology, 25*, 563–570. doi:10.1016/j.tibtech.2007.09.005.
167. Cova, E., et al. (2015). Antibody-engineered nanoparticles selectively inhibit mesenchymal cells isolated from patients with chronic lung allograft dysfunction. *Nanomedicine (London, England), 10*, 9–23. doi:10.2217/nnm.13.208.
168. Salvadori, M., Rosso, G., & Bertoni, E. (2015). Update on ischemia-reperfusion injury in kidney transplantation: Pathogenesis and treatment. *World Journal of Transplantation, 5*, 52–67. doi:10.5500/wjt.v5.i2.52.
169. Menke, J., Sollinger, D., Schamberger, B., Heemann, U., & Lutz, J. (2014). The effect of ischemia/reperfusion on the kidney graft. *Current Opinion in Organ Transplantation, 19*, 395–400. doi:10.1097/mot.0000000000000090.
170. Rogers, N. M., Stephenson, M. D., Kitching, A. R., Horowitz, J. D., & Coates, P. T. (2012). Amelioration of renal ischaemia-reperfusion injury by liposomal delivery of curcumin to renal tubular epithelial and antigen-presenting cells. *British Journal of Pharmacology, 166*, 194–209. doi:10.1111/j.1476-5381.2011.01590.x.
171. Hosgood, S. A., van Heurn, E., & Nicholson, M. L. (2015). Normothermic machine perfusion of the kidney: Better conditioning and repair? *Transplant International : Official Journal of the European Society for Organ Transplantation, 28*, 657–664. doi:10.1111/tri.12319.

172. Brat, A., Pol, R. A., & Leuvenink, H. G. (2015). Novel preservation methods to increase the quality of older kidneys. *Current Opinion in Organ Transplantation, 20*, 438–443. doi:10.1097/mot.0000000000000215.
173. O'Callaghan, J. M., Morgan, R. D., Knight, S. R., & Morris, P. J. (2013). Systematic review and meta-analysis of hypothermic machine perfusion versus static cold storage of kidney allografts on transplant outcomes. *The British Journal of Surgery, 100*, 991–1001. doi:10.1002/bjs.9169.
174. Jiao, B., et al. (2013). Hypothermic machine perfusion reduces delayed graft function and improves one-year graft survival of kidneys from expanded criteria donors: A meta-analysis. *PLoS ONE, 8*, e81826. doi:10.1371/journal.pone.0081826.
175. Brasile, L., Glowacki, P., Castracane, J., & Stubenitsky, B. M. (2010). Pretransplant kidney-specific treatment to eliminate the need for systemic immunosuppression. *Transplantation, 90*, 1294–1298. doi:10.1097/TP.0b013e3181ffba97.
176. Lobb, I., et al. (2015). Hydrogen sulfide treatment mitigates renal allograft ischemia reperfusion injury during cold storage and improves early transplant kidney function and survival following allogeneic renal transplantation. *The Journal of Urology*. doi:10.1016/j.juro.2015.07.096.
177. Thuillier, R., et al. (2014). Cyclodextrin curcumin formulation improves outcome in a preclinical pig model of marginal kidney transplantation. *American Journal of Transplantation: Official Journal of the American Society of Transplantation and the American Society of Transplant Surgeons, 14*, 1073–1083. doi:10.1111/ajt.12661.
178. Pan, N., et al. (2015). Comparison of methods for the reconstruction of the hepatic artery in mouse orthotopic liver transplantation. *PLoS ONE, 10*, e0133030. doi:10.1371/journal.pone.0133030.
179. Zhu, H., Yu, L., He, Y., & Wang, B. (2014). Nonhuman primate models of type 1 diabetes mellitus for islet transplantation. *Journal of Diabetes Research*, 785948. doi:10.1155/2014/785948 (2014).
180. Chandra, P. K., et al. (2012). Inhibition of hepatitis C virus replication by intracellular delivery of multiple siRNAs by nanosomes. *Molecular Therapy : The Journal of the American Society of Gene Therapy, 20*, 1724–1736. doi:10.1038/mt.2012.107.
181. Kesharwani, P., Banerjee, S., Padhye, S., Sarkar, F. H., & Iyer, A. K. (2015). Parenterally administrable nano-micelles of 3,4-difluorobenzylidene curcumin for treating pancreatic cancer. *Colloids and Surfaces. B, Biointerfaces, 132*, 138–145. doi:10.1016/j.colsurfb.2015.05.007.
182. Kepsutlu, B., Nazli, C., Bal, T., & Kizilel, S. (2014). Design of bioartificial pancreas with functional micro/nano-based encapsulation of islets. *Current Pharmaceutical Biotechnology, 15*, 590–608.

Chapter 8
Tissue Bioengineering in Transplantation

Ravi Katari, Lauren Edgar, Kevin Enck, Andrea Peloso, Riccardo Tamburrini, and Giuseppe Orlando

Abstract Organ transplantation has emerged in recent decades as one of the most effective modalities for the treatment of end-stage organ disease. Over 100,000 transplants are performed worldwide each year; however, the supply has not been able to keep up with increasing demand. Furthermore, transplant recipients are committed to a lifelong regimen of brutal immunosuppressive medications that themselves carry significant side effect profiles, influencing clinical outcomes. Recent advances in the fields of tissue engineering and regenerative medicine are beginning to offer alternative solutions that could potentially improve the longevity, functionality, and biocompatibility profiles of transplants. Decellularization technology to produce extracellular matrix scaffolds represents one of the most promising strategies currently under investigation. Such methods can produce bioengineered, transplantable organs using autologous cells that would bypass the need for immunosuppression and its associated side effects. Furthermore, bioengineering strategies in general are not bound by supply constraints imposed by organ donation.

Abbreviations

ECM	Extracellular matrix
ESC	Embryonic stem cell
iPSC	Induced pluripotent stem cell

R. Katari • L. Edgar • K. Enck • R. Tamburrini
Wake Forest School of Medicine, Winston Salem, NC, USA

A. Peloso
General Surgery, Fondazione IRCCS Policlinico San Matteo and University of Pavia, Pavia, Italy

G. Orlando, MD, PhD (✉)
Department of Surgery, Section of Transplantation, Wake Forest School of Medicine, Winston Salem, NC, USA

Wake Forest Baptist Hospital, Medical Center Blvd, Winston Salem, NC 27157, USA
e-mail: gorlando@wakehealth.edu

S.N. Nadig, J.A. Wertheim (eds.), *Technological Advances in Organ Transplantation*, DOI 10.1007/978-3-319-62142-5_8

MSC	Mesenchymal stem cell
PEG	Polyethylene glycol
TE	Tissue engineering

8.1 Introduction

In recent decades, organ transplantation technology has progressed substantially, which has paved the way for its emergence as the gold standard treatment for a myriad of clinical settings characterized by end-stage organ disease [1]. Although medical management has made impressive strides as a treatment modality, surgical transplantation remains the best and only means of permanently restoring the original function of chronically or acutely irreversibly damaged organs. The aim of this chapter is to review relevant concepts in the science and practice of modern transplantation along with related important developments in tissue engineering (TE).

Over 100,000 solid organ transplants are performed every year worldwide, including approximately 70,000 kidney and 20,000 liver transplants [2]. End-stage renal disease, for example, represents an extraordinary public health burden worldwide, which has contributed to the kidney being the most commonly transplanted organ. In 2014, about 15,700 people received kidney transplants in the United States, while over 100,000 candidates remained on the waiting list because of limited supply [3, 4].

Success rates have been improving incrementally over the years, but the challenges of donor organ supply and complications relating to immunosuppressive medications remain major obstacles in the way of further progress for the treatment of irreversible end-organ damage. Temporizing measures, such as dialysis, extends life expectancy, which predictably contributes to a growing demand for donor grafts and an ever-expanding waiting list. But candidates who do ultimately receive a transplanted organ can develop secondary neoplasms or other adverse sequelae associated with prolonged immunosuppression.

In light of these shortcomings, developing methods of tissue engineering (TE) has been very promising as a potential solution to the challenges associated with modern transplantation. TE, often touted as a subfield of regenerative medicine, refers to approaches that attempt to restore physiological or anatomical function by either regenerating cellular material or replacing diseased cells with healthy cells in tissues that might not perform this task spontaneously or sufficiently [5]. TE can be accomplished in experimental and clinical settings by administering cellular material, noncellular biomaterials, or a combination of the two [6]. Biomaterials, naturally derived or synthetic, are often employed, as are anatomically relevant "scaffolds" on which cells can be distributed or that allow for the migration and proliferation of cells in vivo as a way to replace damaged tissue.

The introduction of cellular material for the purpose of regeneration or replacement arguably can be a natural extension of the current donation system underlying modern transplantation. Indeed, it has been recognized that transplantation shares many of the same guiding principles that now guide developing TE methodologies [7]. For example, the idea of replacing diseased tissue by the direct administration of bioactive, cellular material underlies present surgical transplantation. A TE alternative may merely involve seeding cellular material within a biocompatible, organ-shaped scaffold to bioengineer a functional organ for use as a replacement [8]. Further, whether by direct healing, via administration of cellular material and biomaterials, or ex vivo fabrication of whole organs for subsequent transplantation, TE—though still its early stages—enjoys the advantage of not being limited by organ shortage because cell cultures can be expanded in the laboratory and biomaterials can be manufactured on industrial scales without donor morbidity or mortality.

8.2 Major Challenges in Transplantation

8.2.1 The Organ Shortage

The dramatic success of organ transplantation for the treatment of organ failure has stoked an overwhelming demand without a corresponding supply for donor grafts. Kidney transplantation, for example, has proven itself to be superior to dialysis as a modality of renal replacement therapy in terms of both cost-effectiveness and clinical outcomes [9, 10]. Life expectancy for dialysis patients in the United States is approximately 8 years for patients between 40 and 44 years and 4.5 years for patients between 60 and 64 years. Transplantation offers a superior life expectancy profile of 85% at 5 years, 70% at 10 years, and 44% at 20 years postoperatively.

Thus, modern transplantation can be considered a victim of its own success [11]. In 2014, only 15,700 patients received a kidney transplant, while over 100,000 remained on waiting lists at the end of the year [12, 13]. According to the Global Observatory on Donation and Transplantation (GODT), approximately 114,690 solid organs were transplanted worldwide in 2012, representing a modest increase of 1.8% from the previous year. The effort fell short, however, and met less than 10% of the global need [14].

To offset some of the growing demand, transplant operators have widened the donor pool by expanding the criteria for acceptable donors with limited success. For example, the expansion of living and deceased donor acceptance criteria to include donation after cardiac death, paired donations, altruistic donations, immunologically suboptimal donations, and donation from people with comorbidities has helped improve the probability of receiving an organ in the critical time frame while on a waiting list [15, 16]. Nevertheless, the rising incidence of chronic non-communicable diseases such as diabetes and hypertension—leading causes of kidney disease—is projected to limit supply further and exacerbate discrepancy with demand [17].

8.2.2 A Halfway Technology

Though representing one of the greatest achievements of modern medicine, transplantation can be usefully described as a "halfway technology" in light of inherent major obstacles. Lewis Thomas, a renowned pathologist, first articulated this concept by referring to technologies that compensate for the adverse sequelae of various disease processes as opposed to actually curing them [18]. He anticipated the laborious burden of "incomplete" technologies that complicates the overall treatment regimen.

It is readily apparent how modern transplantation in its current state fits Thomas' description of an "incomplete" technology [19]. For example, although replacing native kidneys with donor grafts reconstitutes the filtration and hormonal functions of a healthy kidney, intervention does not eradicate the baseline disease that first led to the procedure in the first place: thus the stage remains set for secondary organ failure. Preventing graft failure, moreover, requires a lifelong commitment to antirejection medications, which also cause acute and/or chronic toxicity, leading to additional clinical syndromes [20]. Consequently, meticulous long-term management is necessary to prevent side effects or graft failure and to optimize quality of life.

A clear example of the incompleteness of transplantation is seen in transplants performed for end-stage kidney disease caused by diabetes mellitus [21]. Although transplantation is a lifesaving intervention in these cases, it does not cure the baseline disease which instead will recur after the operation.

Disease recurrence after liver transplantation is, in fact, virtually inevitable. Histological evidence of infectious damage is detectable often within weeks, and progression to cirrhosis requiring another transplant occurs in 25% of cases within 5–10 years following transplantation. Consequently, the recurrence of hepatitis C presents one of the major challenges to successful modern liver transplantation and is one of the most frequent causes of recurrent disease resulting in graft failure, for all organs [22]. Naturally, the chief exacerbating factor in recurrent hepatitis C cases is immunosuppression secondary to antirejection therapy. There is, indeed, evidence to suggest that weaning from antirejection therapy in hepatitis C liver transplant recipients delays the recurrence of disease and consequently improves overall morbidity and mortality outcomes [23]. Although there are rare cases in which transplantation has been fully curative without the need for immunosuppression, that technology is entirely consistent with Thomas's description.

Thomas's concept of a "halfway" or "incomplete" technology is useful because it provides boundaries and benchmarks by which alternative solutions can be measured. TE, for example, can be compared to transplantation and profitably evaluated using Thomas's conceptual framework because it shares with transplantation overlapping ancestry and methodology [24, 25]. The essential goal to restore tissue and organ function via the introduction of bioactive materials is a shared attribute, albeit with different parameters.

As will be discussed in later sections, TE allows for the use of recipient-derived cellular material to manufacture functional tissues and organs ex vivo for autologous reimplantation thereby obviating the need for toxic immunosuppressive therapy. Without transplantation's associated side effects, a TE-derived graft would be more temporally durable and less susceptible to complications resulting in graft failure. TE methods could move transplantation closer to being a "complete" technology than organs obtained from conventional sources presently allow.

8.2.3 The Burden of Immunosuppression

Although advances in the transplantation sciences have dramatically improved morbidity and mortality outcomes, subsequent immunosuppressive protocols remain a massive burden for patients following surgery because of the need to address associated infectious, neoplastic, and end-organ complications [26, 27]. For this reason, tolerance to organ allografts has traditionally been the ultimate goal for transplant practitioners and researchers. Indeed, achieving an immunosuppression-free state (IFS) that avoids all complications and costs associated with lifelong immunosuppression is often referred to as the "Holy Grail" among affiliated clinicians [28].

Infections are among the most common causes of hospitalization following kidney transplantation and are, in fact, the major cause of hospitalization in pediatric transplant patients postoperatively [29]. Chronic immunosuppression also contributes to adverse cardiovascular events, which represent the chief cause of death in recipients [30]. Solid organ transplant recipients also have a higher risk of developing cancer because of immunosuppression and oncogenic viral infections.

Engels et al. evaluated medical data from over 175,000 transplant recipients from 1987 to 2008 (approximately 40% of all organ recipients in the United States) and found an overall doubling of cancer risk compared with the general population [31]. Standardized incidence ratios were significantly elevated for most infection-related malignancies including non-Hodgkin's lymphoma, Kaposi sarcoma, Hodgkin lymphoma, and cancers of the stomach, oropharynx, anus, vulva, and penis in kidney, liver, heart, and lung recipients. The risk of liver cancer was only elevated among liver recipients, partly because of recurrent hepatitis B or C or diabetes—showing the efficacy of Thomas's concept of the halfway technology.

Ultimately, the duration of immunosuppression along with the intensity of treatment is the most powerful predictor of malignancy [32]. Achieving a balance between the risk profile of antirejection therapy and the need to prevent graft rejection is a supremely difficult task of critical importance. In this context, it is no wonder that the IFS is considered the Holy Grail among transplant clinicians.

To ensure graft survival and function, patients must adhere to long-term immunosuppression. It can be estimated that over 300,000 patients in the United States alone are currently transplant recipients committed to brutal immunosuppressive regimens, with the attendant burden of close and frequent monitoring, which adversely impact health and quality of life in an unusually severe manner. There are over 1000 individuals presently living with an intestinal transplant, approximately 14,000 patients with pancreas grafts, 11,000 with lung transplants, 27,000 with heart transplants, almost 200,000 with kidney transplants, and over 59,000 with liver transplants [33–38].

Organ transplantation is also one of the most expensive medical therapies currently available and has a major impact on hospital and health system expenditures, in addition to a patient's personal finances. Countries in which national health services are overburdened by the increasing medical needs of aging populations resort to strict cost-optimizing policies to limit access to organ transplantation and to limit expenses [39]. Thus, one of the primary challenges associated with transplantation is devising management strategies that lower the cost of the procedure without adversely impacting clinical outcomes [40].

The cost of immunosuppressant drugs and requisite frequent follow-up visits constitute an immense financial burden. For those without long-term insurance coverage, drug costs may represent a significant out-of-pocket expense that is unaffordable. Patients who depend on Medicare for health insurance face a coverage cliff at 36 months after transplant for immunosuppression drugs, a policy widely regarded as shortsighted given the massive expense incurred in the case of graft failure [41, 42]. Evans et al. surveyed kidney transplant programs in the United States and found that 70% reported that their patients have an extremely or very serious problem paying for their medications [43]. Astoundingly, 68% of the programs reported deaths and graft losses attributed to cost-related nonadherence to immunosuppressive medications.

8.3 Tissue Engineering Solutions in Transplantation

TE efforts have resulted in successful manufacture of relatively simple, hollow organs (e.g., bladder, airway, etc.) from autologous cellular sources and subsequent implantation in over 200 patients with various medical conditions without the need for subsequent antirejection therapy. (In contrast, the number of transplant recipients who have been successfully weaned off of immunosuppressive drugs in the postoperative period is much lower.) At present, the production of more complex, modular, solid organs such as the kidney, liver, pancreas, heart, and lung has not yet occurred to a clinically relevant degree. Nevertheless, methods in TE are constantly being refined and updated; consequently, the field offers a promising new arena for transplant research that could ultimately undergo a paradigm shift as it seeks to remove the obstacles described above. Proof of concept has already been established, and the transition from hollow to solid, complex organs should be the next milestone.

8.3.1 Cell-Scaffold Technology

TE approaches typically involve the use of cellular material either alone or in conjunction with a supporting scaffold [44]. The latter approach typically involves the manipulation of biomaterials into geospatially appropriate scaffolds. When seeded with cells and allowed to mature, these constructs are implanted as functional organoids with the bioengineered graft assuming the role of the transplanted organ. The use of autologous cellular material circumvents immunosurveillance, and, thus, antirejection therapy is not warranted: a theoretically supported concept that has been observed empirically [45].

Determining the cell type most suitable for regeneration and/or reseeding onto a scaffold remains a critical objective for investigators in each organ system. Experimental and clinical investigations have commonly employed fully differentiated adult cells, organ-specific progenitor cells, or pluripotent stem cells [46], all of which can be acquired from a recipient.

Natural biomaterial scaffolds can be fabricated by perfusion of detergents through animal- or human-derived organs. This method removes the cellular compartment from the organ and leaves behind an extracellular matrix (ECM). These ECM-based scaffolds are highly biomimetic because they retain their original structural architecture in three dimensions including the intricate, internal vascular networks, adhesion molecules, and cellular signaling proteins [47–51]. Thus, naturally derived scaffolds are particularly suited for cellular seeding.

The idea of fabricating natural ECM scaffolds was pioneered by Harvard researchers Vacanti et al. in the 1980s. In one of the earliest investigations, fetal and adult rat cells, mouse hepatocytes, pancreatic islet cells, and cells from the small intestine were seeded onto synthetically derived scaffolds [52]. The supporting materials consisted of synthetic polymers organized into fiber networks that simulated the intertwining branching networks of the connective tissue present in all organs that allows cells to remain viable by diffusion, promoting vascular ingrowth and encouraging proliferation [53]. After allowing the cells to culture on the synthetic scaffolds for 4 days, these constructs were then implanted into animals of varying species. The investigators recorded six cases of successful engraftment, each of which demonstrated viable cells, mitotic figures, and vascularization of the cell mass. This was the first report of ex vivo manufacture of implantable organoid constructs consisting of cellular material seeded on artificial supporting scaffolds. The report paved the way for further studies that ultimately led to the first clinical application: a tissue-engineered vascular graft that was used to replace an intermediate pulmonary artery in a child with right ventricular and pulmonary atresia [54].

In all successful ECM cases reported to date, autologous cells were isolated and expanded in vitro prior to seeding. In some cases, multipotent stem cells were isolated and differentiated toward specific somatic cell types prior to seeding on scaffolds [55–57]. Macchiarini et al. removed cells and MHC antigens from a human donor trachea and repopulated the ECM scaffold with epithelial cells and stem cell-derived chondrocytes that had been cultured from cells acquired by the recipient. This graft was then used to replace the patient's left main bronchus. At 5-year follow-up, the tissue-engineered trachea remained open over its entire length, was well-vascularized, was completely recellularized with respiratory epithelium, and had normal ciliary function and mucus clearance [58]. Thus, the investigators provided a solid foundation for further research into potentially groundbreaking cell-scaffold technology. Comparable success using biodegradable non-ECM-based scaffolds has been reported in the urethra [59], the bladder [60], and the vagina [61].

8.3.2 ECM-Based Scaffolds

The ideal scaffold in the context of organ transplantation is biomaterial-based and approximates the three-dimensional structure of the organ to be transplanted. It should also lend biological and mechanical signals to induce cell growth and differentiation [62]. Generally, applicable design considerations include minimal immunogenicity, a degradation profile which parallels tissue regeneration, the presence of environmental factors appropriate for the seeded cell type, and appropriate mechanostructural parameters such as stiffness and tensile strength. In fact, ECM-based scaffolds are an attractive option because they include these considerations and support cells in vivo.

In recent decades, it has become increasingly clear that ECMs play a fundamental role in the viability and functionality of cells, tissues, and organs. Indeed, one of the primary goals of TE investigators is to elucidate fully the relationship between ECMs and cells to facilitate the manufacturing of better cell-scaffold grafts. Decellularized tissue matrices obtained via detergent perfusion carry the benefit of preserving the native architecture and mechanical properties of the original tissue, i.e., the ECM remains virtually intact. These frames can be obtained autologously and reseeded with autologous cells. They also have been shown to support cellular regeneration without immunologic rejection: a feature of tremendous importance to transplant investigators [63].

ECM technology is able to drive differentiation of progenitor cells into organ-specific phenotypes, indicating the potential use of stem cells for cell-scaffold technology [64]. Indeed, Remuzzi et al. recently reported on the successful recellularization of acellular rat kidney scaffolds using embryonic stem cells via infusion through the renal artery and subsequent pressure-controlled perfusion with recirculating culture medium [65].

Variations in ECM are likely to parallel the specializations of their corresponding organ systems. As such, the design of a single, all-encompassing biomaterial is not a practical design goal. To design the next generation of advanced biomaterial scaffolds, current investigators must take into consideration that body tissues and organs are highly specialized in structure and function. Decellularization, recellularization, and maturation protocols should, instead, be optimized and recent successes expanded on.

8.4 Organ Bioengineering

8.4.1 Cell Types

There are a number of potential approaches that can be used to fabricate organs ex vivo for eventual implantation into patients. Cell-scaffold technology consists of seeding cellular material on supporting scaffolds. Determining the proper cell type most suitable for regeneration and/or seeding on a scaffold remains a chief objective for each organ system investigated. Experimental and clinical investigations have commonly involved fully differentiated adult cells, organ-specific progenitor cells, or pluripotent stem cells.

Adult somatic cells hold the advantage of being well-defined and readily isolated, but, compared with progenitor and stem cells, their lifespan and regenerative potential are limited. Further, they lose function and differentiability when removed from the native environment. Their use is confined to the specific organ from which they are obtained. For example, cultured hepatocytes or renal tubular epithelial cells can be used only for liver scaffolds or kidney scaffolds, respectively, and they would likely be minimally proliferative.

Stem cells, however, possess the notable capacity of self-renewal and differentiability. Their applicability has a broader range than adult somatic cells. Within the category of stem cells, the two types most commonly utilized are adult stem cells and embryonic stem cells [66].

Embryonic stem cells (ESCs) are pluripotent stem cells derived from the inner cell mass of the embryo during the blastocyst stage that have the ability to differentiate into any cell type when induced with specific environmental cues or stimuli. Furthermore, they can self-renew indefinitely [67]. Researchers have successfully induced their differentiation into cell types from all three germ layers in vitro including cardiac cells [68], endothelial cells [69], neurons [70], insulin-producing cells [71], and renal tubular cells [72]. In the context of abdominal bioengineering, the use of embryonic stem cells would find wide applicability for experimental and clinical purposes. Nevertheless, their use is significantly limited by ethical issues along with potential teratogenicity [73, 74].

Induced pluripotent stem cells (iPSCs) are generated via special reprogramming of adult cells resulting in their dedifferentiation into pluripotent stem cells. This process does not involve manipulation or destruction of primordial embryos, thus avoiding the ethical issues posed by ESCs. iPSCs have been derived from several cell types in experimental investigations, including human keratinocytes and umbilical cord blood mononuclear cells [75, 76].

Adult stem cells, in contrast to ESCs, benefit from wide availability in the adult organism and relative ease of acquisition. They are thought to perform a major role in cellular repopulation and the replenishment process over time [77]. Indeed, adult stem cells residing in anatomical "niches" are responsible for the continual replacement of damaged or dead tissue compartments in response to microenvironmental cues [78]. Wound healing and bone remodeling are familiar examples of tissue systems with high turnover. Adult tissue systems with recognized niche populations of stem cells include the skin, fat, intestine, and kidney, among others [79–82].

Unlike somatic cells, adult stem cells can be induced to differentiate along lineages other than the organ system from which they are derived. For example, fat stem cells have been successfully differentiated into cartilage cells, hematopoietic cells, neurons, bone cells, and skeletal muscle cells [83]. Nevertheless, their differentiation potential is generally recognized to be limited to the germ layer from which they are derived, unlike the pluripotency of the ESC.

Adult cells harvested for regenerative purposes can be heterologous (e.g., porcine), allogeneic (from a donor patient), or autologous (from the patient). Autologous cells are beneficial because they are shielded from the recipient patient's immune system because they are not recognized as nonself. Given their ability to circumvent rejection, autologous cells represent the most desirable cell type in TE investigations. Allogeneic cells are another option; acquired from another individual, their use is complicated by potential immunogenicity [84]. Heterologous cells derived from non-human organisms (e.g., porcine) represent another option under investigation.

Ideally, adult stem cells used for TE purposes would be expanded ex vivo and either (1) introduced directly into the patient to repair the damaged organ system or (2) seeded onto a supporting scaffold for eventual reimplantation in the patient. Isolating and expanding these cells at clinically relevant levels remain a challenge for current investigators.

8.4.2 Gastrointestinal Tract

The intestinal conduits are complex, hollow structures involved in nutrient passage and absorption, digestion, excretion, and the innate immune system. The human tract undergoes continuous turnover and renewal throughout life. This process is governed by the activity of resident stem cells in the gut wall [85].

The preexisting process of regeneration and renewal in the gastrointestinal tract underscores the unique applicability of methods in tissue engineering for repairing and regenerating segments of the GI tract. Given the microanatomical and functional characteristics of the gut, such an endeavor would require regenerating smooth muscle, specialized neuronal tissue, and a mucosal-epithelial bilayer and meeting one of the most challenging tasks—recapitulating the diverse motility patterns that are essential to proper intestinal functioning. Nevertheless, preliminary investigations have been quite promising [86].

Speer et al. isolated organoid units from mouse glandular stomach and seeded them onto biodegradable scaffolds composed of polyglycolic acid (PGA) coated with poly-L-lactic acid (PLA) and type I collagen [87]. The seeded scaffolds were implanted into the omentum of adult mice and harvested at designated time points for analysis. The constructs were found to grow and proliferate as expanding spheres with simple columnar epithelium organized into gastric glands and an adjacent muscularis. Mucous, enteroendocrine, chief, and parietal cells were all expressed in the regenerated epithelium.

Maemura et al. have employed biodegradable polymers seeded with stomach epithelial organoid units obtained from neonatal rats in various investigations [88]. They replaced recipient rat stomachs with tissue-engineered stomachs and discovered no evidence of stenosis or obstruction at the sites of anastomoses. Histological evaluation demonstrated well-developed vascularized tissue and stratified smooth muscle layers [89]. They also used their seeded scaffolds to patch gastric wall defects in rat models with considerable success [90].

The small intestine is the primary segment for nutrient absorption within the gastrointestinal tract. This process depends on the integrity of microvilli structures that line the intestinal epithelium. Early regenerative investigations employed autologous tissue patches to repair defects in the intestinal wall, but later studies underscored the potential use of absorbable biomaterials as a patch scaffold to facilitate tissue ingrowth [91]. Commonly used biomaterials have been collagen scaffolds, PLA, and PGA, among others. Fibrin hydrogels have demonstrated the sufficient mechanical rigidity to allow self-organization of circular sphincteric and intestinal smooth muscles, even in humans [92, 93].

The Vacanti group has employed intestinal organoid units acquired from the small intestine and remodeled them on biocompatible matrices [94]. These implants were able to reduce morbidity associated with massive bowel resection in rat models. Chen et al. implanted ECM-derived submucosal tubes in canine models with small bowel resections without seeding material to ascertain any regenerative potential [95]. The investigation used four dogs and concluded that this method was not efficacious, although patch repair showed potential for wall regeneration.

Numerous studies have reported the use of bone marrow-derived mesenchymal cells (MSCs) for the regeneration of intestinal tissue. Hori et al. were able to regenerate intestinal segments by seeding MSCs onto collagen scaffolds [96]. Their constructs showed a transient distribution of alpha-smooth muscle actin-positive cells but failed to regenerate a muscle layer, which is essential for peristalsis.

ESCs have also been used to regenerate functional or semi-functional intestinal tissue, under proper induction by growth factors [97, 98]. Watson et al. recently reported on the generation of human intestinal organoids procured in vitro from human embryonic stem cells or pluripotent stem cells that they were able to engraft and vascularize in mouse models [99]. In vivo transplantation resulted in marked expansion and maturation of differentiated epithelium and mesenchyme and even demonstrated digestive functions. Wieck et al. showed in a recent report that human-derived tissue-engineered colon can be populated with nervous tissue when cultured with enteric nervous system progenitor cells to reconstitute motility functions impaired in conditions such as Hirschsprung's disease [100].

Detergent-based organ decellularization allows for the manufacture of ECM-based intestinal scaffolds. Notably, the ECM preserves the villus-crypt architecture and vasculature that supports intestinal epithelial regeneration [101]. Though clinical translation has not yet proven feasible, recent findings and ongoing studies are showing great promise.

Given the limited successful results of employing cells or biodegradable scaffolds alone, adult or stem cell seeding on ECM-based scaffolds will likely help to achieve optimal results in future investigations. Identifying the most suitable cell type remains a primary challenge for future studies, along with establishing a scaffold that recaptures the microenvironmental cues of native intestine.

8.4.3 Kidney

The kidney is a complex solid organ with important roles in endocrine, metabolic, and immunologic homeostasis. Functions include filtering the blood, maintaining adequate blood pressure and volume, and excreting toxic metabolic waste products, highlighting the crucial role of the kidney.

At present, dialysis and transplantation represent the gold standard treatment modalities for chronic kidney disease. Dialysis as a modality of renal replacement therapy does not, however, cure kidney damage. Rather, it assumes a portion of the kidney's functions, particularly filtration. As such, dialysis leaves much to be desired.

Replacing the organ with a healthy, functioning graft, on the other hand, is a much more effective treatment. With the worldwide burden of hypertension and diabetes growing, the prevalence of chronic kidney disease is reaching epidemic proportions, and the need for functional kidneys is rising correspondingly [102, 103].

Contemporary investigators agree that significant progress for TE approaches to renal replacement therapy will require further elucidation of the native repair and regeneration processes occurring in vivo at the cellular and molecular levels [104, 105]. Because of the proportionally increased metabolic demand of the kidney itself and its waste and toxin filtration functions, renal tubular cells are constantly under the threat of acute injury and oxidative stress. It has been stipulated that, for this reason, these cell populations feature unique regenerative abilities to compensate for continuous insult. Indeed, surviving renal tubular cells have been observed to give rise to a new population of the cells following physiologic kidney damage [106].

Nagaike et al. observed that unilateral nephrectomy induces mitogenesis and hypertrophy in the contralateral kidney [107]. When Cochrane et al. created ureteral obstructions in murine models of renal injury to induce cortical tubular cell atrophy, tubular dilation, and interstitial macrophage infiltration, they observed a rapid process of reconstruction and interstitial matrix expansion upon reversal of the obstruction which ultimately restored the glomerular filtration rate [108]. However, continuous and supraphysiologic damage characteristic of chronic kidney disease overpowers the regenerative properties of these cells, and the growth of new nephrons, i.e., frank nephrogenesis, has not been shown to occur [109]. But nephron progenitor cells derived from human pluripotent stem cells are currently under investigation for TE purposes given their nephron-forming capacity [110].

Researchers have explored the potential of cell therapy to restore kidney function in the face of widespread damage. Cell-based approaches seek to achieve kidney repair and regeneration in situ upon therapeutic administration. They are based on the observation that exogenously supplied cells can stimulate and/or contribute to repair and proliferative processes [111]. Progenitor cells harvested from the proximal tubules, glomerulus, peritubules, and papillae have all demonstrated some level of regenerative capacity in recent investigations [112, 113]. Stem cells obtained from the urine have also shown some potential to reverse kidney damage and aid in the repair process [114].

A recent study, for example, demonstrated that cultured mesenchymal stem cells injected into mice with ischemia-induced kidney injuries resulted in improved renal function, promoted macrovasculature repair, attenuated kidney peritubular capillary loss, increased the proliferation of parenchymal cells, and significantly reduced overall mortality [115]. Such approaches are attractive because of the ease of isolation and expansion of mesenchymal stem cells and their potentially autologous use to reduce the risk of immunogenic rejection. Furthermore, the use of adult cells circumvents the ethical obstacles encountered when using ESCs. The results of investigations utilizing amniotic fluid-derived stem cells and induced pluripotent stem cells have been similarly encouraging [116–119].

Enhanced understanding of the regenerative properties of renal cells has led to another avenue for treatment of kidney damage: the use of embryonic kidney tissue. These primordial cells have been shown to integrate within adult organ systems, richly vascularize, and form new, mature nephrons (i.e., result in frank nephrogenesis) [120, 121]. Ureteric bud and metanephric mesenchyme cultures retain the capability to form collecting ducts through tubulogenesis and epithelization by virtue of the inherent developmental capacities of the mesonephric duct tissue.

Investigators prepared the kidney tissue in vitro and subsequently implanted it in mice models that survived for over 5 weeks. Microstructural analysis revealed glomerular vascularization in vivo, lending support for the therapeutic potential of these primordial tissues [122]. Imberti et al. implanted renal primordia under the kidney capsule of male rats with kidney injury [123]. The grafts developed glomeruli and tubuli that filtered blood and produced urine in cyst-like structures. Additionally, newly developed metanephroi initiated a process of regeneration in host tissue segments, as indicated by increased cell proliferation and vessel growth.

Preliminary and animal model investigations into kidney bioengineering using cell-scaffold technology also have been highly encouraging. ECM scaffolds produced from animal and human kidneys have been shown to retain their innate biomolecular and biophysical properties in addition to their native external anatomy [124, 125]. Orlando et al. successfully fabricated renal ECM scaffolds from porcine kidneys by pumping an aqueous detergent solution through the renal artery, thereby decellularizing the organ [126]. These scaffolds achieved total cell clearance and retained their essential architectures. The injection of contrast media through the renal artery confirmed preservation and potency of a vascular network with hierarchical branching structures without extravasation into the parenchymal compartment. Seeded endothelial cells demonstrated steady growth and adherence to structures on the bioscaffolds. Finally, unseeded scaffolds were successfully implanted in pigs to assess their in vivo biocompatibility. They were easily reperfused, sustained blood pressure, and were tolerated for 2 weeks.

The same group identified the approximately 2600 human kidneys originally intended for transplant purposes but discarded each year due to anatomical or pathological anomalies as a platform for kidney bioengineering at clinically relevant scales. Given their human origins, the potential for use of these discarded grafts in TE investigations is monumental, at least as a stepping stone to realizable clinical translation. In 2013, they successfully produced ECM scaffolds using these discarded kidneys [127]. These human scaffolds, like their porcine analogs, were completely decellularized and maintained their structural composition. HLA antigen molecules were notably absent: an important indicator of the potential immunocompatibility of these constructs. They were able to induce vessel formation in chick chorioallantoic membranes, suggesting strong angiogenic properties. Lastly, the innate vascular system was able to resist pressure treatment at physiological levels, as a marker of compliance. Using a resin casting methodology to carry out deeper microanatomical analysis, they observed that glomerular shape and capillary width were preserved following decellularization (Fig. 8.1). Furthermore, the branching pattern and vessel integrity were unchanged. Protein analysis confirmed the retention of several growth factors implicated in renal repair and angiogenesis within the scaffolds [128].

Investigators seeded human amniotic fluid-derived stem cells onto discarded kidney ECM scaffolds and observed their attachment and proliferation thereafter [129]. Furthermore, they synthesized and secreted various growth factors and chemokines involved in angiogenesis and matrix remodeling, indicating a dynamic, stimulatory relationship between ECM and seeded cells.

In 2013, Song et al. reported on the novel orthotopic transplantation of bioengineered kidneys in rat models [130]. Acellular rat kidneys were seeded with endothelial and epithelial cells through the renal artery and ureter. Epithelial cells showed engraftment along with organization into tubular structures expressing Na/K--ATPase and aquaporin similar to native proximal tubular epithelium. Electron microscopy showed perfused glomerular capillaries with engrafted podocytes and formation of foot processes. These bioengineered grafts were perfused by recipient circulation and produced rudimentary urine via the ureteral conduit in vivo when transplanted orthotopically.

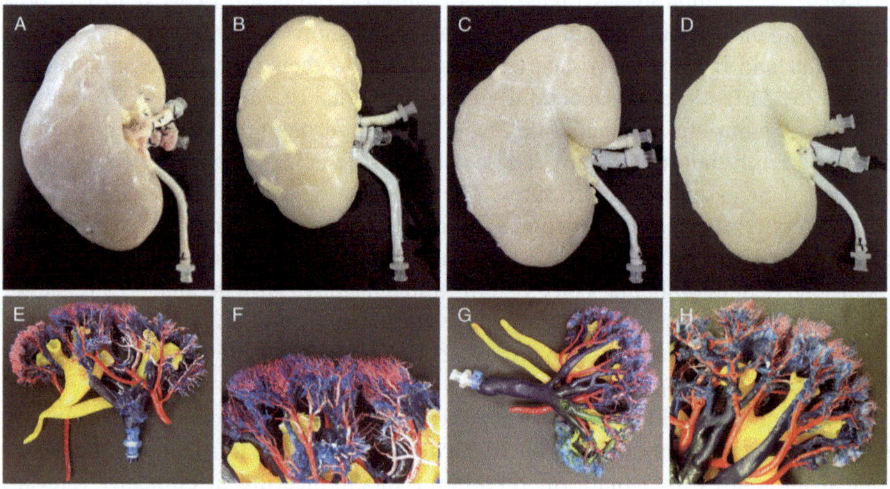

Fig. 8.1 Acellular ECM scaffolds obtained from a discarded human kidney (**a–d**) and related vascular corrosion cast (**e–h**) (From Peloso et al., *Transplantation*, 2015, Reprinted with permission of the American Journal of Transplantation. Copyright © 2017 American Journal of Transplantation) [128]

8.4.4 Pancreas

The pancreas is functionally composed of two parts, the exocrine pancreas and endocrine pancreas. Despite deriving from the same embryologic cells, these components are structurally different and serve important functions for different physiological systems. Despite accounting for 90–95% of the pancreas mass, the exocrine pancreas is not essential for survival because of exogenous pancreatic digestive enzyme replacement therapy [131]. In contrast, the endocrine pancreas only accounts for roughly 2% of the pancreas mass while serving a vital physiological role. The endocrine pancreas is not localized to a specific region of the pancreas but is interspersed around islets of Langerhans (highly vascularized centers) over the entire tissue. Although here are five main types of endocrine pancreatic cells (alpha, beta, delta, gamma, and epsilon), the beta cell plays a particularly vital role in glucose homeostasis and energy metabolism, drawing the most attention from bioengineering research related to endocrine pancreatic deficiencies.

One of the most common pathologic endocrine deficiencies is type 1 diabetes mellitus in which beta cells are destroyed by the host's own immune system. About 1 in 300 individuals in the United States is diagnosed with type 1 diabetes mellitus by the age of 18 [132]. Therapies for type 1 diabetes mellitus focus on the supplementation of exogenous insulin that is normally produced by the beta cells. This method requires strict patient compliance, however, and adequate patient education. Recent innovations remove the need for conscious dosing. Although exogenous insulin therapy is effective at preventing acute decompensation, less than 40% achieve and maintain therapeutic targets [133]. The most commonly accepted treatment regimens lower the incidence of diabetic emergency, but they require lifelong pharmacologic intervention without the possibility of remission. Beta cell replacement, whether by whole pancreas or individual islet cell transplantation, is currently the only method to restore long-term, stable euglycemia in type 1 diabetics.

Pancreatic transplantation is an invasive option that provides the patient an opportunity to be free from strict glycemic control by way of exogenous insulin. It is, however, typically only offered to adults in conjunction with kidney transplants because of the high risk of surgical complications and the subsequent necessity for immunosuppressant medications [134]. Nevertheless, recipients exhibit greater improvements in the micro- and macro-vascular complications of diabetes while enjoying a better quality of life. The 5-year organ survival rate for pancreatic transplantations is roughly 50% [34].

An alternative solution involves the ectopic implantation of purified islet cells and simultaneous depletion of T cells to recapture beta cell function while suppressing the autoimmune component of the disease [135, 136]. Although human islet cell purification and implantation show the same 5-year insulin independence rate as pancreatic transplantation, ectopic implantation has been found to have a lower risk of complication and represents a less invasive solution [137].

Contemporary investigations employ TE solutions for treatment of the disease without the need for lifelong pharmacologic intervention. One of the most heavily researched areas involves the induction of beta cell regeneration within host islets.

There is a current debate in the research community regarding the source material of endogenous beta cell regeneration: it is still unknown whether the existing beta cells or resident stem cells are primarily responsible [138, 139]. Insulin-producing cells derived from ESCs are physiologically closer to beta cells than any adult stem cell-based beta cell. The conversion of human embryonic stem cells toward beta cell phenotypes has been achieved through the initial differentiation into endodermal cells which can be further differentiated into insulin-producing endoderm derivatives [140, 141].

To avoid the controversy of ESC procurement, recent advancements in beta cell replacement have turned toward induced pluripotent stem cells. By using retroviruses to manipulate gene expressions, adult somatic cells can be reprogrammed and individualized to specific patients. In animal models, experimental therapy has already demonstrated long-term correction of hyperglycemia [142]. Though promising, further success with these cells is challenged by premature senescence, subchromosomal abnormalities, and destruction of patient-derived cells due to autoimmune phenomena.

The recellularization of ECM-based scaffolds has also been under investigation for the de novo fabrication of pancreas organs. It has been observed that islets cultured in vitro on ECM increase the longevity of insulin production in response to glucose, likely because of natural growth factors within the ECM that direct cell lines toward beta cells [143]. In addition, recent studies have hinted at the potential use of xenographic transplants. Mirmalek-Sani et al. decellularized the porcine pancreas and showed that the ECM scaffold had a patent vasculature and could be subsequently seeded with human amniotic fluid-derived stem cells [144]. The seeded islets demonstrated the ability to support normal pancreatic function by showing an increased metabolic rate and insulin secretion over isolated islets in culture. Later, similar data were confirmed by the same team on ECM scaffolds obtained from the human pancreas (Fig. 8.2) [145]. Goh et al. successfully decellularized mouse pancreata and recellularized them with acinar and beta cell lines [146]. They showed successful engraftment within the three-dimensional decellularized pancreas along with strong upregulation of insulin gene expression. Their findings supported the utility of whole pancreas ECM for enhancing pancreatic cell functionality. De Carlo et al. reported similar findings [147].

Although all of these technologies may restore the functionality of destroyed islets, the primary pathology of type 1 diabetes is autoimmune, which will remain a persistent obstacle if left unaddressed. Future strategies must incorporate adjuvant immunomodulation or autologous tissue in current technologies to terminate the chronic disease state and preserve endocrine pancreas function.

8.4.5 Liver

Liver disease morbidity and mortality outcomes are increasing worldwide. The total deaths caused by cirrhosis and liver cancer have increased by 50 million per year since 1990 [148]. At present, liver transplantation is the only curative option for patients with end-stage liver disease. Twenty percent of this patient population dies

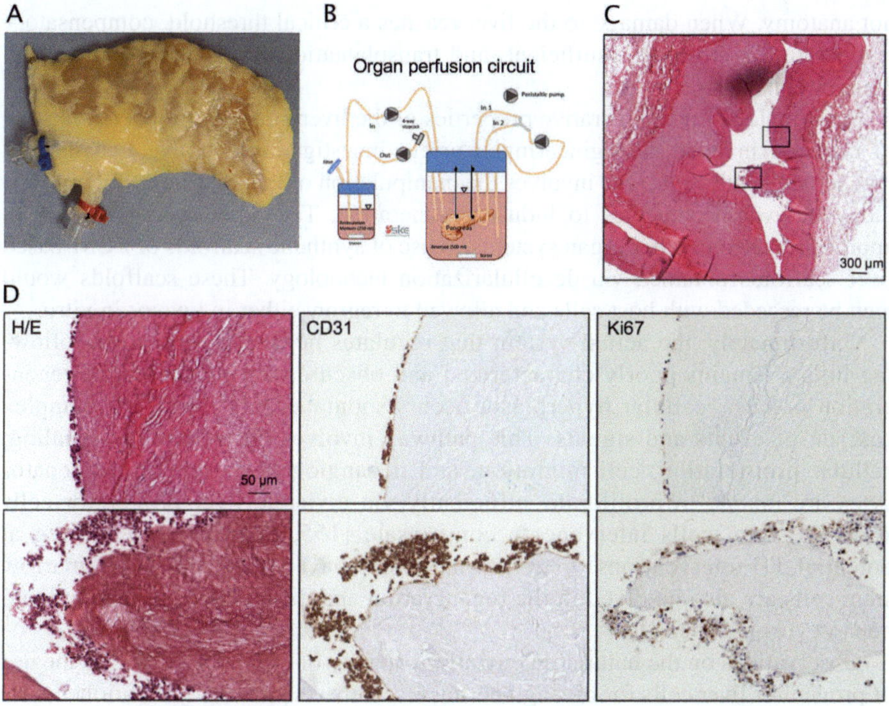

Fig. 8.2 Endothelial cell seeding on acellular ECM scaffolds obtained from the human pancreas. The ECM scaffold was seeded with human pancreatic endothelial cells and cultured for 6 days in a bioreactor, consisting in a closed circuit with one chamber for organ housing, a reservoir for medium oxygenation, and a peristaltic pump (Ismatec), connected by tubing (ID 1/16″, Pharmed BPT). Pancreatic tail was surgically isolated in order to obtain a smaller volume to seed keeping at the same time an inflow (SA – red connector) and an outflow (splenic vein –SV – blue connector) (Panel **a**). Panel **b** indicates a schematic representation of the perfusion circuit for seeded pancreatic scaffold culture. Panel **c** shows a representative image of H&E stain showing localization of infused cells in vessels. *Boxes* indicate areas reported with high magnification in the panel below. Panel **d** illustrates representative images of H&E (*left*), CD31 (*middle*), and Ki67 (*right*) matrix staining (From Peloso et al., *Ann Surg*, 2016, Reprinted with permission of the American Journal of Transplantation. Copyright © 2017 American Journal of Transplantation) [145]

while on a waiting list, however, due to the shortage of transplantable organs [149]. Alternative modalities such as bioartificial livers and hepatocyte transplantation have been investigated, but without successful yield [150–152]. Given the obstacles currently facing end-stage liver patients, TE strategies for liver repair and regeneration are particularly attractive for contemporary investigators.

The liver has an innate capability to regenerate beyond the potential of other abdominal organs. In the event of injury to less than 70% of the organ, the liver can fully regenerate within 6 months of the trauma [153]. Regeneration occurs through cellular hyperplasia of the residual liver, however, and is not considered true regeneration [154]. Instead, there is an observed compensatory enlargement of the remnant liver to meet the functional demands of the body. From an evolutionary perspective, liver hyperplasia is a mechanism of repair designed to restore function, not anatomy. When damage to the liver reaches a critical threshold, compensatory regeneration is no longer sufficient, and transplantation is the usual approach in these circumstances.

Nevertheless, the regenerative properties of the liver make it uniquely amenable to methods in liver bioengineering. Current investigations involve two distinct regenerative pathways. One involves the manipulation of the existing cellular physiology in liver hepatocytes to induce regeneration. The other option, similar to approaches used in other organ systems, is use of synthetic scaffolds or ECM-based liver scaffolds obtained via decellularization technology. These scaffolds would then be reseeded with host cells and allowed to regrow either in vivo or in vitro.

Unfortunately, the actual system that regulates hepatic regeneration following injury remains poorly characterized and obscure. During the native regeneration process, cellular hyperplasia occurs spontaneously through a complex cascade of events and signals. This pathway involves inflammatory signaling, cellular proliferation, cell migration, and neoangiogenesis. If mature hepatocytes are unable to proliferate sufficiently via division, liver progenitor cells known as oval cells intervene to compensate [155, 156], thereby hinting at potential TE interventions. In addition, studies have shown that bone marrow stem cells are also involved in the regeneration and differentiation of functional hepatocytes [157, 158].

Observations on the natural regenerative capacity of injured livers led to the use of progenitor liver cells for bioengineering research. As previously mentioned, cellular hyperplasia replaces the lost volume following liver amputation, and current studies attempt to enhance the innate regenerative ability of the liver through terminal differentiation via resident bone marrow-derived or liver progenitor cells [159–161]. Infusing autologous bone marrow cells into patients with decompensated liver cirrhosis has already been demonstrated to be safe and, in some cases, clinically beneficial [162, 163]. Another avenue is the use of cytokine granulocyte-colony stimulating factor (G-CSF) for the activation of hematopoietic stem cell differentiation and mobilization, increasing the ability of resident progenitor cells to respond to injury [164]. Stem cells isolated from the peripheral bloodstream of liver disease patients infused with G-CSF were isolated and infused back into the patients. These patients showed significant improvements in liver enzyme levels as well as serum bilirubin when compared with the control group [165, 166]. It is thought that G-CSF is able to mobilize bone marrow cells and circulating progenitor cells while increasing hepatocyte growth factor, which is prominently involved in liver regeneration [167]. Despite these findings, researchers are unable to determine the specific progenitor cell type that is actually aiding, or is primarily responsible, in the regeneration process.

Recent work has also highlighted the use of cell-scaffold technology for liver bioengineering [168]. As in other organ systems, the chief objectives include demonstrating successful organ decellularization, successful seeding of acellular scaffolds with hepatocytes or hepatocyte progenitors, and recapitulation of original liver functions including protein synthesis/breakdown, detoxification, and bilirubin metabolism.

Investigators successfully converted the murine liver into acellular scaffolds via perfusion with SDS detergent to remove cellular material and debris. The remaining constructs retained proper collagen structure, laminin basement membranes, and internal vascular networks [169]. In the same study, the scaffold had enough structural integrity to withstand cannulation and perfusion with the reseeding cells. A follow-up study illustrated the ability of the perfused cells to leave the vasculature and distribute among the matrix [170]. Using electron microscopy, it was observed that rodent hepatocytes migrated into decellularized sinusoidal spaces and displayed levels of urea synthesis, albumin synthesis, and cytochrome P450 expression comparable to sandwiched hepatocyte cultures. This is particularly important because the liver requires a wide range of enzymes, cofactors, serum proteins, and acute phase reactants to be fully functional. This graft was able to withstand transplantation for 8 h in a rat host prior to harvesting, which showed that the hepatocytes retained their morphology, position, and integrity.

Yagi et al. recently reported on the successful decellularization of porcine livers, thereby producing ECM scaffolds at scales relevant to humans. These scaffolds were capable of supporting hepatocyte engraftment and reorganization in three dimensions [171]. Baptista et al. also observed that it is possible to decellularize whole liver in mice, rats, ferrets, rabbits, and adult pigs and also that they can be successfully seeded with human hepatocyte progenitors [172]. This study utilized the inferior vena cava for decellularization and used the portal vein, as well as the vena cava, for perfusion of cells. The decellularized vascular network was able to withstand fluid flow that entered through a central inlet vessel, branched into an extensive capillary bed, and coalesced in a single outlet vessel. Discrete populations of human fetal liver and endothelial cells were found throughout the matrix in their putative native locations, suggesting that the retention of glycosaminoglycans and collagen structure provides the necessary environmental signaling to regulate cell differentiation.

Wang et al. demonstrated that progenitor cells seeded onto a scaffold could differentiate into hepatocytes and cholangiocytes, illustrating that the scaffold could dictate the differentiation toward distinct cell lineages [173]. Barakat et al. decellularized porcine livers and successfully repopulated the scaffolds with human cells [174]. Analysis revealed that the cells were induced to differentiate into mature hepatocytes while maintaining active metabolism, the ability to withstand physiologic shear stress from blood flow, and the ability to synthesize albumin.

Most recently, Mazza et al. demonstrated the complete decellularization of discarded whole human liver and lobes to form ECM scaffolds with preserved architecture [175]. These scaffolds were repopulated with human hepatic cells that thereafter showed excellent viability, motility, and proliferation and remodeling of the ECM. The group demonstrated biocompatibility by both omental and subcutaneous xenotransplantation of liver scaffold cubes into immune-competent mice. No foreign body response was observed, dramatically underscoring the clinical potential of bioengineered organs.

8.4.6 Heart

The inability of cardiomyocytes to regenerate following infarction remains the primary cause of congestive heart failure [176]. Infarct scar tissue has long been believed to be acellular and physiologically inert. Recent studies, however, suggest that cardiac scar tissue is composed of phenotypically transformed fibroblast-like cells which behave dynamically and undergo continuous turnover. Such bioactivity suggests possible intervention using TE strategies in order to regenerate functional, healthy cardiac tissue and minimize adverse accumulation of fibrous tissue. At present, there are no available therapies which prevent or reverse cardiac damage and adverse ventricular remodeling following infarction.

The fabrication of engineered heart tissue was demonstrated for the first time over 20 years ago using embryonic chicken cardiac myocytes [177]. Since then, investigators have made enormous strides in the field of stem cell biology suggesting that widespread clinical translation may be rapidly approaching. Recent studies have demonstrated the therapeutic potential of stem cells to regenerate contractile cardiac tissue [178]. Contemporary investigators must determine the most appropriate cell type to use to achieve regeneration and choose the most efficacious engraftment technique. During the early years of cardiac restoration research, a common technique involved a direct bolus of cardiac cells suspended in saline directed into regions of infarcted tissue [179]. Other cell types used with this technique include skeletal myoblasts, neonatal cardiomyocytes, fibroblasts, smooth muscle cells, embryonic stem cells, and bone marrow progenitors with varying improvements in cardiac function [180]. Nevertheless, the proportion of cells successfully engrafting at the infarct site tends to be very low using this method [181].

Recently, investigators have improved engraftment rates by using cultured cardiomyocytes seeded on biomaterial scaffolds to create transplantable cardiac "patches." Scaffolds are not always necessary since cardiac myocytes cultured on standard plastic dishes tend to detach as intact monolayers. Serially stacking multiple sheets on top of one another creates three-dimensional tissues that "beat" and generate force [182]. Cardiac tissue segments thus engineered in vitro demonstrate organized sarcomeres, electrical conductivity, and contractile characteristics resembling native adult myocardium [183]. These "patches" have been shown to improve left ventricular performance when implanted in doxorubicin-treated rats as models of dilated cardiomyopathy [184].

Ott et al. produced acellular ECM-based scaffolds from rodent hearts which were subsequently recellularized using neonatal rat cardiomyocytes and endothelial cells [185]. These constructs were then subjected to perfusion treatment in bioreactors for 28 days. By day 4, investigators observed macroscopic contractions, and by day 8, the maturing organoids were able to sustain pump function under physiological load and electrical stimulation.

Biomaterials used to enhance engraftment and regeneration have also been an important parameter in cardiac tissue engineering. Recent investigators have incorporated solidifying gel polymers such as fibrin glue and PEGylated fibrinogen hydrogels [186, 187]. The advantage of this technique consists in the ability to fabricate any three-dimensional architecture and manipulate the structure prior to implantation. Furthermore, certain biomaterials are able to protect seeded cells from host inflammation and facilitate functional integration within native myocardium at the injury site.

8.4.7 Airway

Approximately 2000 lung transplants were performed in 2013 to treat various diseases including alpha-1 antitrypsin deficiency, primary pulmonary hypertension, cystic fibrosis, emphysema, and others [188]. TE approaches for airway repair could potentially improve patient survival and reduce waiting times for transplantation. Among tissue engineering approaches to transplantation, the airway was one of the first systems in which clinical success was demonstrated. Macchiarini et al. performed the first bioengineered trachea transplantation in humans [55]. The team removed cells and MHC antigens from a human donor trachea and reseeded the resultant scaffold with epithelial cells and MSC-derived chondrocytes which were acquired autologously. The engineered graft was then used to replace the left main bronchus in a 30-year-old woman with end-stage bronchomalacia.

Progress with the bioengineered trachea has alluded to potential inroads for larynx and lung transplantation investigations. At present, only two human laryngeal allotransplants have been reported, and both patients required lifelong immunosuppression regimes thereafter. Baiguera et al. successfully decellularized human larynxes enzymatically to obtain acellular scaffolds [189]. Electron microscopy confirmed that their matrices retained the hierarchical structures of the native larynx. Mechanical testing demonstrated intact biomechanical integrity. Furthermore, chorioallantoic membrane analysis showed that their constructs induced a strong in vivo angiogenic response, underscoring their integration and engraftment potential. Lung transplantation, too, is a field particularly amenable to TE strategies for currently existing problems. Given the limited regenerative potential of the adult lung, investigators have sought to determine whether lung tissue can be regenerated in vitro for potential subsequent implantation. Petersen et al. decellularized the lungs of adult rats to obtain acellular matrix scaffolds [190]. They then used a bioreactor to culture pulmonary epithelium and vascular endothelium on the acellular matrices. Subsequent analysis showed that the seeded epithelium displayed robust hierarchical organization within the matrix. Furthermore, the seeded endothelial cells efficiently repopulated the vascular compartment of the decellularized constructs. In vitro testing demonstrated that the engineered lungs possessed biomechanical parameters similar to those of native lungs. These seeded scaffolds were then implanted in rats for in vivo testing and were observed to participate in gas exchange, remarkably.

Bioengineering strategies for the trachea and larynx differ from those for the lung given their relatively simpler, hollow architectures. Furthermore, their purpose is mainly derived from their structural characteristics (i.e., as air conduits) rather than their cellular functions, unlike complex solid organs. Synthetic scaffolds, such as those involving polyester urethane, polypropylene mesh, alginate gel, or PEGylated hydrogels, were used initially to integrally replace the trachea as they benefit from not requiring a donor and being easily modified to conform to recipient anatomy. Nevertheless their use is limited due to their having different biomechanical properties, being more susceptible to infection, and not vascularizing [191]. Vascularization is critical to maintain the vitality of seeded or recruited cells and subsequent proliferation [192].

Recellularization strategies in these contexts seek to recover the chondrocyte and epithelial compartments of the native airway. Stem cells, appropriately selected, have the potential to differentiate into both of these cell types. Berg et al. repopulated a detergent-decellularized human cadaveric donor trachea with autologous stem cells and implanted the resulting airway construct into a 76-year-old patient with tracheal stenosis [193]. Within a week the graft was compromised by a thick fungal infection which nevertheless controlled with local and systemic antifungal therapy. The patient eventually died after 3 weeks due to cardiac arrest, but the airway graft was found to be patent, open, and stable with intact anastomoses. Histopathological analysis during autopsy showed squamous but not ciliated epithelium, neovascularization, nerve fibers in the submucosa, and intact chondrocytes in the cartilage. Investigators have also experimented with other stem cell types including iPSCs [194], marrow-derived MSCs [195], human ESCs [196], and amniotic fluid-derived stem cells [197] for regeneration of airway tissue.

8.5 Conclusion

The use of autologous cellular material has the potential to obviate the need for lifelong antirejection therapies. De novo organ fabrication using cell-scaffold technology could, theoretically, provide a limitless supply of transplantable organs for waiting list patients, thus circumventing the challenge of organ shortage.

Large-scale clinical translation of abdominal organ bioengineering remains a distant possibility, however, because substantial work remains to be done in the laboratory. The chief obstacles posed by current medical and surgical modalities (particularly transplantation) are quite concerning and have created a pocket of necessity within which TE investigations are being conducted.

Further research is needed to identify the most suitable cell type for regenerative investigations in each organ system. Although adult somatic cells specific to the organ system being treated have shown promise, stem cells have the advantage of being multipotent and renewable, which are particularly attractive qualities in the context of TE. The risks associated with their use, however, need to be further characterized before they can be used safely in patients on a large scale.

The immunogenicity of decellularized scaffolds is another issue that needs to be assessed. Though their tolerability has been recognized both empirically and theoretically, suboptimal decellularization protocols can leave behind residual antigen activity, triggering an immune response in the host.

Investigators must also seek to understand better the interaction between ECM and cells, both endogenous and exogenous. The improved knowledge base would likely result in improved cellular engraftment, improved migration, and improved differentiation within the acellular matrix. The source of new understanding could very well lie in developmental biology and organogenesis, in which ECM-guided differentiation is paramount.

Though TE solutions for abdominal organ engineering show great promise, current researchers must also keep an eye toward research funding limitations and the role of commercial entities in bringing their innovations from bench to bedside. With these last obstacles addressed, TE solutions for transplantation may very well bring about a paradigm shift in the modern era for the treatment of end-stage organ diseases.

References

1. World Health Organization. (2013). *World health statistic*. Geneva: World Health Organization.
2. World Health Organization. *GKT1 activity and practices*. http://www.who.int/transplantation/gkt/statistics/en/. Accessed 24 June 2015.
3. National Kidney Foundation. *Organ donation and transplant statistics*. https://www.kidney.org/news/newsroom/factsheets/Organ-Donation-and-Transplantation-Stats. Accessed 24 June 2015.
4. Organ Procurement and Transplantation Network. http://optn.transplant.hrsa.gov/converge/latestData/rptData.asp. Accessed 24 June 2015.
5. Katari, R., Peloso, A., & Orlando, G. (2014). Tissue engineering and regenerative medicine: Semantic considerations for an evolving paradigm. *Frontiers in Bioengineering and Biotechnology, 2*, 57.
6. Katari, R. S., Peloso, A., & Orlando, G. (2014). Tissue engineering. *Advances in Surgery, 48*, 137–154.
7. Orlando, G., Soker, S., Stratta, R. J., & Atala, A. (2013). Will regenerative medicine replace transplantation? *Cold Spring Harbor Perspectives in Medicine, 3*(8).
8. Peloso, A., Dhal, A., Zambon, J. P., et al. (2015). Current achievements and future perspectives in whole-organ bioengineering. *Stem Cell Research & Therapy, 6*, 107.
9. Wolfe, R. A., Ashby, V. B., Milford, E. L., et al. (1999). Comparison of mortality in all patients on dialysis, patients on dialysis awaiting transplantation, and recipients of a first cadaveric transplant. *The New England Journal of Medicine, 341*(23), 1725–1730.
10. Abecassis, M., Bartlett, S. T., Collins, A. J., et al. (2008). Kidney transplantation as primary therapy for end-stage renal disease: A National Kidney Foundation/Kidney Disease Outcomes Quality Initiative (NKF/KDOQITM) conference. *Clinical Journal of the American Society of Nephrology, 3*(2), 471–480.
11. ERA-EDTA Annual Report. 2012. http://www.era-edta-reg.org/files/annualreports/pdf/AnnRep2012.pdf. Accessed 10 Mar 2016.
12. National Kidney Foundation. *Organ donation and transplant statistics*. https://www.kidney.org/news/newsroom/factsheets/OrganDonation-and-Transplantation-Stats. Accessed 10 Mar 2016.
13. Organ Procurement and Transplantation Network. http://optn.transplant.hrsa.gov/converge/latestData/rptData.asp. Accessed 10 Mar 2016.
14. *Organ donation and transplantation activity report 2012*. http://www.transplant-observatory.org/Documents/dataReports/Basicslides2012.pdf. accessed 10 Mar 2016.

15. Shin, E., Kwon, S. W., Yang, W. S., et al. (2015). Long-term outcomes of ABO-incompatible living donor kidney transplantation: A comparative analysis. *Transplantation Proceedings, 47*(6), 1720–1726.
16. Coilly, A., & Samuel, D. (2015). Pros and Cons: Usage of organs from donors infected with hepatitis C virus – Revision in the direct acting antiviral era. *Journal of Hepatology*.
17. Perico, N., & Remuzzi, G. (2012). Chronic kidney disease: A research and public health priority. *Nephrology, Dialysis, Transplantation, 27*(suppl 3), iii19–iii26.
18. Thomas, L. (1971). Notes of a biology-watcher: The technology of medicine. *The New England Journal of Medicine, 285*, 1366–1368.
19. Brown, E. (1996). Halfway technologies. *Physician Executive, 22*(12), 44–45.
20. Zsom, L., Wagner, L., & Fülöp, T. (2015). Minimization vs tailoring: Where do we stand with personalized immunosuppression during renal transplantation in 2015? *World Journal of Transplantation, 5*(3), 73–80.
21. Nyumura, I., Honda, K., Tanabe, K., Teraoka, S., & Iwamoto, Y. (2012). Early histologic lesions and risk factors for recurrence of diabetic kidney disease after kidney transplantation. *Transplantation, 94*(6), 612–619.
22. Mitchell, O., & Gurakar, A. (2015). Management of hepatitis C post-Liver transplantation: A comprehensive review. *Journal of Clinical and Translational Hepatology, 3*(2), 140–148.
23. Manzia, T. M., Angelico, R., Ciano, P., et al. (2014). Impact of immunosuppression minimization and withdrawal in long-term hepatitis C virus liver transplant recipients. *World Journal of Gastroenterology, 20*(34), 12217–12225.
24. Heidary Rouchi, A., & Mahdavi-Mazdeh, M. (2015). Regenerative medicine in organ and tissue transplantation: Shortly and practically achievable? *International Journal of Organ Transplantation Medicine, 6*(3), 93–98.
25. Dutkowski, P., de Rougemont, O., & Clavien, P. A. (2008 Oct). Alexis Carrel: Genius, innovator and ideologist. *American Journal of Transplantation : Official Journal of the American Society of Transplantation and the American Society of Transplant Surgeons, 8*(10), 1998–2003.
26. Orlando, G., Soker, S., & Wood, K. (2009). Operational tolerance after liver transplantation. *Journal of Hepatology, 50*(6), 1247–1257.
27. Orlando, G. (2010). Finding the right time for weaning off immunosuppression in solid organ transplant recipients. *Expert Review of Clinical Immunology, 6*(6), 879–892.
28. Schroeder RA, Marroquin CE, Kuo PC. Tolerance and the "Holy Grail" of transplantation. The Journal of Surgical Research 2003;111(1):109-119.
29. Dharnidharka, V. R., Stablein, D. M., & Harmon, W. E. (2004). Post-transplant infections now exceed acute rejection as cause for hospitalization: A report of the NAPRTCS. *American Journal of Transplantation, 4*(3), 384–389.
30. U.S. Renal Data System. (2013). USRDS 2013 Annual Data Report: Atlas of CKD & ESRD, Volume 2; Chapter 7: Transplantation. National Institutes of Health, National Institute of Diabetes and Digestive and Kidney Diseases. http://www.usrds.org/atlas.aspx. Accessed 11 Mar 2016.
31. Engels, E. A., Pfeiffer, R. M., Fraumeni, J. F., et al. (2011). Spectrum of cancer risk among US solid organ transplant recipients. *Journal of the American Medical Association, 306*(17), 1891–1901.
32. Bouwes Bavinck, J. N., Hardie, D. R., Green, A., et al. (1996). The risk of skin cancer in renal transplant recipients in Queensland, Australia. A follow-up study. *Transplantation, 61*(5), 715–721.
33. Smith, J. M., Skeans, M. A., Horslen, S. P., et al. (2015). OPTN/SRTR 2013 annual data report: Intestine. *American Journal of Transplantation, 15*(Suppl 2), 1–16.
34. Kandaswamy, R., Skeans, M. A., Gustafson, S. K., et al. (2015). OPTN/SRTR 2013 annual data report: Pancreas. *American Journal of Transplantation, 15*(Suppl 2), 1–20.
35. Valapour, M., Skeans, M. A., Heubner, B. M., Smith, J. M., Hertz, M. I., Edwards, L. B., Cherikh, W. S., Callahan, E. R., Snyder, J. J., Israni, A. K., et al. (2015). OPTN/SRTR 2013 annual data report: Lung. *American Journal of Transplantation, 15*, 1–28.

36. Colvin-Adams, M., Smith, J. M., Heubner, B. M., Skeans, M. A., Edwards, L. B., Waller, C. D., Callahan, E. R., Snyder, J. J., Israni, A. K., & Kasiske, B. L. (2015). OPTN/SRTR 2013 annual data report: Heart. *American Journal of Transplantation, 15*, 1–28.
37. Kim, W. R., Lake, J. R., Smith, J. M., Skeans, M. A., Schladt, D. P., Edwards, E. B., Harper, A. M., Wainright, J. L., Snyder, J. J., Israni, A. K., et al. (2015). OPTN/SRTR 2013 annual data report: Liver. *American Journal of Transplantation, 15*, 1–28.
38. Matas, A. J., Smith, J. M., Skeans, M. A., et al. (2015). OPTN/SRTR 2013 annual data report: Kidney. *American Journal of Transplantation, 15*(Suppl 2), 1–34.
39. Filipponi, F., Pisati, R., Cavicchini, G., Ulivieri, M. I., Ferrara, R., & Mosca, F. (2003). Cost and outcome analysis and cost determinants of liver transplantation in a European National Health Service hospital. *Transplantation, 75*(10), 1731–1736.
40. Irwin, F. D., Wu, C., Bannister, W. M., et al. (2016). A commercial transplant network's perspective of value in solid organ transplantation: Strategizing for value in transplant care. *Transplantation Reviews (Orlando, Fla.)*
41. United States Cong. House. Committee on Energy and Commerce, Subcommittee on Health. (2013, June 28). *Examining reforms to improve the medicare part B drug program for seniors*. Washington D.C. (statement of the American Society of Transplant Surgeons). Available at https://asts.org/docs/default-source/legislative/asts-statement-to-house-committee-on-energy-and-commerce-subcommittee-on-health-june-28-2013.pdf. Accessed 14 Mar 2016.
42. Sack K. 2009, September 13. U.S. cost-saving policy forces new kidney transplant. *The New York Times*.
43. Evans, R. W., Applegate, W. H., Briscoe, D. M., et al. (2010). Cost-related immunosuppressive medication nonadherence among kidney transplant recipients. *Clinical Journal of the American Society of Nephrology, 5*(12), 2323–2328.
44. Marx, V. (2015). Tissue engineering: Organs from the lab. *Nature, 522*(7556), 373–377.
45. He, M., & Callanan, A. (2013). Comparison of methods for whole-organ decellularization in tissue engineering of bioartificial organs. *Tissue Engineering. Part B, Reviews, 19*(3), 194–208.
46. Goldfarb, D. A. (2005). Tissue engineering stem cells, and cloning: Opportunities for regenerative medicine. *The Journal of Urology, 173*(4), 1431.
47. Orlando, G., Soker, S., & Stratta, R. J. (2013). Organ bioengineering and regeneration as the new Holy Grail for organ transplantation. *Annals of Surgery, 258*(2), 221–232.
48. Orlando, G., Wood, K. J., De Coppi, P., Baptista, P. M., Binder, K. W., Bitar, K. N., Breuer, C., Burnett, L., Christ, G., Farney, A., Figliuzzi, M., Holmes, J. H., Koch, K., Macchiarini, P., Mirmalek Sani, S.-H., Opara, E., Remuzzi, A., Rogers, J., Saul, J. M., Seliktar, D., Shapira-Schweitzer, K., Smith, T., Solomon, D., Van Dyke, M., Yoo, J. J., Zhang, Y., Atala, A., Stratta, R. J., & Soker, S. (2012). Regenerative medicine as applied to general surgery. *Annals of Surgery, 255*(5), 867–880.
49. Orlando, G., Wood, K. J., Stratta, R. J., Yoo, J. J., Atala, A., & Soker, S. (2011). Regenerative medicine and organ transplantation: Past, present, and future. *Transplantation, 91*(12), 1310–1317.
50. Orlando, G., Baptista, P., Birchall, M., De Coppi, P., Farney, A., Guimaraes-Souza, N. K., Opara, E., Rogers, J., Seliktar, D., Shapira-Schweitzer, K., Stratta, R. J., Atala, A., Wood, K. J., & Soker, S. (2011). Regenerative medicine as applied to solid organ transplantation: Current status and future challenges. *Transplant International, 24*(3), 223–232.
51. Badylak, S. F., Weiss, D. J., Caplan, A., & Macchiarini, P. (2012). Engineered whole organs and complex tissues. *Lancet, 379*(9819), 943–952.
52. Vacanti, J. P., Morse, M. A., Saltzman, W. M., et al. (1988). Selective cell transplantation using bioabsorbable artificial polymers as matrices. *Journal of Pediatric Surgery, 23*, 3–9.
53. Vacanti, J. P. (1988). Beyond transplantation. Third annual Samuel Jason Mixter lecture. *Archives of Surgery, 123*, 545–549.
54. Shinoka, T., Imai, Y., & Ikada, Y. (2001). Transplantation of a tissue-engineered pulmonary artery. *The New England Journal of Medicine, 344*, 532–533.

55. Macchiarini, P., Jungebluth, P., Go, T., et al. (2008). Clinical transplantation of a tissue-engineered airway. *Lancet, 372*, 2023–2030.
56. Jungebluth, P., Alici, E., Baiguera, S., et al. (2011). Tracheobronchial transplantation with a stem-cell-seeded bioartificial nanocomposite: A proof-of-concept study. *Lancet, 378*, 1997–2004.
57. Elliott, M. J., De Coppi, P., Speggiorin, S., et al. (2012). Stem-cell-based, tissue engineered tracheal replacement in a child: A 2-year follow-up study. *Lancet, 380*, 994–1000.
58. Gonfiotti, A., Jaus, M. O., Barale, D., et al. (2014). The first tissue-engineered airway transplantation: 5-year follow-up results. *Lancet, 383*(9913), 238–244.
59. Raya-rivera, A., Esquiliano, D. R., Yoo, J. J., Lopez-bayghen, E., Soker, S., & Atala, A. (2011). Tissue-engineered autologous urethras for patients who need reconstruction: An observational study. *Lancet, 377*(9772), 1175–1182.
60. Atala, A., Bauer, S. B., Soker, S., Yoo, J. J., & Retik, A. B. (2006). Tissue-engineered autologous bladders for patients needing cystoplasty. *Lancet, 367*(9518), 1241–1246.
61. Raya-rivera, A. M., Esquiliano, D., Fierro-pastrana, R., et al. (2014). Tissue-engineered autologous vaginal organs in patients: A pilot cohort study. *Lancet, 384*(9940), 329–336.
62. Lu, L., Zhu, X., Valenzuela, R. G., Currier, B. L., & Yaszemski, M. J. (2001). Biodegradable polymer scaffolds for cartilage tissue engineering. *Clinical Orthopaedics and Related Research, (391 Suppl)*, S251–S270.
63. Chen, F., Yoo, J. J., & Atala, A. (1999). Acellular collagen matrix as a possible "off the shelf" biomaterial for urethral repair. *Urology, 54*(3), 407–410.
64. Ross, E. A., Williams, M. J., Hamazaki, T., Terada, N., Clapp, W. L., Adin, C., et al. (2009). Embryonic stem cells proliferate and differentiate when seeded into kidney scaffolds. *J Am Soc Nephrol, 20*(11), 2338–2347.
65. Bonandrini, B., Figliuzzi, M., Papadimou, E., et al. (2014). Recellularization of well-preserved acellular kidney scaffold using embryonic stem cells. *Tissue Engineering. Part A, 20*(9–10), 1486–1498.
66. Gjorevski, N., Ranga, A., & Lutolf, M. P. (2014). Bioengineering approaches to guide stem cell-based organogenesis. *Development, 141*(9), 1794–17804.
67. Brivanlou, A. H., Gage, F. H., Jaenisch, R., Jessell, T., Melton, D., & Rossant, J. (2003). Stem cells. Setting standards for human embryonic stem cells. *Science, 300*(5621), 913–916.
68. Kehat, I., Kenyagin-karsenti, D., Snir, M., et al. (2001). Human embryonic stem cells can differentiate into myocytes with structural and functional properties of cardiomyocytes. *The Journal of Clinical Investigation, 108*(3), 407–414.
69. Levenberg, S., Golub, J. S., Amit, M., Itskovitz-eldor, J., & Langer, R. (2002). Endothelial cells derived from human embryonic stem cells. *Proceedings of the National Academy of Sciences of the United States of America, 99*(7), 4391–4396.
70. Reubinoff, B. E., Itsykson, P., Turetsky, T., et al. (2001). Neural progenitors from human embryonic stem cells. *Nature Biotechnology, 19*(12), 1134–1140.
71. Assady, S., Maor, G., Amit, M., Itskovitz-eldor, J., Skorecki, K. L., & Tzukerman, M. (2001). Insulin production by human embryonic stem cells. *Diabetes, 50*(8), 1691–1697.
72. Narayanan, K., Schumacher, K. M., Tasnim, F., et al. (2013). Human embryonic stem cells differentiate into functional renal proximal tubular-like cells. *Kidney International, 83*(4), 593–603.
73. Denker, H. W. (2006). Potentiality of embryonic stem cells: An ethical problem even with stem cell source. *Journal of Medical Ethics, 32*, 665–671.
74. Aldahmash, A., Atteya, M., Elsafadi, M., Al-Nbaheen, M., Al-Mubarak, H. A., Vishnubalaji, R., Al-Roalle, A., Al-Harbi, S., Manikandan, M., Matthaei, K. I., & Mahmood, A. (2013). Teratoma formation in immunocompetent mice after syngeneic and allogeneic implantation of germline capable mouse embryonic stem cells. *Asian Pacific Journal of Cancer Prevention, 14*, 5705–5711.
75. Aasen, T., Raya, A., Barrero, M. J., et al. (2008). Efficient and rapid generation of induced pluripotent stem cells from human keratinocytes. *Nature Biotechnology, 26*(11), 1276–1284.

76. Wang, J., Gu, Q., Hao, J., et al. (2013). Generation of induced pluripotent stem cells with high efficiency from human umbilical cord blood mononuclear cells. *Genomics, Proteomics & Bioinformatics, 11*(5), 304–131.
77. Spradling, A., Drummond-barbosa, D., & Kai, T. (2001). Stem cells find their niche. *Nature, 414*(6859), 98–104.
78. Li, L., & Xie, T. (2005). Stem cell niche: Structure and function. *Annual Review of Cell and Developmental Biology, 21*, 605–631.
79. Fuchs, E. (2008). Skin stem cells: Rising to the surface. *The Journal of Cell Biology, 180*(2), 273–284.
80. Sousa, B. R., Parreira, R. C., Fonseca, E. A., Amaya, M. J., Tonelli, F. M., Lacerda, S. M., Lalwani, P., Santos, A. K., Gomes, K. N., Ulrich, H., Kihara, A. H., & Resende, R. R. (2014). Human adult stem cells from diverse origins: An overview from multiparametric immunophenotyping to clinical applications. *Cytometry. Part A, 85*, 43–77.
81. Van der flier, L. G., & Clevers, H. (2009). Stem cells, self-renewal, and differentiation in the intestinal epithelium. *Annual Review of Physiology, 71*, 241–260.
82. Bussolati, B., Bruno, S., Grange, C., et al. (2005). Isolation of renal progenitor cells from adult human kidney. *The American Journal of Pathology, 166*(2), 545–555.
83. Lindroos, B., Suuronen, R., & Miettinen, S. (2011). The potential of adipose stem cells in regenerative medicine. *Stem Cell Reviews, 7*(2), 269–291.
84. Strom, T. B., Field, L. J., & Ruediger, M. (2002). Allogeneic stem cells, clinical transplantation and the origins of regenerative medicine. *Current Opinion in Immunology, 14*(5), 601–605.
85. Simons, B. D., & Clevers, H. (2011). Stem cell self-renewal in intestinal crypt. *Experimental Cell Research, 317*(19), 2719–2724.
86. Bitar, K. N., & Raghavan, S. (2012). Intestinal tissue engineering: Current concepts and future vision of regenerative medicine in the gut. *Neurogastroenterology and Motility, 24*(1), 7–19.
87. Speer, A. L., Sala, F. G., Matthews, J. A., & Grikscheit, T. C. (2011). Murine tissue-engineered stomach demonstrates epithelial differentiation. *The Journal of Surgical Research, 171*(1), 6–14.
88. Maemura, T., Shin, M., Kinoshita, M., et al. (2008). A tissue-engineered stomach shows presence of proton pump and G-cells in a rat model, resulting in improved anemia following total gastrectomy. *Artificial Organs, 32*(3), 234–239.
89. Maemura, T., Shin, M., Sato, M., Mochizuki, H., & Vacanti, J. P. (2003). A tissue-engineered stomach as a replacement of the native stomach. *Transplantation, 76*(1), 61–65.
90. Maemura, T., Kinoshita, M., Shin, M., et al. (2012). Assessment of a tissue-engineered gastric wall patch in a rat model. *Artificial Organs, 36*(4), 409–417.
91. Chen, M. K., & Beierle, E. A. (2004). Animal models for intestinal tissue engineering. *Biomaterials, 25*(9), 1675–1681.
92. Hecker, L., Baar, K., Dennis, R. G., & Bitar, K. N. (2005). Development of a three-dimensional physiological model of the internal anal sphincter bioengineered *in vitro* from isolated smooth muscle cells. *American Journal of Physiology. Gastrointestinal and Liver Physiology, 289*(2), G188–G196.
93. Somara, S., Gilmont, R. R., Dennis, R. G., & Bitar, K. N. (2009). Bioengineered internal anal sphincter derived from isolated human internal anal sphincter smooth muscle cells. *Gastroenterology, 137*(1), 53–61.
94. Grikscheit, T. C., Siddique, A., Ochoa, E. R., et al. (2004). Tissue-engineered small intestine improves recovery after massive small bowel resection. *Annals of Surgery, 240*(5), 748–754.
95. Chen, M. K., & Badylak, S. F. (2001). Small bowel tissue engineering using small intestinal submucosa as a scaffold. *The Journal of Surgical Research, 99*(2), 352–358.
96. Hori, Y., Nakamura, T., Kimura, D., et al. (2002). Experimental study on tissue engineering of the small intestine by mesenchymal stem cell seeding. *The Journal of Surgical Research, 102*(2), 156–160.

97. Finkbeiner, S. R., & Spence, J. R. (2013). A gutsy task: Generating intestinal tissue from human pluripotent stem cells. *Digestive Diseases and Sciences, 58*(5), 1176–1184.
98. Spence, J. R., Mayhew, C. N., Rankin, S. A., et al. (2011). Directed differentiation of human pluripotent stem cells into intestinal tissue *in vitro*. *Nature, 470*(7332), 105–109.
99. Watson, C. L., Mahe, M. M., Múnera, J., et al. (2014). An *in vivo* model of human small intestine using pluripotent stem cells. *Nature Medicine, 20*(11), 1310–1314.
100. Wieck, M. M., El-nachef, W. N., Hou, X., et al. (2016). Human and murine tissue-engineered colon exhibit diverse neuronal subtypes and can be populated by enteric nervous system progenitor cells when donor colon is aganglionic. *Tissue Engineering. Part A, 22*(1–2), 53–64.
101. Resende, R. R., Fonseca, E. A., Tonelli, F. M. P., Sousa, B. R., Santos, A. K., Gomes, K. N., Guatimosim, S., Kihara, A. H., & Ladeira, L. O. (2014). Scale/topography of substrates surface resembling extracellular matrix for tissue engineering. *Journal of Biomedical Nanotechnology, 10*, 1157–1193.
102. Eckardt, K. U., Coresh, J., Devuyst, O., et al. (2013). Evolving importance of kidney disease: From subspecialty to global health burden. *Lancet, 382*(9887), 158–169.
103. Hoerger, T. J., Simpson, S. A., Yarnoff, B. O., Pavkov, M. E., Ríos Burrows, N., Saydah, S. H., Williams, D. E., & Zhuo, X. (2015). The future burden of CKD in the United States: A simulation model for the CDC CKD initiative. *American Journal of Kidney Diseases, 65*, 403–411.
104. Zambon, J. P., Magalhaes, R. S., Ko, I., et al. (2014). Kidney regeneration: Where we are and future perspectives. *World Journal of Nephrology, 3*(3), 24–30.
105. Salvatori, M., Peloso, A., Katari, R., & Orlando, G. (2014). Regeneration and bioengineering of the kidney: Current status and future challenges. *Current Urology Reports, 15*(1), 379.
106. Rinkevich, Y., Montoro, D. T., Contreras-trujillo, H., et al. (2014). Vivo clonal analysis reveals lineage-restricted progenitor characteristics in mammalian kidney development, maintenance, and regeneration. *Cell Reports, 7*(4), 1270–1283.
107. Nagaike, M., Hirao, S., Tajima, H., et al. (1991). Renotropic functions of hepatocyte growth factor in renal regeneration after unilateral nephrectomy. *The Journal of Biological Chemistry, 266*(34), 22781–22784.
108. Cochrane, A. L., Kett, M. M., Samuel, C. S., et al. (2005). Renal structural and functional repair in a mouse model of reversal of ureteral obstruction. *Journal of the American Society of Nephrology, 16*(12), 3623–3630.
109. Davidson, A. J. (2011). Uncharted waters: Nephrogenesis and renal regeneration in fish and mammals. *Pediatric Nephrology, 26*(9), 1435–1443.
110. Lam, A. Q., & Bonventre, J. V. (2015). Regenerating the nephron with human pluripotent stem cells. *Current Opinion in Organ Transplantation, 20*(2), 187–192.
111. Little, M. H. (2006). Regrow or repair: Potential regenerative therapies for the kidney. *Journal of the American Society of Nephrology, 17*(9), 2390–2401.
112. Pleniceanu, O., Harari-Steinberg, O., & Dekel, B. (2010). Concise review: Kidney stem/progenitor cells: Differentiate, sort out, or reprogram? *Stem Cells, 28*(9), 1649–1660.
113. Romagnani, P., Lasagni, L., & Remuzzi, G. (2013). Renal progenitors: An evolutionary conserved strategy for kidney regeneration. *Nature Reviews. Nephrology, 9*(3), 137–146.
114. Zhang, Y., McNeill, E., Tian, H., Soker, S., Andersson, K.-E., Yoo, J. J., & Atala, A. (2008). Urine derived cells are a potential source for urological tissue reconstruction. *The Journal of Urology, 180*, 2226–2233.
115. Xing, L., Cui, R., Peng, L., et al. (2014). Mesenchymal stem cells, not conditioned medium, contribute to kidney repair after ischemia-reperfusion injury. *Stem Cell Research & Therapy, 5*(4), 101.
116. Hauser, P. V., de Fazio, R., Bruno, S., et al. (2010). Stem cells derived from human amniotic fluid contribute to acute kidney injury recovery. *The American Journal of Pathology, 177*, 2011–2021.
117. Sedrakyan, S., da Sacco, S., Milanesi, A., et al. (2012). Injection of amniotic fluid stem cells delays progression of renal fibrosis. *Journal of the American Society of Nephrology, 23*, 661–673.

118. Lee, P. Y., Chien, Y., Chiou, G. Y., Lin, C. H., Chiou, C. H., & Tarng, D. C. (2012). Induced pluripotent stem cells without c-Myc attenuate acute kidney injury via downregulating the signaling of oxidative stress and inflammation in ischemia-reperfusion rats. *Cell Transplantation, 21*, 2569–2585.
119. Taguchi, A., Kaku, Y., Ohmori, T., et al. (2014). Redefining the in vivo origin of metanephric nephron progenitors enables generation of complex kidney structures from pluripotent stem cells. *Cell Stem Cell, 14*(1), 53–67.
120. Abrahamson, D. R., St John, P. L., Pillion, D. J., & Tucker, D. C. (1991). Glomerular development in intraocular and intrarenal grafts of fetal kidneys. *Laboratory Investigation, 64*(5), 629–639.
121. Woolf, A. S., Palmer, S. J., Snow, M. L., & Fine, L. G. (1990). Creation of a functioning chimeric mammalian kidney. *Kidney International, 38*(5), 991–997.
122. Rosines, E., Johkura, K., Zhang, X., Schmidt, H. J., De Cambre, M., Bush, K. T., & Nigam, S. K. (2010). Constructing kidney-like tissues from cells based on programs for organ development: Toward a method of i*n vitro* tissue engineering of the kidney. *Tissue Engineering Part A, 16*, 2441–2455.
123. Imberti, B., Corna, D., Rizzo, P., et al. (2015). Renal primordia activate kidney regenerative events in a rat model of progressive renal disease. *PLoS ONE, 10*(3), e0120235.
124. Nakayama, K. H., Batchelder, C. A., Lee, C. I., & Tarantal, A. F. (2010). Decellularized rhesus monkey kidney as a three-dimensional scaffold for renal tissue engineering. *Tissue Engineering. Part A, 16*(7), 2207–2216.
125. Sullivan, D. C., Mirmalek-sani, S. H., Deegan, D. B., et al. (2012). Decellularization methods of porcine kidneys for whole organ engineering using a high-throughput system. *Biomaterials, 33*(31), 7756–7764.
126. Orlando, G., Farney, A. C., Iskandar, S. S., et al. (2012). Production and implantation of renal extracellular matrix scaffolds from porcine kidneys as a platform for renal bioengineering investigations. *Annals of Surgery, 256*(2), 363–370.
127. Orlando, G., Booth, C., Wang, Z., et al. (2013). Discarded human kidneys as a source of ECM scaffold for kidney regeneration technologies. *Biomaterials, 34*, 5915–5925.
128. Peloso, A., Petrosyan, A., Da Sacco, S., et al. (2015). Renal extracellular matrix scaffolds from discarded kidneys maintain Glomerular Morphometry and vascular resilience and retains critical growth factors. *Transplantation, 99*(9), 1807–1816.
129. Petrosyan, A., Orlando, G., Peloso, A., Wang, Z., Farney, A. C., Rogers, G., Katari, R., Da Sacco, S., Sedrakyan, S., De Filippo, R. E., Stratta, R. J., Soker, S., & Perin, L. (2015). Understanding the bioactivity of stem cells seeded on extracellular matrix scaffolds produced from discarded human kidneys: A critical step towards a new generation bio-artificial kidney. *CellR4., 3*(1), e1401.
130. Song, J. J., Guyette, J. P., Gilpin, S. E., Gonzalez, G., Vacanti, J. P., & Ott, H. C. (2013). Regeneration and experimental orthotopic transplantation of a bioengineered kidney. *Nature Medicine, 19*(5), 646–651.
131. D'Haese, J. G., Ceyhan, G. O., Demir, I. E., Layer, P., Uhl, W., Löhr, M., Rychlik, R., Pirilis, K., Zöllner, Y., Gradl, B., et al. (2014). Pancreatic enzyme replacement therapy in patients with exocrine pancreatic insufficiency due to chronic pancreatitis: A 1-year disease management study on symptom control and quality of life. *Pancreas, 43*, 834–841.
132. Tao, Z., Shi, A., & Zhao, J. (2015). Epidemiological perspectives of diabetes. *Cell Biochemistry and Biophysics*, 1–5.
133. Orlando, G., Stratta, R. J., & Light, J. (2011). Pancreas transplantation for type 2 diabetes mellitus. *Current Opinion in Organ Transplantation, 16*(1), 110–115.
134. Bottino, R., & Trucco, M. (2015). Clinical implementation of islet transplantation: A current assessment. *Pediatric Diabetes, 16*(6), 393–401.
135. Bellin, M. D., Kandaswamy, R., Parkey, J., Zhang, H.-J., Liu, B., Ihm, S. H., Ansite, J. D., Witson, J., Bansal-Pakala, P., Balamurugan, A. N., et al. (2008). Prolonged insulin independence after islet allotransplants in recipients with type 1 diabetes. *American Journal of Transplantation, 8*, 2463–2470.

136. Ricordi, C., Lacy, P. E., Finke, E. H., Olack, B. J., & Scharp, D. W. (1988). Automated method for isolation of human pancreatic islets. *Diabetes, 37*, 413–420.
137. Bellin, M. D., Barton, F. B., Heitman, A., Harmon, J. V., Kandaswamy, R., Balamurugan, A. N., Sutherland, D. E. R., Alejandro, R., & Hering, B. J. (2012). Potent induction immunotherapy promotes long-term insulin independence after islet transplantation in type 1 diabetes. *American Journal of Transplantation, 12*, 1576–1583.
138. Dor, Y., Brown, J., Martinez, O. I., & Melton, D. A. (2004). Adult pancreatic beta-cells are formed by self-duplication rather than stem-cell differentiation. *Nature, 429*, 41–46.
139. Xu, X., D'Hoker, J., Stangé, G., Bonné, S., De Leu, N., Xiao, X., Van De Casteele, M., Mellitzer, G., Ling, Z., Pipeleers, D., et al. (2008). β cells can be generated from endogenous progenitors in injured adult mouse pancreas. *Cell, 132*, 197–207.
140. D'Amour, K. A., Agulnick, A. D., Eliazer, S., Kelly, O. G., Kroon, E., & Baetge, E. E. (2005). Efficient differentiation of human embryonic stem cells to definitive endoderm. *Nature Biotechnology, 23*, 1534–1541.
141. D'Amour, K. A., Bang, A. G., Eliazer, S., Kelly, O. G., Agulnick, A. D., Smart, N. G., Moorman, M. A., Kroon, E., Carpenter, M. K., & Baetge, E. E. (2006). Production of pancreatic hormone-expressing endocrine cells from human embryonic stem cells. *Nature Biotechnology, 24*, 1392–1401.
142. Alipio, Z., Liao, W., Roemer, E. J., et al. (2010). Reversal of hyperglycemia in diabetic mouse models using induced-pluripotent stem (iPS)-derived pancreatic beta-like cells. *Proceedings of the National Academy of Sciences of the United States of America, 107*(30), 13426–13431.
143. De Carlo, E., Baiguera, S., Conconi, M. T., Vigolo, S., Grandi, C., Lora, S., Martini, C., Maffei, P., Tamagno, G., Vettor, R., et al. (2010). Pancreatic acellular matrix supports islet survival and function in a synthetic tubular device: In vitro and in vivo studies. *International Journal of Molecular Medicine, 25*, 195–202.
144. Mirmalek-Sani, S. H., Orlando, G., McQuilling, J., Pareta, R., Mack, D., Salvatori, M., Farney, A. C., Stratta, R. J., Atala, A., Opara, E. C., & Soker, S. (2013). Porcine pancreas extracellular matrix as a platform for endocrine pancreas bioengineering. *Biomaterials, 34*(22), 5488–5495.
145. Peloso, A., Urbani, L., Katari, R., Maghsoudlou, P., Alvarez Fallas, M. E., Sordi, V., Citro, A., Niu, G., Zambon, J. P., Farney, A. C., Iskandar, S. S., Rogers, J., Stratta, R. J., Opara, E. C., Piemonti, L., Furdui, C. M., Soker, S., De Coppi, P., & Orlando, G. (2016). The human pancreas as a source of ECM scaffold for a new generation bio-artificial endocrine pancreas. *Annals of Surgery, 264*(1), 169–179.
146. Goh, S. K., Bertera, S., Olsen, P., Candiello, J., Halfter, W., Uechi, G., Balasubramani, M., Johnson, S., Sicari, B., Kollar, E., Badylak, S. T., & Banerjee, I. (2013). Perfusion-decellularized pancreas as a natural 3D scaffold for pancreatic tissue and whole organ engineering. *Biomaterials, 34*(28), 6760–6772.
147. De Carlo, E., Baiguera, S., Conconi, M. T., et al. (2010). Pancreatic acellular matrix supports islet survival and function in a synthetic tubular device: *In vitro* and *in vivo* studies. *International Journal of Molecular Medicine, 25*(2), 195–202.
148. Murray, C. J., & Lopez, A. D. (2013). Measuring the global burden of disease. *The New England Journal of Medicine, 369*(5), 448–457.
149. Dutkowski, P., Oberkofler, C. E., Béchir, M., et al. (2011). The model for end-stage liver disease allocation system for liver transplantation saves lives, but increases morbidity and cost: A prospective outcome analysis. *Liver Transplantation, 17*(6), 674–684.
150. Fox, I. J., Chowdhury, J. R., Kaufman, S. S., et al. (1998). Treatment of the Crigler-Najjar syndrome type I with hepatocyte transplantation. *The New England Journal of Medicine, 338*(20), 1422–1426.
151. Dhawan, A., Mitry, R. R., Hughes, R. D., et al. (2004). Hepatocyte transplantation for inherited factor VII deficiency. *Transplantation, 78*(12), 1812–1814.
152. Struecker, B., Raschzok, N., & Sauer, I. M. (2014). Liver support strategies: Cutting-edge technologies. *Nature Reviews. Gastroenterology & Hepatology, 11*(3), 166–176.

153. Minuk, G. Y. (2003). Hepatic regeneration: If it ain't broke, don't fix it. *Canadian Journal of Gastroenterology, 17*, 418–424.
154. Kay, M. A., & Fausto, N. (1997). Liver regeneration: Prospects for therapy based on new technologies. *Molecular Medicine Today, 3*, 108–115.
155. Meng, F., Francis, H., Glaser, S., et al. (2012). Role of stem cell factor and granulocyte colony-stimulating factor in remodeling during liver regeneration. *Hepatology, 55*(1), 209–221.
156. Cardinale, V., Wang, Y., Carpino, G., et al. (2011). Multipotent stem/progenitor cells in human biliary tree give rise to hepatocytes, cholangiocytes, and pancreatic islets. *Hepatology, 54*(6), 2159–2172.
157. Petersen, B. E., Bowen, W. C., Patrene, K. D., et al. (1999). Bone marrow as a potential source of hepatic oval cells. *Science, 284*(5417), 1168–1170.
158. Alison, M. R., Poulsom, R., Jeffery, R., et al. (2000). Hepatocytes from non-hepatic adult stem cells. *Nature, 406*(6793), 257.
159. Schmelzer, E., Wauthier, E., & Reid, L. M. (2006). The phenotypes of pluripotent human hepatic progenitors. *Stem Cells, 24*, 1852–1858.
160. Sugimoto, S., Harada, K., Shiotani, T., Ikeda, S., Katsura, N., Ikai, I., Mizuguchi, T., Hirata, K., Yamaoka, Y., & Mitaka, T. (2005). Hepatic organoid formation in collagen sponge of cells isolated from human liver tissues. *Tissue Engineering, 11*, 626–633.
161. Zhang, L., Theise, N., Chua, M., & Reid, L. M. (2008). The stem cell niche of human livers: Symmetry between development and regeneration. *Hepatology, 48*, 1598–1607.
162. Terai, S., Takami, T., Yamamoto, N., et al. (2014). Status and prospects of liver cirrhosis treatment by using bone marrow-derived cells and mesenchymal cells. *Tissue Engineering. Part B, Reviews, 20*(3), 206–210.
163. Pan, X. N., Zheng, L. Q., & Lai, X. H. (2014). Bone marrow-derived mesenchymal stem cell therapy for decompensated liver cirrhosis: A meta-analysis. *World Journal of Gastroenterology, 20*(38), 14051–14057.
164. Liu, F., Pan, X., Chen, G., et al. (2006). Hematopoietic stem cells mobilized by granulocyte colony-stimulating factor partly contribute to liver graft regeneration after partial orthotopic liver transplantation. *Liver Transplantation, 12*(7), 1129–1137.
165. Duan, X. Z., Liu, F. F., Tong, J. J., et al. (2013). Granulocyte-colony stimulating factor therapy improves survival in patients with hepatitis B virus-associated acute-on-chronic liver failure. *World Journal of Gastroenterology, 19*(7), 1104–1110.
166. Tsolaki, E., Athanasiou, E., Gounari, E., et al. (2014). Hematopoietic stem cells and liver regeneration: Differentially acting hematopoietic stem cell mobilization agents reverse induced chronic liver injury. *Blood Cells, Molecules & Diseases, 53*(3), 124–132.
167. Piscaglia, A. C., Shupe, T. D., SH, O., Gasbarrini, A., & Petersen, B. E. (2007). Granulocyte-colony stimulating factor promotes liver repair and induces oval cell migration and proliferation in rats. *Gastroenterology, 133*(2), 619–631.
168. Soto-Gutierrez, A., Zhang, L., Medberry, C., Fukumitsu, K., Faulk, D., Jiang, H., Reing, J., Gramignoli, R., Komori, J., Ross, M., et al. (2011). A whole-organ regenerative medicine approach for liver replacement. *Tissue Engineering Part C: Methods, 17*, 677–686.
169. Shupe, T., Williams, M., Willenberg, B., & Petersen, B. (2010). Method for the decellularization of intact rat liver. *Organogenesis, 6*, 134–136.
170. Uygun, B. E., Soto-Gutierrez, A., Yagi, H., Izamis, M. L., Guzzardi, M. A., Shulman, C., Milwid, J., Kobayashi, N., Tilles, A., Berthiaume, F., Hertl, M., Nahmias, Y., Yarmush, M. L., & Uygun, K. (2010). Organ reengineering through development of a transplantable recellularized liver graft using decellularized liver matrix. *Nature Medicine, 16*, 814–820.
171. Yagi, H., Fukumitsu, K., Fukuda, K., et al. (2013). Human-scale whole-organ bioengineering for liver transplantation: A regenerative medicine approach. *Cell Transplantation, 22*(2), 231–242.
172. Baptista, P. M., Siddiqui, M. M., Lozier, G., Rodriguez, S. R., Atala, A., & Soker, S. (2011). The use of whole organ decellularization for the generation of a vascularized liver organoid. *Hepatology, 53*(2), 604–617.

173. Wang, Y., Cui, C. B., Yamauchi, M., Miguez, P., Roach, M., Malavarca, R., Costello, M. J., Cardinale, V., Wauthier, E., Barbier, C., Gerber, D. A., Alvaro, D., & Reid, L. M. (2011). Lineage restriction of human hepatic stem cells to mature fates is made efficient by tissue-specific biomatrix scaffolds. *Hepatology, 53*, 293–305.
174. Barakat, O., Abbasi, S., Rodriguez, G., et al. (2012). Use of decellularized porcine liver for engineering humanized liver organ. *The Journal of Surgical Research, 173*(1), e11–e25.
175. Mazza, G., Rombouts, K., Rennie Hall, A., et al. (2015). Decellularized human liver as a natural 3D-scaffold for liver bioengineering and transplantation. *Scientific Reports, 5*, 13079.
176. Sun, Y., & Weber, K. T. (2000). Infarct scar: A dynamic tissue. *Cardiovascular Research, 46*(2), 250–256.
177. Eschenhagen, T., Wakatsuki, T., & Elson, E. L. (1995, August 19–22). *A new method to measure isometric force of contraction in embryonic cardiac myocytes*. Report No. 95-17 on the second international conference on Cellular Engineering, La Jolla (Abstract).
178. Kehat, I., Kenyagin-karsenti, D., Snir, M., et al. (2001). Human embryonic stem cells can differentiate into myocytes with structural and functional properties of cardiomyocytes. *The Journal of Clinical Investigation, 108*(3), 407–414.
179. Fazel, S., Tang, G. H., Angoulvant, D., et al. (2005). Current status of cellular therapy for ischemic heart disease. *The Annals of Thoracic Surgery, 79*(6), S2238–S2247.
180. Murry, C. E., Field, L. J., & Menasché, P. (2005). Cell-based cardiac repair: Reflections at the 10-year point. *Circulation, 112*(20), 3174–3183.
181. Qian, H., Yang, Y., Huang, J., et al. (2007). Intracoronary delivery of autologous bone marrow mononuclear cells radiolabeled by 18F-fluoro-deoxy-glucose: Tissue distribution and impact on post-infarct swine hearts. *Journal of Cellular Biochemistry, 102*(1), 64–74.
182. Hirt, M. N., Hansen, A., & Eschenhagen, T. (2014). Cardiac tissue engineering: State of the art. *Circulation Research, 114*(2), 354–367.
183. Zimmermann, W. H., Schneiderbanger, K., Schubert, P., et al. (2002). Tissue engineering of a differentiated cardiac muscle construct. *Circulation Research, 90*(2), 223–230.
184. Leontyev, S., Schlegel, F., Spath, C., et al. (2013). Transplantation of engineered heart tissue as a biological cardiac assist device for treatment of dilated cardiomyopathy. *European Journal of Heart Failure, 15*(1), 23–35.
185. Ott, H. C., Matthiesen, T. S., Goh, S. K., et al. (2008). Perfusion-decellularized matrix: Using nature's platform to engineer a bioartificial heart. *Nature Medicine, 14*(2), 213–221.
186. Christman, K. L., Vardanian, A. J., Fang, Q., Sievers, R. E., Fok, H. H., & Lee, R. J. (2004). Injectable fibrin scaffold improves cell transplant survival, reduces infarct expansion, and induces neovasculature formation in ischemic myocardium. *Journal of the American College of Cardiology, 44*(3), 654–660.
187. Habib, M., Shapira-schweitzer, K., Caspi, O., et al. (2011). A combined cell therapy and in-situ tissue-engineering approach for myocardial repair. *Biomaterials, 32*(30), 7514–7523.
188. Organ Procurement and Transplantation Network. http://optn.transplant.hrsa.gov. Accessed April 2016.
189. Baiguera, S., Gonfiotti, A., Jaus, M., et al. (2011). Development of bioengineered human larynx. *Biomaterials, 32*(19), 4433–4442.
190. Petersen, T. H., Calle, E. A., Zhao, L., et al. (2010). Tissue-engineered lungs for in vivo implantation. *Science, 329*(5991), 538–541.
191. Grillo, H. C. (2002). Tracheal replacement: A critical review. *The Annals of Thoracic Surgery, 73*(6), 1995–2004.
192. Novosel, E. C., Kleinhans, C., & Kluger, P. J. (2011). Vascularization is the key challenge in tissue engineering. *Advanced Drug Delivery Reviews, 63*(4–5), 300–311.
193. Berg, M., Ejnell, H., Kovács, A., et al. (2014). Replacement of a tracheal stenosis with a tissue-engineered human trachea using autologous stem cells: A case report. *Tissue Engineering. Part A, 20*(1–2), 389–397.

194. Wong, A. P., Bear, C. E., Chin, S., et al. (2012). Directed differentiation of human pluripotent stem cells into mature airway epithelia expressing functional CFTR protein. *Nature Biotechnology, 30*(9), 876–882.
195. Seguin, A., Baccari, S., Holder-espinasse, M., et al. (2013). Tracheal regeneration: Evidence of bone marrow mesenchymal stem cell involvement. *The Journal of Thoracic and Cardiovascular Surgery, 145*(5), 1297–1304.e2.
196. Mcintyre, B. A., Alev, C., Mechael, R., et al. (2014). Expansive generation of functional airway epithelium from human embryonic stem cells. *Stem Cells Translational Medicine, 3*(1), 7–17.
197. Li, Y., Xu, W., Yan, J., et al. (2014). Differentiation of human amniotic fluid-derived mesenchymal stem cells into type II alveolar epithelial cells in vitro. *International Journal of Molecular Medicine, 33*(6), 1507–1513.

Chapter 9
New Strategies in Composite Tissue Allotransplantation

Gerald Brandacher, Maria Chicco, and W.P. Andrew Lee

Abstract Over the last two decades, composite tissue allotransplantation (CTA), more recently referred to as vascularized composite allotransplantation (VCA), has become a valid therapeutic and restorative option for patients with severe tissue defects. While functional outcomes are such that recipients' quality of life is significantly improved and in some cases even dramatically enhanced, long-term risks are still difficult to evaluate in a precise fashion. Strategies have been devised, and new ones are being explored, in order to minimize or, if possible, eliminate the need for immunosuppression while keeping rejection under control. Success in these efforts would ultimately tilt the risk versus benefit balance, so that VCA may become widely adopted and a standard therapeutic option. This chapter introduces a brief historical perspective as well as a description of the current state of play within the discipline. It examines emerging challenges and then focuses on novel research avenues and clinical perspectives for the future.

9.1 Introduction and Historical Perspective

Vascularized composite allotransplantation (VCA) is the transplantation of entities, such as the hand, face, abdominal wall, or urogenital organs such as the uterus and penis, which consist of different types of tissues, e.g., skin, muscle, bone, nerves, and blood vessels with the goal to restore both form and function by replacing "like-with-like" tissue.

This discipline clinically evolved from attempts at skin transplantation to treat severely burnt soldiers, in the wake of World War II. However, negative outcomes led Medawar to the conclusion that allogeneic skin transplants could never be permanently accepted [48, 49]. This initial observation resulted in the long-lived assumption that no skin-bearing structure could be successfully transplanted.

G. Brandacher (✉) • M. Chicco • W.P. Andrew Lee
Department of Plastic and Reconstructive Surgery, Vascularized Composite Allotransplantation (VCA) Laboratory, Johns Hopkins University School of Medicine, Baltimore, MD, USA
e-mail: gbranda2@jhmi.edu

S.N. Nadig, J.A. Wertheim (eds.), *Technological Advances in Organ Transplantation*, DOI 10.1007/978-3-319-62142-5_9

With the advent in the 1960s of the first immunosuppressive drugs, VCA was attempted: hand transplantation was performed in 1964 by Gilbert in Ecuador. Despite the availability of these new medications, the graft was lost 3 weeks after surgery due to acute rejection [27].

While these disappointing experiences led to the temporary abandonment of VCA, the second half of the twentieth century saw the rise of solid organ transplantation (SOT). The development of novel more potent and targeted immunosuppressive agents in the 1980s and 1990s brought about excellent short-term survival in SOT, and as a consequence, VCA was reattempted: the first successful hand transplant was performed in 1998 [18]. Since then, more than 60 and 35 patients received upper limb/hand and face transplants, respectively, and new types of VCA such as the larynx, abdominal wall, knee, uterus, and penis were also developed [51].

9.2 Risk Versus Benefit Balance

The first successful cases of VCA generated considerable enthusiasm, but this resurgence was also met with strong criticism. The transplants are life-enhancing, but not lifesaving per se; this fact gives a unique dimension to the fundamental question in transplantation: can benefits compensate the side effects caused by immunosuppressive drugs required to prevent rejection? While in SOT the answer is clear, as benefits mean life over death, in VCA, the answer comes with a high dose of uncertainty [33].

Supporters of the practice argue that VCA bestows functional and esthetic recovery that cannot be achieved with other conventional techniques. Critics adduce that the unavoidable risks far outweigh the uncertain benefits. It is clear therefore that the risk versus benefit balance constitutes the central issue [5]. As a consequence, VCA is currently a clinical reality only for the most severe impairments. In this chapter, we briefly outline the context in which novel challenges are emerging and then focus on new therapeutic strategies, developments, and indications aimed at tilting the risk versus benefit balance in favor of establishing VCA as a standard procedure.

9.3 Immunosuppression and Side Effects

The skin is considered a potent immunogenic entity, to the point that, as we have seen, skin-bearing structures were deemed to be not permanently transplantable. Early evidence from animal models suggested, however, that a weaker immune response is elicited by the skin when transplanted as part of a VCA graft [41]. This could explain why, contrary to initial expectations, most VCA recipients are successfully maintained on the same immunosuppressive regimens as used in kidney transplantation.

These regimens consist of a triple therapy combining a calcineurin inhibitor, a corticosteroid, and an antiproliferative agent. Experience in SOT showed that immunosuppression involves substantial morbidity including infections, malignancies, metabolic complications, and nephrotoxicity; the International Registry on Hand and Composite Tissue Transplantation (IRHCTT) confirms that VCA recipients share these side effects [57].

9.4 Acute Rejection

Within a VCA graft, the skin elicits the strongest immune response. This is illustrated by the phenomenon of split tolerance observed in animal models, whereby the skin component is rejected, while other elements are accepted long term [47]. In humans, a similar finding was evidenced in post-explantation specimens from the firsthand transplant recipient, who required amputation after voluntary cessation of immunosuppression. Histopathological analysis revealed severe immune-mediated damage to the skin, with relative sparing of other structures [35]. Furthermore, in patients maintaining their graft, the skin proves the focus of acute rejection episodes.

More than 85% of VCA patients experience at least one episode of acute rejection within the first year posttransplant [57]; this rate is substantially higher than the one registered for SOT, e.g., 10% in kidney transplantation [62]. These episodes can usually be reversed so that few, if any, grafts have been lost to acute rejection in patients adherent to immunosuppressive treatment [56]. However, the possible long-term consequences of high acute rejection rates in VCA represent an important concern, since experience from SOT demonstrates that acute rejection increases risks of chronic graft dysfunction and rejection [55].

9.5 Chronic Rejection

Chronic rejection is a well-described phenomenon in SOT and currently represents the leading cause of late graft loss [55]. It consists in a clinical picture of progressive graft dysfunction, combined with specific histopathological changes: parenchymal tissue fibrosis and graft vasculopathy, characterized by myointimal hyperplasia and vascular narrowing. Some authors suggested that VCA might enjoy immune privilege from chronic rejection, in view of an initial absence of case reports [58]. However, these optimistic conjectures must be considered with a grain of salt, given the small patient cohort and the relatively short follow-up. In fact, as VCA patients increase in number and monitoring periods lengthen, case reports suggestive of chronic rejection are emerging.

Graft vasculopathy was identified in animal models of chronic rejection [52, 53, 74], as well as in clinical cases of vascularized knee and hand transplantation [17, 36]. The Louisville group examined six hand transplant recipients by means of high-resolution ultrasound and deep tissue biopsy: all patients presented a degree of myointimal hyperplasia, in the absence of suggestive clinical signs and histopathological features on standard skin biopsy. The authors therefore concluded that current VCA monitoring practices may not be adequate for detecting early signs of chronic rejection [37].

The Lyon group recently reported an intriguing case of suspected chronic rejection in face transplantation. Histopathological analysis revealed extensive epidermal and adnexal atrophy with diffuse dermal sclerosis; however, graft vasculopathy was not observed [60].

9.5.1 Antibody-Mediated Rejection

Antibody-mediated rejection (AMR) is a diagnosis characterized by high donor-specific antibodies (DSAs) and evidence of complement activation, in the context of graft dysfunction. AMR could constitute a concern in VCA: in fact, the limited donor pool available, combined with restrictions posed by blood group and by esthetic considerations such as donor age, gender, and ethnicity, renders donor-recipient HLA matching arduous. Transplantation therefore occurs across important HLA barriers, a circumstance which in theory would favor the development of DSA [77]. Until recently, DSA and C4d deposition were only sporadically detected in VCA, prompting some authors to suggest that AMR may not play a significant role [50]. In 2013, however, the first AMR case was reported by the Innsbruck group: a double hand transplant recipient presented, at 9 years postsurgery, substantial de novo DSA production, C4d deposition, and cutaneous lymphoid nodules strongly positive for B cells. The episode proved refractory to conventional therapy, but resolved with rituximab administration [75].

Formation of DSA can also be induced by so-called sensitizing events, such as blood transfusions and temporary skin allografts. The initial management of major trauma such as burns often involves sensitizing events: many patients who could benefit from VCA are therefore sensitized and considered ineligible by most centers [42]. In fact, sensitized patients are at risk of hyperacute AMR, whereby preexisting DSAs cause graft failure within minutes to hours posttransplantation.

In 2014, the Harvard group performed the first fullface transplant in a highly sensitized patient with a positive donor-recipient crossmatch. Desensitization was started the day after transplantation. However, the patient developed high DSA titers and C4d deposition; treatment combining plasmapheresis, intravenous immunoglobulin, eculizumab, bortezomib, and alemtuzumab eventually reversed the episode [11].

Incidence of rejection remains as yet difficult to evaluate and to weigh in the risk versus benefit balance. In 2007, the VCA Banff working group acknowledged that not enough information was available to draw precise conclusions and to set criteria for chronic rejection and AMR [9]. As new cases emerge and understanding of mechanisms increases, our appreciation of the risks will improve. In the meantime, novel strategies are being explored as a means to reduce and, ideally, remove both the need for immunosuppression and the risks of rejection.

9.6 Immune Tolerance in VCA

Medawar's seminal experiments with skin transplantation laid the foundation for transplant immunology; subsequent experimentation in animals led to the discovery of "acquired immune tolerance" [4]. This phenomenon was defined as "a state of donor-specific hyporesponsiveness, in the absence of immunosuppression," while maintaining adequate immune responses to third-party antigens [78]. In other words, the recipient's immune system does not recognize donor antigen as foreign and therefore does not mount a response against the graft while still identifying other antigens, such as those derived from pathogens, as foreign and therefore mounting an adequate response against these immune challenges.

This initial discovery was later complemented by the observation that some transplant patients, who for various reasons had ceased immunosuppressive therapy, survived with intact graft function – this phenomenon was designated as "operational immune tolerance" [44]. This finding paved the way for what is now unanimously recognized as the "holy grail" of transplantation: a method to iatrogenically induce such a state of immune tolerance. Recently, Kawai et al. compared long-term outcomes between tolerant patients and patients treated with conventional immunosuppression: they conclusively demonstrated the decisive benefits of immune tolerance [39]. Induction of immune tolerance in VCA would therefore obviate concerns about immunosuppression side effects as well as rejection: it would ultimately tilt the risk versus benefit balance. Consequently, it represents one of the major research avenues in VCA.

9.7 Mixed Chimerism

To this day, immune tolerance was successfully induced in humans only via "mixed chimerism." Mixed chimerism refers to a state in which donor and recipient cells coexist within the recipient: the term was coined as an analogy with the chimera, a mythical figure often depicted in classical literature as a lion with a goat's head. This state is established in clinical practice via donor hematopoietic stem cells (HSCs), mobilized from the donor bone marrow, collected from peripheral blood, and then administered to the recipient as an intravenous infusion. Donor HSCs engraft into the recipient's bone marrow and, in conjunction with recipient HSCs, guarantee the constant production of myeloid and lymphoid cell lineages: the recipient's immune system is now composed of cells both of donor and of recipient origin.

The specific mechanisms through which mixed chimerism leads to immune tolerance are an active area of research in transplant immunology. Central to this research is the notion of "clonal deletion." Clonal deletion is an established mechanism of tolerance to self: immature T cells are exposed to self-antigen by antigen-presenting dendritic cells within the thymus and, if they exhibit a strong response, are selectively eliminated. In a similar fashion, following the establishment of mixed chimerism, donor-derived dendritic cells travel to the recipient's thymus and conduct the specific elimination of immature T cells that exhibit strong response to donor antigen. This entire process renders the recipient specifically tolerant to donor antigen and therefore to the graft.

Immune tolerance was achieved via mixed chimerism in a variety of animal models; clonal deletion was demonstrated as an important mechanism in this process. In humans, mixed chimerism was achieved, in the vast majority of cases, only as a transient state; however, in certain patients, immune tolerance persists long after the loss of mixed chimerism: this indicates that mechanisms other than clonal deletion might be at play [64]. Peripheral mechanisms, mediated in particular by the regulatory function of various cell types, were demonstrated to be of importance in both induction and maintenance of immune tolerance in various SOT and VCA experimental models [22].

9.8 Conditioning Regimens

Mixed chimerism was established in humans via donor HSC administration only in the context of a conditioning regimen. Conditioning is necessary in order to immunosuppress the recipient and prevent immune-mediated destruction of donor HSCs, as well as to free niche space within the recipient bone marrow and allow for donor HSC engraftment [13]. So far, successful conditioning regimens involve the use of irradiation, either thymic or systemic, as well as of a T cell-depleting agent [38, 43, 66].

Clinical experience with conditioning regimens derives from the field of hematological oncology; while the use of these highly toxic regimens can be justified in the context of hematological malignancies and, more recently, for the purpose of tolerance induction in kidney transplantation [38], they cannot be applied for a similar purpose in VCA. Strategies to reduce conditioning regimens' toxicity are an active area of research in both SOT and VCA: they include, among others, the use of co-stimulation blockade [76] and the development of monoclonal antibodies which allow for depletion of bone marrow progenitors [13].

9.9 Delayed Tolerance

Beyond the development of an irradiation-free regimen, another obstacle remains: current tolerance-inducing regimens span a period from days to weeks prior to transplantation. While this is appropriate in the context of live donor transplantation, it would not be applicable to the deceased donor context imposed on VCA. Efforts to circumvent this issue have led to the emergence of "delayed tolerance," whereby the patient receives a transplant, followed later on by conditioning and donor HSC administration. A theoretical downside to this strategy is possible sensitization and in particular the generation of donor-specific T memory cells (Tmem cells), in the interval between transplantation and donor HSC administration [79]: Tmem cells are increasingly recognized as a barrier to successful donor HSC engraftment [70]. However, several studies in SOT animal models have demonstrated that delayed tolerance via mixed chimerism induction can be achieved [10, 73, 80].

Recently, a delayed tolerance induction protocol was tested in a rat model of heterotopic osteomyocutaneous transplant. VCA recipients, maintained on cyclosporine, received donor bone marrow cells with a conditioning regimen at 2 months posttransplant. The authors reported successful engraftment and induction of transient mixed chimerism in all animals, except those receiving minimal irradiation. Immunosuppression was discontinued after donor bone marrow cell administration, and a majority of recipients accepted the graft for more than 6 months [12]. This study still required the use of irradiation, but it provides proof of principle that delayed tolerance induction is a promising strategy in VCA: patients would have time to recover after surgery, and establishment of mixed chimerism could be undertaken as an elective procedure by means of cryopreserved donor HSCs.

9.10 Vascularized Bone Marrow

Some VCA grafts present a unique characteristic: they include a vascularized bone component, which contains bone marrow – rendering them equivalent to a vascularized bone marrow transplant and prone to mixed chimerism induction. Stromal cells present within the bone marrow play an essential role in maintaining the appropriate microenvironment for hematopoiesis [23]. Thus, transferring donor HSCs within their physiological milieu offers an evident advantage and constitutes an important strategy in order to unleash the immunomodulatory potential of donor HSCs, even in the absence of a conditioning regimen.

To investigate this potential, Barth et al. compared outcomes between groups of nonhuman primates receiving a face transplant, either with or without a donor mandible segment. The recipients were not conditioned and were maintained on standard immunosuppressive treatment. The four animals receiving a transplant with a bone segment experienced prolonged graft survival, while those who did not receive a bone component suffered early graft loss. In the surviving grafts, viable donor bone marrow was present and contained both donor and recipient cell populations. Three of the four animals developed varying degrees of mixed chimerism; however, they all rejected the graft after cessation of immunosuppression [3].

In a more recent study, mice received a heterotopic hind limb, either with or without a bone component, as well as co-stimulation blockade and rapamycin. Eighty percent of animals receiving a bone component developed long-term graft acceptance after treatment cessation. Prolonged graft survival was associated with both in vivo and in vitro evidence of donor-specific hyporesponsiveness. Tolerant mice presented higher donor cell levels within both graft and native femur bone marrow compartments as well as within the thymus: more intense cell trafficking facilitated by the vascularized bone marrow may promote, given the right conditions, the establishment of immune tolerance [45]. These studies confirm the immunomodulatory potential of the vascularized bone marrow. In healthy adults, however, hematopoiesis is largely restricted to the pelvis, vertebrae, sternum, and cranium. It remains uncertain, at present, to what extent the vascularized bone marrow contained in some VCA grafts exerts an immunomodulatory effect.

9.11 Donor Hematopoietic Stem Cells

The immunomodulatory potential of various cellular therapies constitutes a promising strategy in order to minimize the need for immunosuppression in VCA. The first face transplant patient received two donor HSC infusions, harvested from the donor's iliac crest, at postoperative days 4 and 11 [19]. Initial monitoring revealed a low level of mixed chimerism at 2 months: 0.1% of CD34+ cells were of donor origin [29]. However, the patient was not weaned from immunosuppression, and mixed chimerism was lost on subsequent follow-up [59]. Retrospectively, the team expressed their suspicion that the loss of chimerism was caused by poor engraftment, as a consequence of low donor HSC quality [29].

More recently, the Johns Hopkins/Pittsburgh protocol allowed for successful maintenance of five upper limb/hand transplant recipients on low-dose tacrolimus monotherapy [68]. The protocol consists of alemtuzumab and methylprednisolone induction as well as unprocessed donor bone marrow cell infusion. The cells were harvested from donor vertebral bodies and infused at 2 weeks posttransplantation [28]. Mixed chimerism was not evidenced in the peripheral blood; nonetheless, the authors suggest that the immunomodulatory properties of the protocol may account for the patients' stable condition while on minimal immunosuppression. Recipients experienced similar rates of acute rejection as patients on standard immunosuppressive regimens, successfully treated with conventional therapy and did not have any infectious side effects of complications [68].

9.12 Mesenchymal Stem Cells

Mesenchymal stem cells (MSCs) are multipotent stem cells which hold potential to differentiate into various cell types such as osteoblasts, chondrocytes, and myocytes. MSCs modulate immune responses via a variety of mechanisms, including cytokine production and expansion of T regulatory cells [61]. They can be derived from bone marrow or adipose tissue and harvested from either the recipient (autologous) or the donor (allogeneic) [69]. While the former presents the advantage of avoiding an immune response, the latter might be necessary in order to facilitate immunomodulation toward donor antigen and therefore the graft. In 2013, the Milan/Monza group reported the first clinical case of MSC administration in VCA: the patient, a double hand amputee, received an infusion of autologous MSCs at 12 h posttransplantation. Over the following year, steroid therapy was weaned, and she remained on low-dose double therapy [15].

9.13 T Regulatory Cells

T regulatory cells (Treg cells) are a subpopulation of T cells which possess a distinctive ability to modulate the immune system: they are characterized by expression of the transcription factor forkhead box protein3 (Foxp3), as well as of the surface markers CD4 and CD25. Treg cells play an important role in the maintenance of tolerance to self and thus in the prevention of autoimmunity [63]. Their potential to modulate the alloimmune response has long been recognized and has formed the object of numerous studies. Eljaafari et al. identified Foxp3 expression among graft-infiltrating cells in a clinically stable hand allograft, 6 years posttransplantation [20]; since then, several animal studies suggested an association between Treg cell presence and long-term graft survival and acceptance [32].

Adoptive transfer of Treg cells could prove an effective means to prevent rejection and allow for maintenance on reduced-intensity immunosuppression. Various methods are available for the isolation and in vitro expansion of human Treg cells. Two clinical trials are at present examining the safety and feasibility of Treg cell adoptive transfer: the ONE and ThRIL studies evaluate its impact as an adjunctive treatment in kidney and liver transplantation, respectively [26, 65]. The results of the phase I and II clinical trials, if positive, might prompt similar efforts in VCA.

9.14 A Caveat

The potential benefits of sharp reduction in immunosuppressive therapy are exciting: experience from SOT informs that side effects of immunosuppression are strongly correlated with both dose and length of exposure. VCA patients are, on average, younger, and these new strategies represent an important means for decreasing cumulative exposure and therefore side effect incidence – however, potential long-term consequences are unknown. This dilemma was stressed by the Louisville group: conventional immunosuppression, while far from ideal, delivers a consistent performance, which risks being jeopardized by attempts to withdraw medication [62]. Moreover, as noted by Morelon et al., "given the very small patient cohorts, it is impossible to perform randomized studies, and our ability to rationally optimize patient immunosuppression is therefore limited"; it is currently "impossible to establish the optimum benefit-risk ratio for these new immunosuppressive regimens" [51].

9.15 Current Outcomes

There is a consensus among many patients that VCA delivers considerable improvements in function, independence, and quality of life (Fig. 9.1). Functional and esthetic gains allow these patients to perform a variety of daily activities while, in some cases, dramatically improving their social life [51].

Fig. 9.1 Bilateral upper extremity transplant recipient 6 years postsurgery. The left hand was transplanted at the level of the wrist, while on the right side, an above elbow-level transplant was performed. The patient has regained a high level of both motor and sensory functions and is completely independent and able to perform all tasks of daily life

The hand and face are the most common transplants performed to date, and detailed functional outcomes are available: they demonstrate consistent restoration of protective sensation, adequate recovery of temperature, and vibration senses as well as proprioception, with more modest results for discriminative sensation [57]. However, restoration of muscle function has proved more challenging and currently constitutes an important hurdle to overcome [8]. Success depends on reinnervation of donor muscles via peripheral nerve regeneration. Experiences with limb replantation demonstrate that, when regeneration must occur over longer distances, outcomes correspondingly worsen [34]. Enhancing peripheral nerve regeneration would maximize VCA functional benefits, as well as expand its indications to include, e.g., full-arm transplants at the shoulder level or the transplantation of lower limbs.

9.16 Enhancing Nerve Regeneration

VCA involves end-to-end coaptation of donor and recipient peripheral nerves. This can be achieved without tension, as additional donor nerve length can be easily harvested. Nerve regeneration occurs at a rate of approximately 1–3 mm per day: new strategies aim at increasing both speed and adequacy of reinnervation. These include pharmacotherapy such as administration of hormones and growth factors, as well as cell-based therapies [21]. After nerve injury, Schwann cells (SCs) convert to a regenerative phenotype and upregulate various molecules, thus forming a permissive environment for axonal regeneration. This characteristic accounts, in part, for the superior regenerative potential of the peripheral as compared to the central nervous system [67]. SCs can be harvested, expanded ex vivo, and readministered by local delivery. This approach generated encouraging results in various experimental models; however, SCs have proven difficult to substantially expand in vitro.

Embryonic, induced pluripotent and mesenchymal stem cells all possess the ability to differentiate into SC-like precursors and can be easily expanded in vitro. Furthermore, they can be genetically engineered so as to express specific growth factors and further improve their regenerative potential [21]. The use of embryonic and induced pluripotent stem cells raises, however, safety concerns over undesirable cell differentiation.

MSCs display a narrower differentiation potential, which is nonetheless adequate for the purpose of nerve regeneration. In various experimental models, both systemic and local MSC administrations improved neuronal growth [14, 46]. Furthermore, as we have seen, MSCs can be harvested from accessible sources such as adipose tissue and also hold potential for immunomodulation: the use of adipose-derived MSCs is considered one of the most promising strategies to reduce risks and maximize benefits in VCA.

9.17 Uterine Transplantation

Given the highly encouraging outcomes seen in upper limb/hand and face transplantation, indications for VCA to reconstruct other devastating tissue defects or organ losses are rapidly expanding. Two isolated cases of uterine transplantation were reported in 2000 and 2011 by teams in Saudi Arabia and in Turkey. While both surgeries were initially successful, the first patient lost the graft at 3 months due to uterine artery thrombosis, and the second patient suffered two early miscarriages after embryo transfer [6].

In 2013, a team in Sweden reported the first birth of a child to a uterine transplant recipient, who had undergone single embryo transfer 1 year posttransplantation [7]. To date, 12 uterine transplants have been performed in five countries: four grafts failed, while five successful pregnancies were reported, all of them in Sweden. Teams in various other countries are developing dedicated programs [2].

Uterine transplantation represents the first "ephemeral" type of transplantation. Since hysterectomy could be performed after successful pregnancy, concerns about side effects from long-term immunosuppression would not apply [6]. Uterine transplantation achieved, in some cases, full functionality and holds therefore clear therapeutic potential; it also raises a series of ethical questions. For instance, Huet et al. reviewed potential candidates for transplantation: some presented with complete androgen insensitivity syndrome [31]. These patients are phenotypically and psychosocially female but genetically male – the question that has been posed was: should they be considered eligible? This question can further be extended to transgender individuals and will require further sensitive discussion.

9.18 Penile Transplantation

The first penile transplant was performed in 2006 by a team in China: a patient who sustained traumatic amputation received a transplant from a deceased donor. The surgery involved microsurgical anastomoses of the urethra, deep and superficial dorsal veins, dorsal arteries, and dorsal nerves, as well as of the corpus spongiosum and corpora cavernosa. Despite encouraging early results, the graft was explanted because of "severe psychological problem of the recipient and his wife" [30]. The second and third cases were performed in 2014 and 2016 in South Africa and in the USA, respectively [1]. Extensive coverage was given by the lay press, but scientific reports on outcomes have not yet been published.

The three transplants performed to date provide proof of principle that penile transplantation is feasible and constitutes a promising option for repair of extensive urogenital tissue loss. Nevertheless, it remains uncertain at present to what extent functions such as micturition, sexual intercourse, and reproduction can be achieved. Several animal models have been developed; however, they do not allow for functional evaluation [40, 71]. Recently, the Johns Hopkins group designed an ex vivo model aimed at investigating the impact of immune responses on erectile tissue function. Contractile response was assessed by electrical field stimulation: rejected corpora cavernosa tissues demonstrated impaired contraction and relaxation. Interestingly, addition of cyclosporine prevented rejection, while impaired relaxation persisted. This effect was specific to cyclosporine, but absent with tacrolimus [72].

9.19 VCA from Living Donors and the "Aggregate Risk Versus Benefit Balance"

In 2013, the US Department of Health and Human Services (DHHS) adopted a final rule modifying its definition of "covered human organ" so as to include VCA grafts, defined in a broad manner on the basis of eight functional and structural criteria [16]. It also entrusted the Organ Procurement and Transplantation Network (OPTN) with the oversight of VCA donations, transferring this competence from the Food and Drug Administration (FDA).

With the development of new types of VCA, new perspectives arise: e.g., uterine transplantation opens the possibility of live donation. In its guidelines, the OPTN remarked the success of upper limb/hand and face transplantation, as well as "the use of different, less extensive types of VCA"; it considered therefore that "the possibility of utilizing a living VCA donor is now medically feasible and may be ethically appropriate for both reconstructive and non-reconstructive (e.g. reproductive organ) types of VCA transplantation" [54].

The OPTN recognized, however, that the use of VCA grafts from living donors posed a series of specific ethical issues, with regard in particular to the protection of donors, and concluded that "the concept of living VCA donation can be supported only as long as an unacceptably severe permanent disability or disfigurement to the donor does not occur." Therefore, the analysis of the risk-benefit balance should be extended to both donor and recipient: VCA from living donors should only be conducted "when the aggregated benefits to the donor-recipient pair out-weigh the risks to the donor-recipient pair" [54], for instance, in the setting of uterus transplantation or other free tissue transfers, such as a vascularized fibula graft for extremity reconstruction, from a parent to their child.

9.20 Novel Technologies and Future Outlook

Novel technologies such as tissue engineering and regenerative medicine will also play a significant role in the future development of VCA. Regenerated nerves, vessels, bone, etc. will allow to eventually perform transplants in a more rapid, safer, and successful fashion. Along those lines, 3D bioprinting could help us further optimize outcomes by enhancing the graft or advancing rehabilitation strategies and approaches currently in use for upper extremity transplantation. Ultimately bioengineering technologies, which are currently developed in advanced prosthetic programs, could further help to improve outcome efforts and even augment or enhance function after VCA in the future.

Another exciting aspect of the future of VCA is the potential for noninvasive immune monitoring technologies (Fig. 9.2). The skin component of these types of transplants lends itself as a perfect scenario to advance such novel noninvasive monitoring technologies in the form of multimodal imaging technologies including IR, near IR, or 3CCD camera devices [24].

Finally, novel drug delivery routes and systems such as the use of hydrogels and nano- or microparticles will be another technology that will significantly shape the future of the field of VCA. First studies, such as the one from the group in Bern, Switzerland, in collaboration with bioengineers at MIT, have shown proof of concept that these technologies could replace the need for systemic immunosuppression and thereby significantly contribute to favor the risk versus benefit balance [25].

9.21 Conclusion

Basic science and translational and clinical research efforts are at present focused on tilting the risk versus benefit balance in VCA, so that it may become a standard therapeutic option. In this chapter, we have reviewed new strategies, developments, and indications explored to this effect.

As fittingly enunciated by Carty, "the boundaries of VCA continue to be both defined and refined" [8]: more and more transplants are being carried out, and new types of VCA are being performed. However, it seems likely that, in the foreseeable future, VCA will remain a procedure based on individual choice, in which subjective factors play a decisive role.

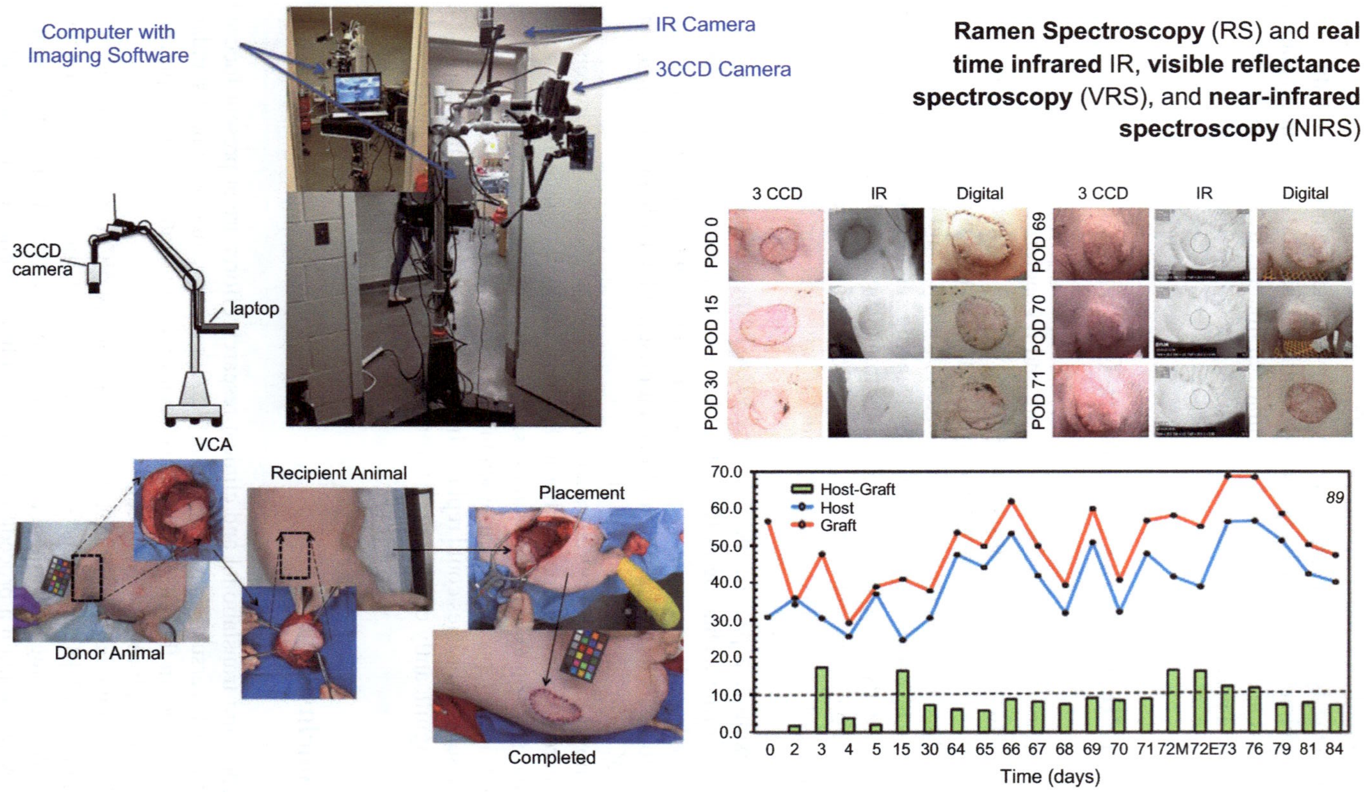

Fig. 9.2 Noninvasive optical techniques and multimodal imaging modalities such as Raman spectroscopy (*RS*) and real-time infrared IR, visible reflectance spectroscopy (*VRS*), and near-infrared spectroscopy (*NIRS*) applied in a large animal VCA model (heterotopic hind limb transplantation) for skin rejection diagnosis

Well-informed patients who opt for this procedure may have in mind the quote attributed to the Roman philosopher Seneca: "the part of life we really live is small. For, all the rest of existence is not life, but merely time."

References

1. Albersen, M. (2016). Getting ready for penile transplantation. *European Urology, 71*(4), 594–595. Available online at: http://dx.doi.org/10.1016/j.eururo.2016.10.025. Accessed on 31st Oct 2016.
2. Balayla, J. (2016). Uterine transplants in the Canadian setting: A theoretical framework. *Journal of Obstetrics and Gynaecology Canada, 38*, 955–960.
3. Barth, R. N., Rodriguez, E. D., Mundinger, G. S., Nam, A. J., Ha, J. S., Hui-Chou, H., Jones, L. S., Panda, A., Shipley, S. T., Drachenberg, C. B., Kukuruga, D., & Bartlett, S. T. (2011). Vascularized bone marrow-based immunosuppression inhibits rejection of vascularized composite allografts in nonhuman primates. *American Journal of Transplantation, 11*, 1407–1416.
4. Billingham, R. E., Brent, L., & Medawar, P. B. (1953). Activity acquired tolerance of foreign cells. *Nature, 172*, 603–606.
5. Brandacher, G., Lee, W. P., & Schneeberger, S. (2012). Minimizing immunosuppression in hand transplantation. *Expert Review of Clinical Immunology, 8*, 673–683.
6. Brännström, M., Johannesson, L., Dahm-Kähler, P., Enskog, A., Mölne, J., Kvarnström, N., Diaz-Garcia, C., Hanafy, A., Lundmark, C., Marcickiewicz, J., Gäbel, M., Groth, K., Akouri, R., Eklind, S., Holgersson, J., Tzakis, A., & Olausson, M. (2014). First clinical uterus transplantation trial: A six-month report. *Fertility and Sterility, 101*, 1228–1236.
7. Brännström, M., Johannesson, L., Bokström, H., Kvarnström, N., Mölne, J., Dahm-Kähler, P., Enskog, A., Milenkovic, M., Ekberg, J., Diaz-Garcia, C., Gäbel, M., Hanafy, A., Hagberg, H., Olausson, M., & Nilsson, L. (2015). Livebirth after uterus transplantation. *Lancet, 385*, 607–616.
8. Carty, M. J., Zuker, R., Cavadas, P., Pribaz, J. J., Talbot, S. G., & Pohamac, B. (2013). The case for lower extremity allotransplantation. *Plastic and Reconstructive Surgery, 131*, 1272–1277.
9. Cendales, L. C., Kanitakis, J., Schneeberger, S., Burns, C., Ruiz, P., Landin, L., Remmelink, M., Hewitt, C. W., Landgren, T., Lyon, B., Drachenberg, C. B., Solez, K., Kirk, A. D., Kleiner, D. E., & Racusen, L. (2008). The Banff 2007 working classification of skin-containing composite tissue allograft pathology. *American Journal of Transplantation, 8*, 1396–1400.
10. Cetrulo, C. L., Jr., Drijkoningen, T., & Sachs, D. H. (2015). Tolerance induction via mixed chimerism in vascularized composite allotransplantation: Is it time for clinical application? *Current Opinion in Organ Transplantation, 20*, 602–607.
11. Chandraker, A., Arscott, R., Murphy, G. F., Lian, C. G., Bueno, E. M., Marty, F. M., Rennke, H. G., Milford, E., Tullius, S. G., & Pomahac, B. (2014). The management of antibody-mediated rejection in the first presensitized recipient of a full-face allotransplant. *American Journal of Transplantation, 14*, 1446–1452.
12. Chen, B., Hong, X., Corbin, D. R., & Ildstad, S. T. (2012). A clinically feasible approach to induce delayed tolerance in recipients of prior kidney or vascularized composite allotransplants. *Transplantation, 94*, 671–678.
13. Chhabra, A., Ring, A. M., Weiskopf, K., Schnorr, P. G., Gordon, S., Le, A. C., Kwon, H. S., Ring, N. G., Volkmer, J., Ho, P. H., Tseng, S., Weissman, I. L., & Shizuru, J. A. (2016). Hematopoietic stem cell transplantation in immunocompetent hosts without radiation or chemotherapy. *Science Translational Medicine, 8*, 351ra105.
14. Cooney, D. S., Wimmers, E. G., Ibrahim, Z., Grahammer, J., Christensen, J. M., Brat, G. A., Wu, L. W., Sarhane, K. A., Lopez, J., Wallner, C., Furtmüller, G. J., Yuan, N., Pang, J., Sarkar, K., Lee, W. P., & Brandacher, G. (2016). Mesenchymal stem cells enhance nerve regenera-

tion in a rat sciatic nerve repair and hindlimb transplant model. *Scientific Reports, 6*, 31306. Available online at: doi: 10.1038/srep31306. Accessed on 31st Oct 2016.
15. Del Bene, M., Peri Di Caprio, A., Melzi, M. L., Pioltelli, P. E., & Bonomi, S. (2013). Autologous mesenchymal stem cells as a new strategy in immunosuppressant therapy in double hand allotransplantation. *Plastic and Reconstructive Surgery, 131*, 305–307.
16. Department of Health and Human Services. (2013). Final rule: Organ procurement and transplantation network. Published online July 3rd 2013. Available at: https://www.gpo.gov/fdsys/pkg/FR-2013-07-03/html/2013-15731.htm. Accessed on 31st Oct 2016.
17. Diefenbeck, M., Nerlich, A., Schneeberger, S., Wagner, F., & Hofmann, G. O. (2011). Allograft vasculopathy after allogeneic vascularized knee transplantation. *Transplant International, 24*, 1–5.
18. Dubernard, J. M., Owen, E., Lefrançois, N., Petruzzo, P., Martin, X., Dawahre, M., Jullien, D., Kanitakis, J., Frances, C., Preville, X., Gebuhrer, L., Hakim, N., Lanzetta, M., Kapila, H., Herzberg, G., & Revillard, J. (2000). First human hand transplantation. *Transplant International, 13*, 521–524.
19. Dubernard, J. M., Lengelé, B., Morelon, E., Testelin, S., Badet, L., Moure, C., Beziat, J. L., Dakpé, S., Kanitakis, J., D'Hauthuille, C., El Jaafari, A., Petruzzo, P., Lefrancois, N., Taha, F., Sirigu, A., Di Marco, G., Carmi, E., Bachmann, D., Cremades, S., Giraux, P., Burloux, G., Hequet, O., Parquet, N., Francès, C., Michallet, M., Martin, X., & Devauchelle, B. (2007). Outcomes 18 months after the first human partial face transplantation. *The New England Journal of Medicine, 357*, 2451–2460.
20. Eljaafari, A., Badet, L., Kanitakis, J., Ferrand, C., Farre, A., Petruzzo, P., Morelon, E., Dubosson, M., Tiberghien, P., Dubois, V., Martin, X., Miossec, P., & Dubernard, J. M. (2006). Isolation of regulatory T cells in the skin of a human hand-allograft, up to six years posttransplantation. *Transplantation, 82*, 1764–1768.
21. Faroni, A., Mobasseri, S. A., Kingham, P. J., & Reid, A. J. (2015). Peripheral nerve regeneration: experimental strategies and future perspectives. *Advanced Drug Delivery Reviews, 82*, 160–167.
22. Fryer, M., Grahammer, J., Khalifian, S., Furtmüller, G. J., Lee, W. P., Raimondi, G., & Brandacher, G. (2015). Exploring cell-based tolerance strategies for hand and face transplantation. *Expert Review of Clinical Immunology, 11*, 1189–1204.
23. Fukushima, N., & Ohkawa, H. (1995). Hematopoietic stem cells and microenvironment: The proliferation and differentiation of stromal cells. *Critical Reviews in Oncology/Hematology, 20*, 255–270.
24. Furtmüller, G., Crane, N., Luthra, R., Swanson, E., Cheng, H, T., Oh, B., Fryer, M., Elster, E., Lee, W., Brandacher, G. (2015). Non-invasive imaging modalities for immune monitoring of vascularized composite allografts using a translational large animal model. Abstract presented at 2015. American Transplant Congress. Available online at: http://www.atcmeetingabstracts.com/abstract/non-invasive-imaging-modalities-for-immune-monitoring-of-vascularized-composite-allografts-using-a-translational-large-animal-model/. Accessed on 31st Oct 2016.
25. Gajanayake, T., Olariu, R., Leclère, F. M., Dhayani, A., Yang, Z., Bongoni, A. K., Banz, Y., Constantinescu, M. A., Karp, J. M., Vemula, P. K., Rieben, R., & Vögelin, E. (2014). A single localized dose of enzyme-responsive hydrogel improves long-term survival of a vascularized composite allograft. *Science Translational Medicine, 6*, 249ra110.
26. Geissler, E. K. (2012). The ONE Study compares cell therapy products in organ transplantation: introduction to a review series on suppressive monocyte-derived cells. *Transplantation Research, 1*, 10–12.
27. Gilbert, R. (1964). Hand transplanted from a cadaver is reamputated. *Medical Tribune and Medical News, 5*, 23.
28. Gorantla, V. S., Brandacher, G., Schneeberger, S., Zheng, X. X., Donnenberg, A. D., Losee, J. E., & Lee, W. P. (2011). Favoring the risk-benefit balance for upper extremity transplantation – The Pittsburgh Protocol. *Hand Clinics, 27*, 511–520.

29. Hequet, O., Morelon, E., Bourgeot, J. P., Dubois, V., Devauchelle, B., Dubernard, J. M., & Michallet, M. (2008). Allogeneic donor bone marrow cells recovery and infusion after allogeneic face transplantation from the same donor. *Bone Marrow Transplantation, 41*, 1059–1061.
30. Hu, W., Lu, J., Zhang, L., Wu, W., Nie, H., Zhu, Y., Deng, Z., Zhao, Y., Sheng, W., Chao, Q., Qiu, X., Yang, J., & Bai, Y. (2006). A preliminary report of penile transplantation. *European Urology, 50*, 851–853.
31. Huet, S., Tardieu, A., Filloux, M., Essig, M., Pichon, N., Therme, J. F., Piver, P., Aubard, Y., Ayoubi, J. M., Garbin, O., Collinet, P., Agostini, A., Lavoue, V., Piccardo, A., & Gauthier, T. (2016). Uterus transplantation in France: for which patients? *European Journal of Obstetrics & Gynecology and Reproductive Biology, 205*, 7–10.
32. Issa, F., & Wood, K. J. (2014). The potential role for regulatory T-cell therapy in vascularized composite allograft transplantation. *Current Opinion in Organ Transplantation, 19*, 558–565.
33. Jones, N. F. (2002). Concerns about human hand transplantation in the 21st century. *The Journal of Hand Surgery, 27*, 771–787.
34. Jones, N. F., & Schneeberger, S. (2009). Arm transplantation: Prospects and visions. *Transplantation Proceedings, 41*, 476–480.
35. Kanitakis, J., Jullien, D., Petruzzo, P., Hakim, N., Claudy, A., Revillard, J. P., Owen, E., & Dubernard, J. M. (2003). Clinicopathological features of graft rejection of the first hand allograft. *Transplantation, 76*, 688–693.
36. Kanitakis, J., Petruzzo, P., Badet, L., Gazarian, A., Thaunat, O., Testelin, S., Devauchelle, B., Dubernard, J. M., & Morelon, E. (2016). Chronic rejection in human vascularized composite allotransplantation (hand and face recipients): An update. *Transplantation, 10*, 2053–2061.
37. Kaufman, C. L., Ouseph, R., Blair, B., Kutz, J. E., Tsai, T. M., Scheker, L. R., Tien, H. Y., Moreno, R., Ozyurekoglu, T., Banegas, R., Murphy, E., Burns, C. B., Zaring, R., Cook, D. F., & Marvin, M. R. (2012). Graft vasculopathy in clinical hand transplantation. *American Journal of Transplantation, 12*, 1004–1016.
38. Kawai, T., Cosimi, A. B., Spitzer, T. R., Tolkoff-Rubin, N., Suthanthiran, M., Saidman, S. L., Shaffer, J., Preffer, F. I., Ding, R., Sharma, V., Fishman, J. A., Dey, B., Ko, D. S., Hertl, M., Goes, N. B., Wong, W., Williams, W. W., Jr., Colvin, R. B., Sykes, M., & Sachs, D. H. (2008). HLA-mismatched renal transplantation without maintenance immunosuppression. *The New England Journal of Medicine, 358*, 353–361.
39. Kawai, T., Sachs, D. H., Sprangers, B., Spitzer, T. R., Saidman, S. L., Zorn, E., Tolkoff-Rubin, N., Preffer, F., Crisalli, K., Gao, B., Wong, W., Morris, H., LoCascio, S. A., Sayre, P., Shonts, B., Williams, W. W., Smith, R. N., Colvin, R. B., Sykes, M., & Cosimi, A. B. (2014). Long-term results in recipients of combined HLA-mismatched kidney and bone marrow transplantation without maintenance immunosuppression. *American Journal of Transplantation, 14*, 1599–1611.
40. Koga, H., Yamataka, A., Wang, K., Kato, Y., Lane, G. J., Kobayashi, H., Sueyoshi, N., & Miyano, T. (2003). Experimental allogenic penile transplantation. *Journal of Pediatric Surgery, 38*, 1802–1805.
41. Lee, W. P., Yaremchuk, M. J., Pan, Y. C., Randolph, M. A., Tan, C. M., & Weiland, A. J. (1991). Relative antigenicity of components of a vascularized limb allograft. *Plastic and Reconstructive Surgery, 87*, 401–411.
42. Leto Barone, A. A., Sun, Z., Montgomery, R. A., Lee, W. P., & Brandacher, G. (2013). Impact of donor-specific antibodies in reconstructive transplantation. *Expert Review of Clinical Immunology, 9*, 835–844.
43. Leventhal, J., Abecassis, M., Miller, J., Gallon, L., Tollerud, D., Elliott, M. J., Bozulic, L. D., Houston, C., Sustento-Reodica, N., & Ildstad, S. T. (2013). Tolerance induction in HLA disparate living donor kidney transplantation by donor stem cell infusion: Durable chimerism predicts outcome. *Transplantation, 95*, 169–176.
44. Levitsky, J. (2011). Operational tolerance: Past lessons and future prospects. *Liver Transplantation, 17*, 222–232.

45. Lin, C. H., Wang, Y. L., Anggelia, M. R., Chuang, W. Y., Cheng, H. Y., Mao, Q., Zelken, J. A., Lin, C. H., Zheng, X. X., Lee, W. P. A., & Brandacher, G. (2016). Combined Anti-CD154/CTLA4Ig costimulation blockade-based therapy induces donor-specific tolerance to vascularized osteomyocutaneous allografts. *American Journal of Transplantation, 16*, 2030–2041.
46. Liu, G., Cheng, Y., Guo, S., Feng, Y., Li, Q., Jia, H., Wang, Y., Tong, L., & Tong, X. (2011). Transplantation of adipose-derived stem cells for peripheral nerve repair. *International Journal of Molecular Medicine, 28*, 565–572.
47. Mathes, D. W., Randolph, M. A., Solari, M. G., Nazzal, J. A., Nielsen, G. P., Arn, J. S., Sachs, D. H., & Lee, W. P. (2003). Split tolerance to a composite tissue allograft in a swine model. *Transplantation, 15*, 25–31.
48. Medawar, P. B. (1944). The behaviour and fate of skin autografts and skin homografts in rabbits: A report to the War Wounds Committee of the Medical Research Council. *Journal of Anatomy, 78*, 176–199.
49. Medawar, P. B. (1945). A second study of the behaviour and fate of skin homografts in rabbits: A report to the War Wounds Committee of the Medical Research Council. *Journal of Anatomy, 79*, 157–176.
50. Morelon, E., Kanitakis, J., & Petruzzo, P. (2012). Immunological issues in clinical composite tissue allotransplantation: Where do we stand today? *Transplantation, 93*, 855–859.
51. Morelon, E., Kanitakis, J., Petruzzo, P., Badet, L., & Thaunat, O. (2015). Immunological challenges in vascularised composite allotransplantation. *Current Transplantation Reports, 2*, 276–283.
52. Mundinger, G. S., Nam, A. J., Hui-Chou, H. G., Stanwix, M. G., Jones, L. S., Drachenberg, C. B., Kukuruga, D., Shipley, S. T., Dorafshar, A. H., Panda, A., Bartlett, S. T., Bart, R. N., & Rodriguez, E. D. (2011). Nonhuman primate model of fibula vascularized composite tissue allotransplantation demonstrates donor-recipient bony union. *Plastic and Reconstructive Surgery, 128*, 1193–1204.
53. Mundinger, G. S., Munivenkatappa, R., Drachenberg, C. B., Ha, J. S., Vaca, E. E., Shipley, S. T., Papadimitriou, J. C., Bartlett, S. T., Rodriguez, E. D., & Barth, R. N. (2013). Histopathology of chronic rejection in a nonhuman primate model of vascularized composite allotransplantation. *Transplantation, 95*, 1204–1210.
54. Organ Procurement and Transplantation Network. (2015). VCA from living donors. Published online on 2nd June 2015. Available at: https://optn.transplant.hrsa.gov/resources/by-organ/vascular-composite-allograft/vcas-from-living-donors/. Accessed on 31st Oct 2016.
55. Pascual, M., Theruvath, T., Kawai, T., Tolkoff-Rubin, N., & Cosimi, A. B. (2002). Strategies to improve long-term outcomes after renal transplantation. *The New England Journal of Medicine, 346*, 580–590.
56. Petruzzo, P., Lanzetta, M., Dubernard, J. M., Landin, L., Cavadas, P., Margreiter, R., Schneeberger, S., Breidenbach, W., Kaufman, C., Jablecki, J., Schuind, D., & Dumontier, C. (2010). The international registry on hand and composite tissue transplantation. *Transplantation, 90*, 1590–1594.
57. Petruzzo, P., & Dubernard, J. M. (2011). The international registry on hand and composite tissue allotransplantation. *Clinical Transplants*, 247–253.
58. Petruzzo, P., Kanitakis, J., Badet, L., Pialat, J. B., Boutroy, S., Charpulat, R., Mouly, J., Gazarian, A., Lanzetta, M., Brunet, M., Devauchelle, B., Testelin, S., Martin, X., Dubernard, J. M., & Morelon, E. (2011). Long-term follow-up in composite tissue allotransplantation: In-depth study of five (hand and face) recipients. *American Journal of Transplantation, 11*, 808–816.
59. Petruzzo, P., Testelin, S., Kanitakis, J., Badet, L., Lengelé, B., Girbon, J. P., Parmentier, H., Malcus, C., Morelon, E., Devauchelle, B., & Dubernard, J. M. (2012). First human face transplantation: 5 years outcomes. *Transplantation, 93*, 236–240.
60. Petruzzo, P., Kanitakis, J., Testelin, S., Pialat, J. B., Buron, F., Badet, L., Thaunat, O., Devauchelle, B., & Morelon, E. (2015). Clinicopathological findings of chronic rejection in a face grafted patient. *Transplantation, 99*, 2644–2650.

61. Plock, J. A., Schnider, J. T., Solari, M. G., Zheng, X. X., & Gorantla, V. S. (2013). Perspectives on the use of mesenchymal stem cells in vascularized composite allotransplantation. *Frontiers in Immunology, 4*, 1–5.
62. Ravindra, K. V., & Ildstad, S. T. (2011). Immunosuppressive protocols and immunological challenges related to hand transplantation. *Hand Clinics, 27*, 467–479.
63. Rudensky, A. Y. (2011). Regulatory T cells and Foxp3. *Immunological Reviews, 241*, 260–268.
64. Sachs, D. H., Kawai, T., & Sykes, M. (2014). Induction of tolerance through mixed chimerism. *Cold Spring Harbor Perspectives in Medicine, 4*, 1–19.
65. Safinia, N., Vaikunthanathan, T., Fraser, H., Thirkell, S., Lowe, K., Blackmore, L., Whitehouse, G., Martinez-Llordella, M., Jassem, W., Sanchez-Fueyo, A., Lechler, R. I., & Lombardi, G. (2016). Successful expansion of functional and stable regulatory T cells for immunotherapy in liver transplantation. *Oncotarget, 7*, 7563–7577.
66. Scandling, J. D., Busquet, S., Dejbakhsh-Jones, S., Benike, C., Sarwal, M., Millan, M. T., Shizuru, J. A., Lowsky, R., Engleman, E. G., & Strober, S. (2012). Tolerance and withdrawal of immunosuppressive drugs in patients given kidney and hematopoietic cell transplants. *American Journal of Transplantation, 12*, 1133–1145.
67. Scheib, J., & Höke, A. (2013). Advances in peripheral nerve regeneration. *Nature Reviews. Neurology, 9*, 668–676.
68. Schneeberger, S., Gorantla, V. S., Brandacher, G., Zeevi, A., Demetris, A. J., Lunz, J. G., Metes, D. M., Donnenberg, A. D., Shores, J. T., Dimartini, A. F., Kiss, J. E., Imbriglia, J. E., Azari, K., Goitz, R. J., Manders, E. K., Nguyen, V. T., Cooney, D. S., Wachtman, G. S., Keith, J. D., Fletcher, D. R., Macedo, C., Planinsic, R., Losee, J. E., Shapiro, R., Starzl, T. E., & Lee, A. W. P. (2013). Upper-extremity transplantation using a cell-based protocol to minimize immunosuppression. *Annals of Surgery, 257*, 345–351.
69. Schweizer, R., Gorantla, V. S., & Plock, J. A. (2015). Premise and promise of mesenchymal stem cell-based therapies in clinical vascularized composite allotransplantation. *Current Opinion in Organ Transplantation, 20*, 608–614.
70. Setoguchi, K., Kishimoto, H., Kobayashi, S., Shimmura, H., Ishida, H., Toki, D., Suzuki, T., Ohnuki, K., Tate, Y., Fujioka, S., Abe, R., & Tanabe, K. (2010). Potential role of host effector memory CD8+ T cells in marrow rejection after mixed chimerism induction in cynomolgus monkeys. *Transplant Immunology, 23*, 194–203.
71. Seyam, R. M., Kattan, S. A., Assad, L. W., El-Sayed, R. M., & Almohanna, F. H. (2013). Penile autotransplantation in rats: An animal model. *Urology Annals, 5*, 255–258.
72. Sopko, N. A., Matsui, H., Lough, D. M., Miller, D., Harris, K., Kates, M., Liu, X., Billups, K., Redett, R., Burnett, A. L., Brandacher, G., & Bivalacqua, T. J. (2017). Ex vivo model of human penile transplantation and rejection: Implications for erectile tissue physiology. *European Urology, 71*(4), 584–593. Available at: http://dx.doi.org/10.1016/j.eururo.2016.07.006. Aaccessed on 31st Oct 2016.
73. Tonsho, M., Lee, S., Aoyama, A., Boskovic, S., Nadazdin, O., Capetta, K., Smith, R. N., Colvin, R. B., Sachs, D. H., Cosimi, A. B., Kawai, T., Madsen, J. C., Benichou, G., & Allan, J. S. (2015). Tolerance of lung allografts achieved in nonhuman primates via mixed hematopoietic chimerism. *American Journal of Transplantation, 15*, 2231–2239.
74. Unadkat, J. V., Schneeberger, S., Horibe, E. H., Goldbach, C., Solari, M. G., Washington, K. M., Gorantla, V. S., Cooper, G. M., Thomson, A. W., & Lee, A. W. P. (2010). Composite tissue vasculopathy and degeneration following multiple episodes of acute rejection in reconstructive transplantation. *American Journal of Transplantation, 10*, 251–261.
75. Weissenbacher, A., Hautz, T., Zelger, B., Zelger, B. G., Mayr, V., Brandacher, G., Pratschke, J., & Schneeberger, S. (2013). Antibody-mediated rejection in hand transplantation. *Transplant International, 27*, 13–17.
76. Wekerle, T., Kurtz, J., Ito, H., Ronquillo, J. V., Dong, V., Zhao, G., Shaffer, J., Sayegh, M. H., & Sykes, M. (2000). Allogeneic bone marrow transplantation with co-stimulatory blockade induces macrochimerism and tolerance without cytoreductive host treatment. *Nature Medicine, 6*, 464–469.

77. Wiebe, C., & Nickerson, P. (2013). Posttransplant monitoring of de novo human leukocyte antigen donor-specific antibodies in kidney transplantation. *Current Opinion in Organ Transplantation, 18*, 470–477.
78. Wu, S., Xu, H., Ravindra, K., & Ildstad, S. T. (2009). Composite tissue allotransplantation: past, present and future – The history and expanding applications of CTA as a new frontier in transplantation. *Transplantation Proceedings, 41*, 463–465.
79. Yamada, Y., Benichou, G., Cosimi, A. B., & Kawai, T. (2012). Tolerance induction after organ transplantation, 'delayed tolerance', via the mixed chimerism approach. *Chimerism, 3*, 24–28.
80. Yamada, Y., Boskovic, S., Aoyama, A., Murakami, T., Putheti, P., Smith, R. N., Ochiai, T., Nadazdin, O., Koyama, I., Boenisch, O., Najafian, N., Bhasin, M. K., Colvin, R. B., Madsen, J. C., Strom, T. B., Sachs, D. H., Benichou, G., Cosimi, A. B., & Kawai, T. (2012). Overcoming memory T-cell responses for induction of delayed tolerance in nonhuman primates. *American Journal of Transplantation, 12*, 330–340.

Chapter 10
Ex Vivo Organ Repair (Drug and Gene Delivery)

Daisuke Nakajima, Marcelo Cypel, and Shaf Keshavjee

Abstract Lung transplantation is a life-saving intervention for patients with end-stage lung diseases. However, ultimately, the success of lung transplantation is dependent on the quality and function of the transplanted donor lungs, which are frequently subjected to multiple different injuries. Recent innovations, in particular the development of ex vivo lung perfusion (EVLP) in which donor lungs are preserved under normothermic conditions outside the body, have enabled clinicians to more accurately evaluate the donor lung function prior to transplantation and have significantly impacted donor lung assessment and utilization worldwide. The advancement of EVLP from organ assessment to organ repair will be the next important and challenging step not only to expand the donor lung pool but also to improve graft survival and long-term outcomes after transplantation. The application of enhanced EVLP techniques combined with targeted repairs and molecular therapeutic strategies, including gene and cell-based therapy, will result in improved rehabilitation of injured donor lungs and provide a framework for the application of a personalized medicine approach in lung transplantation.

10.1 Introduction

Lung transplantation is a life-saving therapy for patients with end-stage lung diseases. Organ donor rates still remain low, and subsequent shortfalls in organ donation have led to shortages in the number of transplants performed [42]. This issue is further exacerbated by the susceptibility of donor lungs to intensive care unit (ICU)-related complications and injuries acquired during brain death, which have resulted in a further obstacle for a clinical use [72]. Worldwide donor lung utilization rates are estimated at only 20% since many lungs are ultimately declined for transplantation due to multiple different injuries or uncertain lung function [42, 72]. A number

D. Nakajima • M. Cypel • S. Keshavjee (✉)
Toronto Lung Transplant Program, University Health Network, University of Toronto, 200 Elizabeth Street, 9N-946, Toronto, ON M5G 2CS, Canada
e-mail: daink@kuhp.kyoto-u.ac.jp; marcelo.cypel@uhn.ca; shaf.keshavjee@uhn.ca

S.N. Nadig, J.A. Wertheim (eds.), *Technological Advances in Organ Transplantation*, DOI 10.1007/978-3-319-62142-5_10

of events occurring in donor lungs prior to retrieval, such as brain death, trauma, mechanical ventilation, infection, and aspiration, could amplify the ischemia-induced lung injury following lung transplantation [72]. Severe ischemia-reperfusion injury is a major cause of primary graft dysfunction (PGD) which affects 15–20% of lung transplant patients and is associated with the early posttransplant morbidity and mortality and an increased risk of acute rejection and chronic lung allograft dysfunction [18].

Ex vivo lung perfusion (EVLP) has recently been advanced to be able to address these issues. EVLP supplies donor lungs with oxygen and nutrients under normothermic conditions (37 °C), which enables lungs to maintain active aerobic cellular metabolism ex vivo, keeping them functional for periods as long as at least 12 h [14, 15]. This achievement has set EVLP as an excellent platform to assess lung function, treat lung injury, and facilitate lung repair before transplantation [13, 80]. This chapter will focus on the recent enhanced EVLP techniques combined with targeted pharmacologic and molecular therapeutic strategies, including gene and cell-based therapy.

10.2 Development of EVLP for Donor Lung Assessment and Preservation

Since the original concept of ex vivo organ perfusion emerged as early as 1935 by Alexis Carrel and Charles Lindbergh [60], isolated lung perfusion systems have been adopted to study pulmonary physiology for most of the past century [27, 44, 46, 47, 93]. In 2000, Steen successfully transplanted lungs donated after cardiac death (DCD), following a short-period assessment of lung function ex vivo with a low-potassium extracellular solution including dextran 40 and human albumin (Steen solution, XVIVO Perfusion, Sweden) [94]. Dextran 40 as a colloid component can protect endothelium from interaction with leukocytes and prevent both the coagulation cascade and platelet aggregation [54, 104]. Albumin helps maintain optimal colloid osmotic pressure and inhibits pulmonary edema formation [7, 50]. In 2008, we developed a successful prolonged EVLP technique, using acellular perfusion and protective mechanical ventilation and perfusion strategy with a low tidal volume (7 mL/kg) and low perfusion flow rate (40% of estimated cardiac output) in a preclinical pig model, where 12 h of extended normothermic EVLP resulted in superior lung function during EVLP and after transplantation and preserved lung histology [15].

A critical contributing factor to keep the lungs functional on the EVLP circuit as long as possible is to maintain a positive physiological left atrial pressure during EVLP [15, 59]. Several lung transplant centers however have employed a simpler strategy of leaving the left atrium open with the left atrial pressure being 0 mmHg during EVLP [94, 98]. It has been demonstrated in an isolated rat lung perfusion model that such lack of pulmonary venous pressure, caused by a simple open atrium system, could lead to unstable microvascular circulation around the alveolar wall [84]. The Toronto EVLP technique uses a closed left atrial circuit, keeping the left

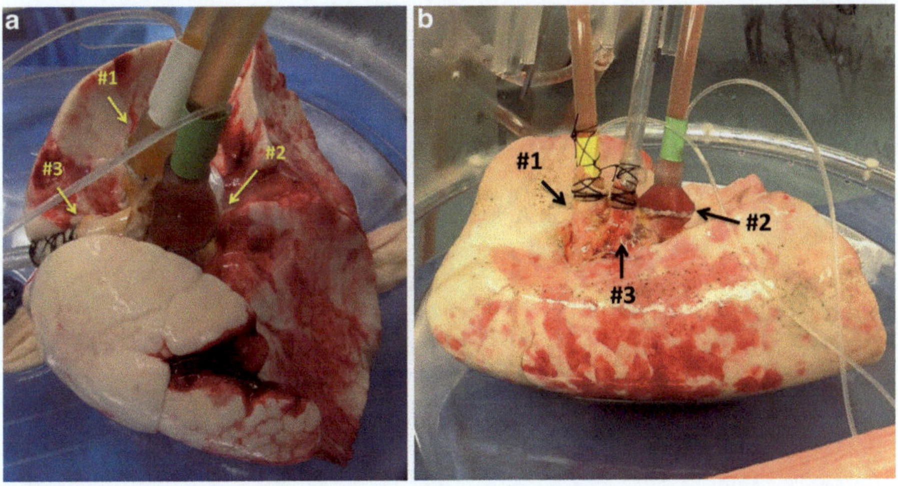

Fig. 10.1 A straight cannula is inserted into the pulmonary artery (PA) and secured with a tie suture, or a cone-shaped cannula is sutured to the PA cuff in cases when the main PA is cut very short (*Arrow #1*). A cone-shaped cannula is attached to the left atrial cuff with a running 4-0 Prolene suture. (*Arrow #2*). A standard endotracheal tube is inserted into the trachea and secured with a tie suture (*Arrow #3*). (**a**) Bilateral lung ex vivo lung perfusion (EVLP) of injured human lungs; (**b**) Single-lung EVLP (left in this case)

atrial pressure physiologically positive between 3 and 5 mmHg, which theoretically avoids the cyclical collapse of alveolar capillaries during the inspiratory and expiratory phases of respiration (Fig. 10.1).

We recently studied this in detail by comparing pig lung performance during 12 h of EVLP between a simple open left atrium (left atrial pressure 0 mmHg) and closed (left atrial pressure 3–5 mmHg). Although both strategies showed the similar lung function during the first 6 h of EVLP, the physiological parameters, including airway pressure, pulmonary compliance, and pulmonary vascular resistance, strikingly deteriorated due to severe pulmonary edema after 7 h in the lungs with open left atrium. This finding suggests that the open left atrium technique leads to injury that can only be tolerated by the short-term perfused donor lung. The closed atrium strategy and the closed EVLP system with physiological normal left atrial pressure can not only evaluate function of injured donor lungs for a longer time period but also provide the opportunity to treat and repair the lungs before transplantation [59].

10.3 Clinical Application of EVLP Protocols

There have currently been published three major EVLP techniques applied to various clinical trials: the Toronto (Fig. 10.2), Lund, and portable organ care system (OCS) (TransMedics, Andover, MA, USA) methods (Table 10.1).

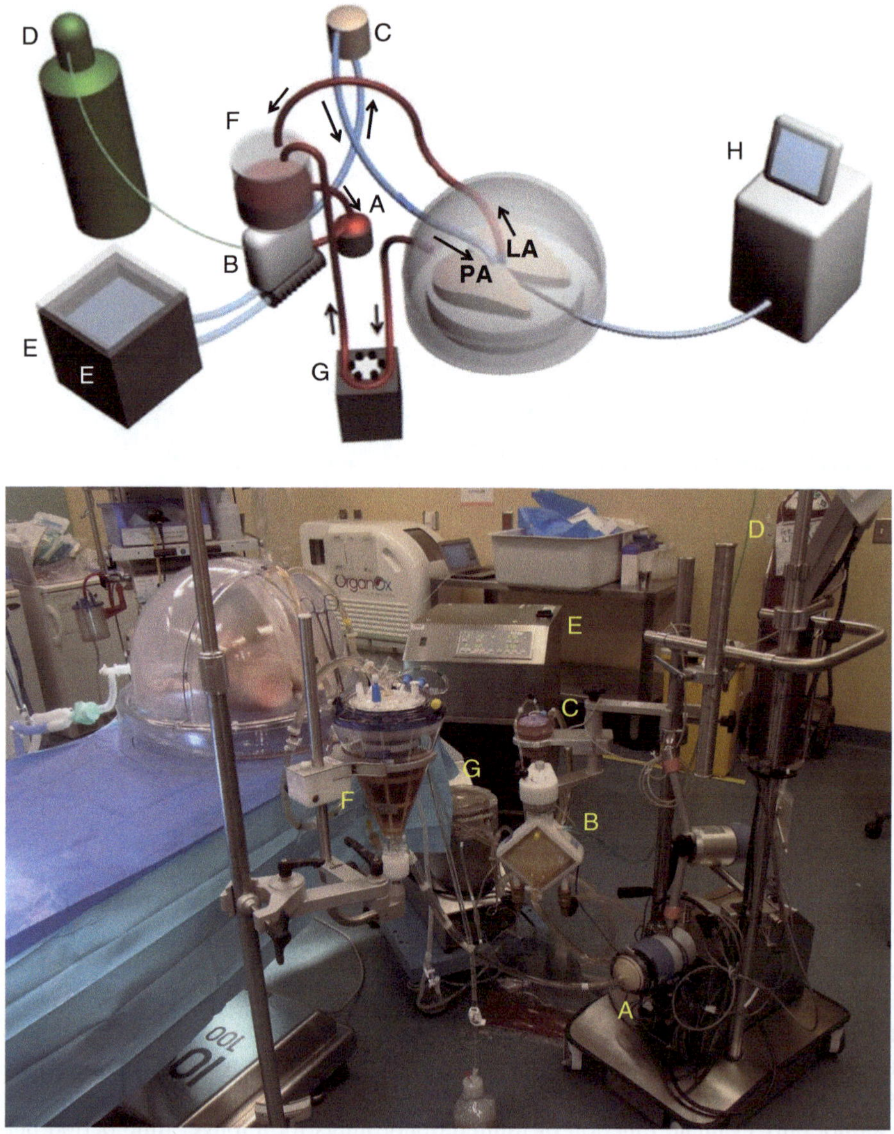

Fig. 10.2 The basic components of the Toronto ex vivo lung perfusion (EVLP) system. In this completely closed system, the perfusate is circulated by a lung protective centrifugal pump (*A*), passing through the gas exchange membrane (*B*) and leukocyte filter (*C*), and enters the lung through the pulmonary arterial (*PA*) cannula. The gas exchange membrane is connected to a gas tank which has a deoxygenating gas mixture with added CO_2 (*D*) and heat exchanger (*E*). The perfusate flows out of the lung through the left atrial (*LA*) cannula and returns to the hard shell reservoir (*F*). The minimal perfusate that might leak from the lung is salvaged from the chamber to the hard shell reservoir by a roller pump (*G*). The lung is ventilated with a state-of-the-art lung protective clinical ICU ventilator (*H*)

Table 10.1 Comparison between the clinically used EVLP protocols

Variable	Toronto	Lund	OCS
EVLP time	4–6 h; up to 12 h	2 h	Transport time
Perfusion			
Perfusate	Steen solution acellular	Steen solution + RBC (Hct 14%)	OCS solution + blood (Hct 15–25%)
Flow characteristic	Centrifugal pump	Roller pump	Roller pump
Target flow	40% CO	100% CO	2.0–2.5 L/min
PA pressure (mmHg)	≤15	≤20	≤20
LA pressure (mmHg)	Atrium closed, 3–5	Atrium open, 0	Atrium open, 0
Ventilation	ICU ventilator	ICU ventilator	Bellows pump
Temperature at start (°C)	32	32	34
Tidal volume (mL/kg)	7	5–7	6
Respiratory rate (bpm)	7	20	10
FiO_2	0.21	0.50	0.21
PEEP (cmH_2O)	5	5	5–7

OCS organ care system, *RBC* red blood cells, *Hct* hematocrit, *CO* cardiac output, *PA* pulmonary artery, *LA* left atrium, *bpm* breaths per minute, *PEEP* positive end-expiratory pressure

10.4 The Toronto EVLP Protocol

The first successful clinical trial in Toronto was reported in the *New England Journal of Medicine* in 2011, demonstrating that there was no significant difference in either the primary end point (primary graft dysfunction, PGD, grade 2 or 3 at 72 h) or the secondary end points (PGD grade 2 or 3 at ICU arrival, 24 and 48 h; extracorporeal membrane oxygen (ECMO) requirement; duration of mechanical ventilation, ICU stay, and hospital stay; 30-day mortality) between 20 EVLP-treated marginal donor lungs and 116 non-EVLP standard donor lungs [16]. In 2012, these results were updated with the first 50 consecutive transplants following EVLP (EVLP lung utilization rate for transplant of 86%) showing that EVLP-treated cases showed posttransplant outcomes equivalent to 367 conventional non-EVLP cases [17]. Notably, the incidence of PGD 3 at 72 h was lower in the EVLP-treated cases (2.0%) than the conventional non-EVLP cases (8.5%), indicating that reparative processes could take place during EVLP. A more recent report focused on the long-term outcomes of 63 EVLP-treated allografts, showing that the allograft survival was 79%, 71%, and 58% at 1, 3, and 5 years after transplantation, respectively, which was also similar to standard non-EVLP lungs [95].

In 2012, the Vienna group reported 13 clinical EVLP cases with initially unacceptable donor lungs for transplantation [1]. EVLP lung function deteriorated in four cases, which were subsequently declined for transplantation. The other nine transplanted cases (utilization rate: 69%) had no incidence of PGD 2 or 3 within 72 h after transplantation and showed early posttransplant outcomes (duration of mechanical ventilation, ICU and hospital stay, and 30-day mortality) comparable to 119 standard non-EVLP cases. The Harefield hospital also presented their experience in 13 EVLP cases, initially considered unsuitable for transplantation [105]. Six cases were eventually transplanted (utilization rate: 46%) and showed hospital length of stay and 3- and 6-month survival comparable to 86 standard non-EVLP cases. Of note, in this study, two cases were turned down for transplantation, because of technical issues, including too short left atrial cuff for transplant and misconnection of the EVLP circuit.

In 2014, the Torino group compared early posttransplant outcomes between the 8 EVLP-treated and 28 conventional non-EVLP cases [5]. EVLP was conducted for 11 marginal donor lungs, and 3 cases were rejected for transplantation due to infection ($n = 2$) or emphysema (utilization rate: 73%). The early posttransplant outcomes, including incidence of PGD 3, ECMO requirement, duration of mechanical ventilation and ICU stay, and 30-day mortality, did not differ between the groups. The Hospital Foch group in Paris, France, has also reported their experience with 32 EVLP-treated cases, initially rejected for transplant by all the French lung transplant centers [91]. Among 32 EVLP-treated cases, 31 lungs were used for bilateral lung transplant with their favorable lung function during 4 h of EVLP (utilization rate: 96%) and demonstrated similar incidence of PGD 3 at 72 h to 81 conventionally transplanted lungs without EVLP. Furthermore, there was no significant difference in duration of mechanical ventilation, ICU and hospital stay, and 30-day mortality between the groups.

10.5 The Lund EVLP Protocol

The Lund EVLP method was developed and used for the first time to evaluate ex vivo lung function of a Maastricht category II DCD donor in 2001 by Steen [94]. The same group adapted this technique to recondition the nine marginal donor lungs, primary deemed unsuitable for clinical use, and reported that six lungs improved their performance toward transplantable quality during short-period EVLP (median EVLP time 1 h 29 min), which contributed to the significant increase over the previous transplant activity in this group [41].

Using the same Lund method, the University of Gothenburg has compared early clinical outcome between 11 EVLP-treated cases for marginal donor lungs and 47 regularly transplanted non-EVLP (control) cases [97]. The incidence of PGD 2 or 3 at 72 h was 18% in EVLP-treated cases vs. 12% in control. Although duration of mechanical ventilation and ICU stay was longer in the EVLP-treated group than the non-EVLP control, length of hospital stay was similar between the groups.

10.6 The Portable Organ Care System (OCS) Protocol

The two lung transplant centers in Hanover and Madrid have reported their joint pilot study in a series of 12 patients, where they preserved the donor lungs using the portable device shortly after donor lung retrieval of standard donor lungs [98].

The early posttransplant outcomes (PGD grade, ECMO requirement, duration of mechanical ventilation, and length of ICU stay) were favorable. Although EVLP has generally been adopted to expand donor lung pool through the assessment and reconditioning of marginal or extended criteria donor lungs ex vivo, this group has focused on the feasibility and safety of the preservation method using OCS for donor lungs meeting the current standard criteria.

10.7 Mechanisms of Action in Lung Repair: Biomarker Discovery

Under our current clinical decision-making algorithm, based on the physiological lung function during 4–6 h of EVLP, approximately 20% of donor lungs have been turned down for transplantation [17]. Furthermore, donor lungs, showing favorable lung function during EVLP, still have an incidence of PGD after transplantation in a small percentage of cases. Recognizing that some components of PGD can be attributable to recipient factors, we believe that it should be possible to abrogate most, if not all, donor-related factors leading to posttransplant graft dysfunction.

EVLP can be utilized not only to facilitate the ex vivo *assessment* of physiological lung function, but if we understand the underlying mechanisms of injury, EVLP can be used as a platform for the ex vivo *treatment and repair* of the organ. Taken a step further, it can be used to *pre-prepare* the organ in anticipation of the transplant procedure. It is therefore important to elucidate the underlying mechanisms of actions of the associated injuries to develop specifically targeted therapies to treat the injured lungs. This is critical to fully realize the therapeutic potential of EVLP. Biomarkers in EVLP perfusate and donor blood or donor lung tissue could be used to fine-tune diagnostic accuracy by objectively revealing donor lungs with poor prognosis to avoid high-risk transplants or identify severely injured lungs that will require further advanced ex vivo therapy before transplantation.

Pro-inflammatory cytokines and chemokines play a crucial role in severe ischemia-reperfusion-induced lung injury which is involved in the pathogenesis of PGD [18]. Elevated expression of tumor necrosis factor (TNF)-α, interleukin (IL)-1β, IL-6, and IL-8 in donor lung tissues, collected at the end of cold preservation, has been shown to be associated with 30-day mortality after transplantation or non-utilization of donor lungs for transplantation [10, 16, 45]. We have recently reported the correlation between EVLP perfusate levels of cytokines, chemokines, and growth factors and graft performance either during perfusion or after transplantation in 50 clinical EVLP cases [63]. IL-8 and growth-regulated oncogene (GRO)-α levels at 4 h of EVLP were identified to predict PGD 3 within 72 h. Stem cell growth factor (SCGF)-β was significantly higher in the declined lungs than the non-PGD lungs at both 1 and 4 h, suggesting that SCGF-β could be used as a biomarker to select poor prognostic lungs which will require advanced therapeutic intervention early in the EVLP course to save the lungs.

Ischemia followed by reperfusion induces apoptosis rapidly, and extracellular release of adenosine triphosphate (ATP) from apoptotic cells acts as one of the "danger signals" or damage-associated molecular patterns (DAMPs) that augment the inflammatory response through toll-like receptor (TLR) signaling [24, 25, 96]. M30 is a major marker of epithelial cell apoptosis, which reflects soluble caspase-cleaved keratin 18 [51]. We have recently described a significant correlation between EVLP perfusate M30 levels and PGD 3 within 72 h, as well as long-term survival, in 79 clinical EVLP lung transplant cases [34]. We have previously demonstrated in animal studies that caspase inhibitors reduced the caspase activity that is increased during ischemia, which contributed to a significant decrease of apoptotic cell death and an improvement in lung function after transplantation [86]. Therefore, the detection of cell death signals during EVLP could be used to implement targeted anti-apoptotic therapy pretransplantation during EVLP to improve posttransplant outcomes.

Endothelin-1 (ET-1) is a potent vasoconstrictor polypeptide, well known for its pathogenetic involvement in pulmonary hypertension and acute lung injury [9, 30, 53, 74]. In clinical lung transplantation, increased ET-1 expression in donor lung tissues just before reperfusion is strongly correlated with severe PGD after transplantation [92]. We have demonstrated the association of ET-1 and big ET-1, a precursor of ET-1, circulating in the EVLP perfusate with lung performance during EVLP and posttransplant outcomes in 48 clinical EVLP cases [64]. ET-1 and big ET-1 levels were significantly increased in the declined lungs both at 1 and 4 h of EVLP, compared to the lungs with good posttransplant outcome. This finding also indicates that these mediators could be useful biomarkers to point out, early in the EVLP process, the severely injured donor lungs which would either be rejected for transplant or subjected to an advanced targeted therapeutic intervention.

We have used microarrays to determine differential mRNA expression in human donor lung tissues taken at the end of cold preservation. A number of different genes were identified in the lungs failing to complete 12 h of EVLP that differentiated them from the lungs which stably completed 12 h of EVLP and the lungs which showed good posttransplant outcome after 4 h of clinical EVLP [103]. The validation of these mRNA biomarkers will be required to clarify the role of the genes in lung performance to direct the development of future therapeutic targets during EVLP.

10.8 Ex Vivo Personalized Treatment and Repair

The advancement of EVLP from organ assessment to organ repair is the next step toward realizing what we have termed a "personalized medicine approach to management of the donor organ." There are several benefits of treating injured donor lungs ex vivo: (1) EVLP allows us to perform invasive and time-consuming procedures, which might be difficult in vivo in multi-organ donors with severe deterioration in lung function, (2) EVLP allows us to specifically treat and repair lungs in isolation without collateral toxicity of the treatment to other organs, and (3) EVLP allows us to confirm the effect of treatment and ultimately to transplant an organ with a known and predictable good function.

The donor lung is subject to multiple different injuries, caused by brain death, trauma, mechanical ventilation, infection, aspiration, etc. Thus, each lung needs to be diagnosed and subsequently treated individually in a personalized approach to significantly increase the donor lung pool. Preclinical studies have started to target and treat different forms of donor lung injury with favorable results. This strategy will ultimately yield an arsenal of ex vivo therapy strategies tailored to each uniquely injured donor lung. This process will lead to clinicians being able to apply a personalized medicine approach, just as we do with our patients, to the management of each injured donor lung during EVLP.

10.9 Infection

Lungs from multi-organ donors managed in the ICU are exposed to the risk of ventilator-associated pneumonia (VAP) [8, 85]. Several clinical studies have reported that 46 to 89% of donor lungs had positive bacterial cultures in bronchoalveolar lavage (BAL) and that frequent bacterial transmission from donor to recipient may put the recipients at higher risk of subsequent lung infection and poor outcomes after transplantation [3, 6, 90]. In fact, concern about infection is one of the main reasons for which donor lungs are turned down for transplantation when evaluated. Prolonged EVLP is potentially an ideal platform for the administration of high-dose, broad-spectrum antibiotics to treat infected donor lungs. Broad-spectrum antibiotic therapy during 3–6 h of EVLP was associated with a decrease in the microbial load in human marginal donor lungs intended for clinical use [2].

We have recently demonstrated the therapeutic benefit of broad-spectrum antibiotics during 12 h of extended EVLP in human donor lungs rejected for transplant due to infectious concerns [80]. Three types of high-dose antibiotics were added to the 1500 mL of perfusate: ciprofloxacin 400 mg or azithromycin 500 mg, vancomycin 15 mg/kg, and meropenem 2 g. The high-dose antibiotic therapy during 12 h EVLP significantly reduced the BAL bacterial count (Fig. 10.3a), suppressed endotoxin release, and improved lung function. Further investigation is required to fine-tune the assessment and treatment of infected donor lungs for safe future clinical translation.

10.10 Gastric Acid Aspiration

Gastric acid aspiration is a common event in organ donors who suffer from a neurologic insult and DCD donors after withdrawal of life support [72]. Aspiration of gastric content results in an intense chemical pneumonitis, which can subsequently lead to surfactant dysfunction with impairment of endogenous surfactant [66, 87]. Thus, gastric acid aspiration or even suspicion that it has occurred is a very common reason to decline donor lungs for transplantation. The presence of surfactant dysfunction in

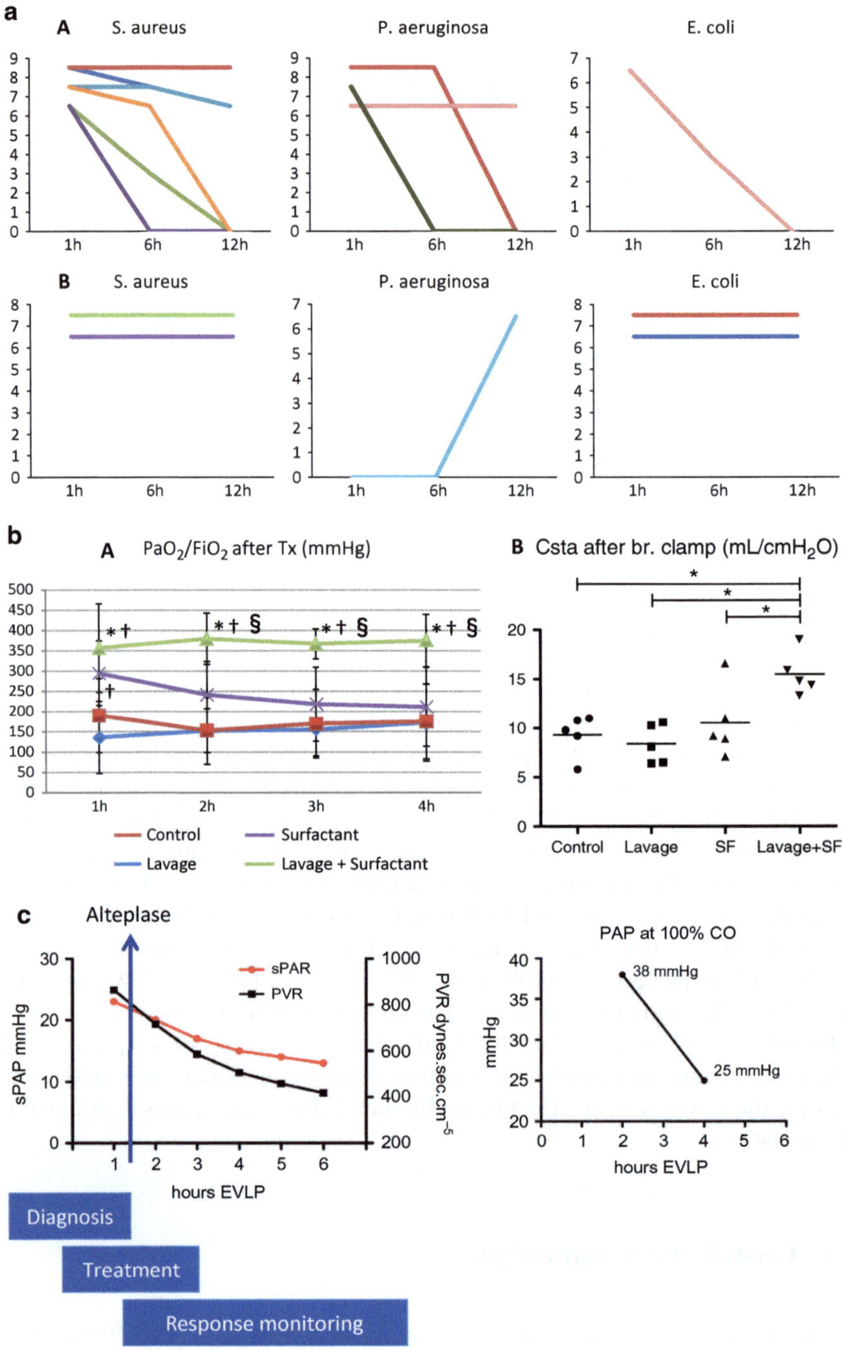

Fig. 10.3 Examples of effective donor lung treatment and repair using ex vivo lung perfusion (EVLP) as a platform for advanced targeted therapies. (**a**) Antibiotic therapy – decreasing bronchoalveolar lavage (BAL) bacterial counts (log10[cfu/L]) over time. *A*, antibiotic group; *B*, control group without antibiotics [80]. (**b**) Treatment of gastric acid aspiration with lavage + surfactant, showing superior posttransplant pulmonary oxygenation (*A*) and static pulmonary compliance (Cstat) (*B*). *A*: $^{*}P < 0.05$ vs. control group; $^{\dagger}P < 0.05$ vs. lavage group; $^{\S}P < 0.05$ vs. surfactant group. *B*: $^{*}P < 0.05$ by post hoc test between the groups. *SF* surfactant [81]. (**c**) Fibrinolytic treatment of pulmonary

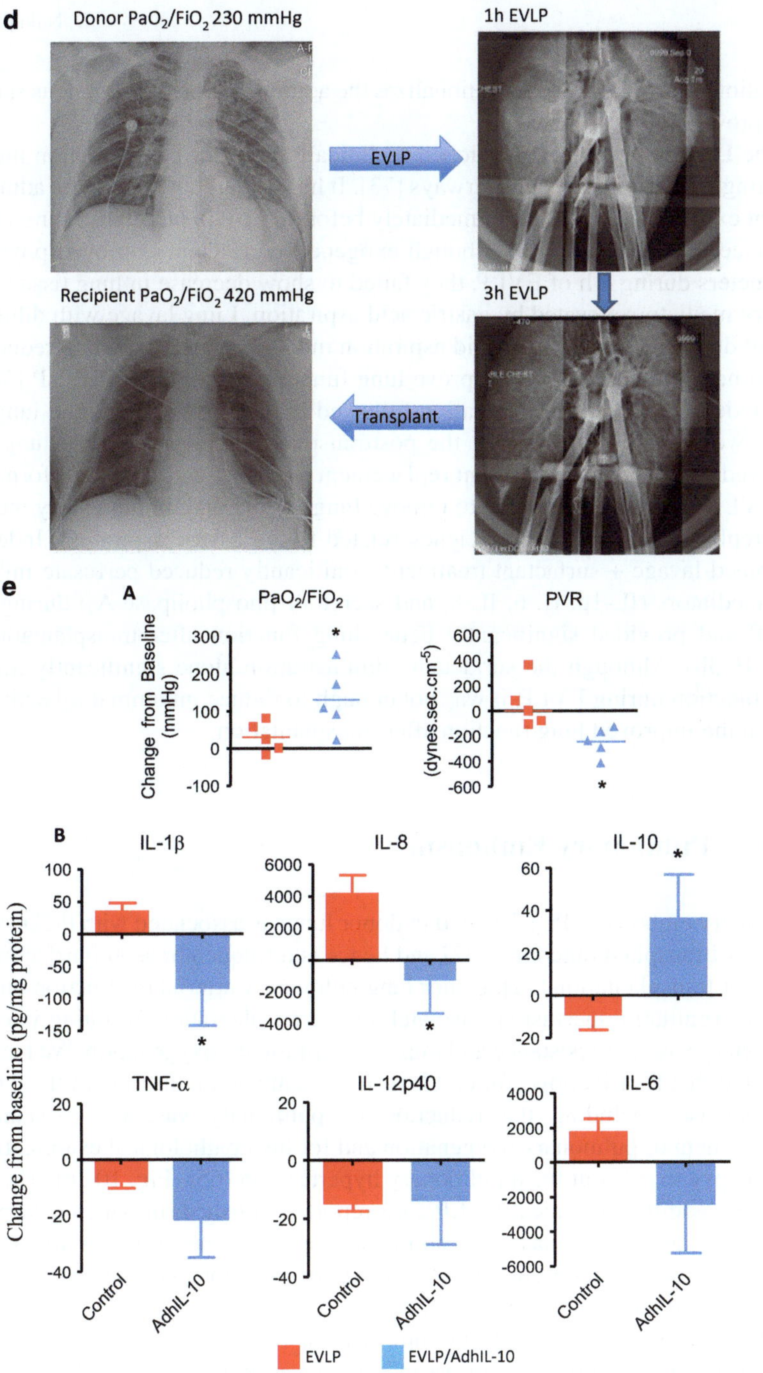

Fig. 10.3 (continued) embolism. Decreasing pulmonary vascular resistance (PVR) over time on EVLP. *PAP* pulmonary arterial pressure, *CO* cardiac output [65]. (Reprinted with permission of the American Thoracic Society. Copyright © 2018 American Thoracic Society. The American Journal of Respiratory and Critical Care Medicine is an official journal of the American Thoracic Society). (**d**) Treatment of neurogenic pulmonary edema in the donor lung with EVLP, reducing congestion on chest X-ray [11]. (**e**) Human lung IL-10 gene therapy, demonstrating improved pulmonary oxygenation and pulmonary vascular resistance (PVR) (*A*) and inflammatory cytokine profile (*B*). $*P < 0.05$ [13]

aspiration injured donor lungs rationalizes the administration of exogenous surfactant to improve their performance.

The Leuven group has developed a clinically relevant pig aspiration model by injecting gastric juice into the airways [73]. It has been reported that the administration of exogenous surfactant immediately before EVLP in pig lungs injured by gastric juice is beneficial [48]. Although exogenous surfactant improved physiologic parameters during 4 h of EVLP, they failed to show decrease in lung tissue inflammatory mediators elevated by gastric acid aspiration. Lung lavage with diluted surfactant during EVLP in a pig acid aspiration model has been shown to reduce BAL inflammatory mediators and improve lung function during 2 h of EVLP [38]. In a pig model of gastric juice aspiration followed by EVLP treatment and lung transplant, we have recently proven the posttransplant efficacy of whole lung lavage followed by exogenous surfactant replacement (lavage + surfactant) performed during EVLP [81]. The idea being to remove lung soiling and inflammatory mediators then replace the surfactant deficiency related to the gastric aspiration. Indeed, the combined lavage + surfactant treatment significantly reduced perfusate inflammatory mediators (IL-1β, IL-6, IL-8, and secretory phospholipase A_2) during 6 h of EVLP and provided significantly better lung function after transplantation [81] (Fig. 10.3b). Although the surfactant administration alone significantly improved lung function during EVLP, it was not enough to reduce inflammatory activity and sustain the improved lung function after transplantation.

10.11 Pulmonary Embolism

Pulmonary embolism (PE) detected in donor lungs is associated with PGD and poor early posttransplant outcomes [82] and hence is a frequent reason for declining the lungs for transplantation [72]. Donor lung pulmonary arterial obstruction due to PE leads to ventilation-perfusion mismatch after transplantation, resulting in elevated pulmonary vascular resistance and impaired pulmonary oxygenation. We have demonstrated that EVLP allows for clot lysis treatment and monitoring of the therapeutic response, including the reduction of pulmonary vascular resistance and improvement of pulmonary oxygenation and for histopathological examination of a lung biopsy to rule out fixed pulmonary hypertension [65] (Fig. 10.3c). There have been two additional clinical EVLP case reports published, in which a fibrinolytic agent (alteplase or urokinase) was administered into the perfusate for donor PE during EVLP, followed by successful lung transplantation without any fibrinolytic complications such as bleeding ([39, 62]).

DCD donors carry a risk of the intravascular formation of microthrombi during prolonged agonal hypotension or warm ischemia after cardiac arrest. Several studies have demonstrated the efficacy of fibrinolytic treatment during EVLP in DCD animal models [40, 78]. Although most clinical case reports and animal studies used a plasminogen activator and showed its therapeutic effects during EVLP and after transplant, some suggest the use of a direct fibrinolytic agent, such as plasmin, during the ex vivo acellular perfusion with less plasminogen in the perfusate [78].

Computed tomography (CT) scanning during EVLP has recently been demonstrated to add useful evaluation information of donor lungs. Furthermore, the recent advances in multidetector CT scanners provide simultaneous acquisition of images at two different kilovoltages for dual-energy scanning. Dual-energy CT (DECT), thus, can provide both anatomic and functional information about the lungs in a variety of pulmonary diseases based on a single contrast-enhanced CT examination [61]. In fact, this quality has been shown to improve the assessment of acute PE [88]. Hence, DECT could be a useful tool to evaluate impaired regional lung function due to PE and the fibrinolytic therapeutic effects during ex vivo perfusion treatment of donor lungs with PE.

10.12 Pulmonary Edema

Pulmonary edema is a condition very frequently encountered in donor lungs secondary to acute lung injury or neurogenic pulmonary edema in brain death donors (Fig. 10.3d). β_2-Adrenergic receptors are present in the alveolar walls and play an important role on regulating the alveolar fluid clearance (AFC) [79]. β_2-Adrenergic agonists can enhance the cyclic adenosine monophosphate (cAMP)-dependent AFC through the upregulation of epithelial Na^+ channel and cystic fibrosis transmembrane conductance regulator (CFTR) and increased Na,K-ATPase activity [71].

A significant increase in the AFC with the β_2-adrenergic agonist (terbutaline) instilled into the airways during EVLP has been demonstrated in 70% of cases in a study of human lungs declined for lung transplantation [28]. The efficacy of β_2-adrenergic agonist (procaterol) inhalation during EVLP has also been demonstrated in an animal DCD model [49]. The inhaled β_2-adrenergic agonist significantly elevated lung tissue cAMP levels and CFTR gene expression, which contributed to a significant decrease in the wet to dry lung weight ratio (a pulmonary edema index) and improved lung function during 4 h of EVLP.

10.13 Donation After Cardiac Death (DCD)

DCD donors fall into two general categories, according to the Maastricht classification: controlled DCD (Maastricht category 3 and 4) and uncontrolled DCD (Maastricht category 1 and 2). Although the early and long-term posttransplant outcomes are generally equivalent between the controlled DCD donors and donation after brain death (DBD) donors [12], controlled DCD donors still have a potential risk of injuries related to prolonged agonal hypotension and hypoxia until cardiac arrest (shock lung) or gastric acid aspiration after the withdrawal of life support. Thus, EVLP could offer the opportunity to assess and treat the lungs from such controlled DCD donors to improve the safety and predictability of outcomes. Uncontrolled DCD donors are clearly considered to be extended donors in that they are definitely subjected to a longer warm ischemia after cardiac arrest, which could contribute to the high incidence of severe PGD after transplantation without the assessment and treatment by EVLP [32].

The lung has the advantage of being able to have therapeutic agents delivered not only through the vascular route but also through the airway. Several therapeutic gases, including nitric oxide (NO), carbon monoxide (CO), and hydrogen (H_2), have been administered directly into the airways for the ex vivo prevention of ischemia-reperfusion-induced lung injury in animal DCD models [22, 23, 33]. NO stimulates soluble guanylate cyclase and increases cyclic guanosine monophosphate (cGMP) in vascular smooth muscle, which leads to the relaxation of vascular smooth muscle [25]. NO also has significant anti-inflammatory properties which inhibit neutrophil and platelet activation [31, 52]. It has been shown in a rat EVLP-transplant model that posttransplant pulmonary vascular resistance and pulmonary oxygenation were improved by the 40 ppm NO inhalation during warm ischemia, during EVLP, and after transplant [23].

CO has anti-inflammatory and cytoprotective effects mediated through the hypoxia-inducible factor (HIF) stabilization and activation of HIF-dependent transcriptional response [25]. CO inhalation (500 ppm) has been shown to improve posttransplant pulmonary oxygenation and the decrease in the expression of pro-inflammatory cytokines in a rat lung EVLP-transplant model [22].

Hydrogen gas (H_2) can bond with hydroxyl radicals to produce water, acting as a potent free radical scavenger [25]. Oxidative stress is an important contributing mechanism of ischemia-reperfusion injury [18], and the antioxidant effects of H_2 have been elucidated in the ischemia-reperfusion injury in many organs [37]. It has been reported in a pig DCD model that the ventilation with 2% H_2 significantly improved lung function and reduced inflammatory mediators during 4 h of EVLP [33].

10.14 Gene Therapy

Donor organs cannot avoid undergoing the process of ischemia and reperfusion, which is associated with an early and rapid release of pro-inflammatory cytokines and chemokines, such as IL-6 and IL-8 [19, 68, 77]. The balance between pro-inflammatory and anti-inflammatory cytokines can predict the posttransplant outcomes [45]. Ischemia-reperfusion injury elicits a strong inflammatory innate immune response mediated through various pathways that manifest as acute posttransplant lung injury or PGD. The acquired immune (alloimmune) response is also activated and likely further fueled by innate immune-related inflammation to worsen chronic lung rejection leading to chronic lung allograft dysfunction (CLAD).

IL-10 is a key cytokine central to the regulation of the innate and acquired immune responses – both of which play a central role in the transplant process and transplant-related lung injury (Fig. 10.4). IL-10 acts as an immune suppressor through the downregulation and inactivation of inflammatory and antigen-presenting cells, and it also inhibits the secretion of pro-inflammatory cytokines [75]. We have previously demonstrated significant success and provided proof of concept that

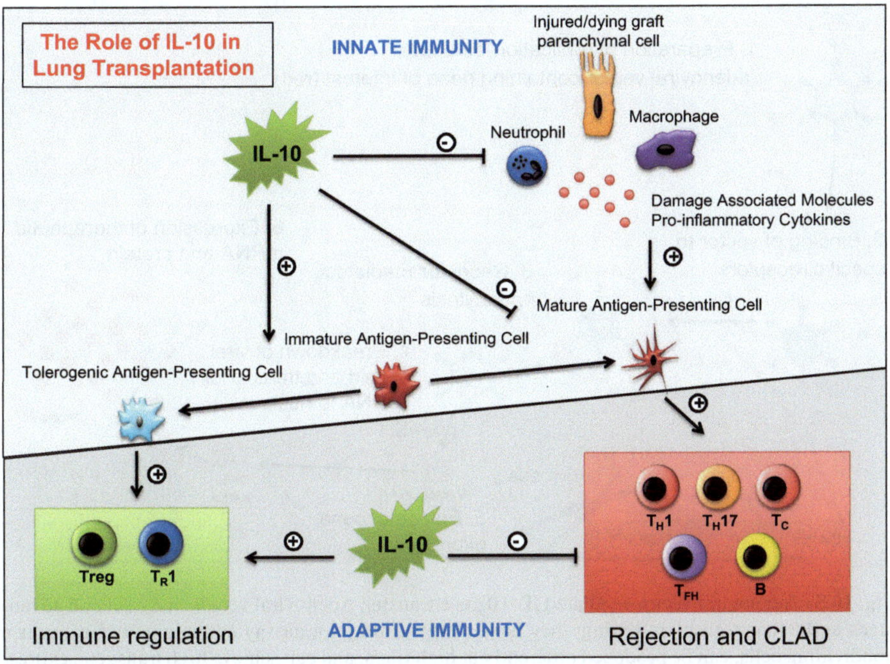

Fig. 10.4 IL-10 is a pleiotropic cytokine which shows anti-inflammatory and immunomodulatory effects in lung transplantation. T_H helper T cell, T_C cytotoxic T cell, *Treg* regulatory T cell, T_R1 Type 1 regulatory T cell

adenoviral-mediated IL-10 (AdIL-10) gene therapy to upregulate IL-10 expression in the donor lung prior to implantation can prevent the inflammatory response related to ischemia-reperfusion-induced lung injury in both small and large animal transplant models [26, 43, 67] (Fig. 10.5). However, gene delivery in vivo in a donor before organ retrieval is not always practical in the clinical setting, since we do not often have sufficient time for the transgene to be expressed in the donor lung in sufficient quantities prior to retrieval, implantation, and reperfusion. Hence, we developed our prolonged EVLP technique, which can maintain the lungs for over 12 h at normothermia without adding injury, to create the platform to effectively genetically modify the lung prior to transplantation.

We have reported the efficacy of AdIL-10 gene therapy intrabronchially delivered during 12 h of EVLP in human lungs rejected for transplantation and in a pig EVLP-transplant model [13]. Administered IL-10 gene was expressed in human lung tissue in a time-dependent manner and required at least 6–9 h to be fully expressed in human lungs. The AdIL-10 gene-treated human lungs showed significant improvement in lung function, including pulmonary oxygenation and pulmonary vascular resistance, a favorable shift from pro-inflammatory to anti-inflammatory cytokine expression, and recovery of alveolar-blood barrier integrity during EVLP (Fig. 10.3e). We also found similar functional and biochemical effects of IL-10 gene therapy after transplantation in a pig transplant model. Although vector-associated

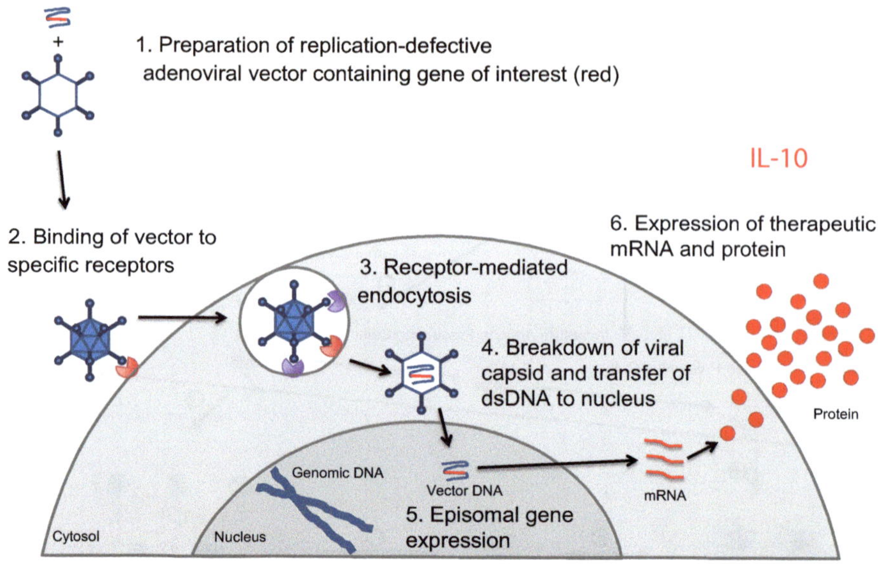

Fig. 10.5 Adenoviral vector-mediated IL-10 gene transfer. Adenoviral vectors have several advantages in the lung transplant setting: they have a high affinity for airway epithelium, can transduce nondividing cells, can be produced efficiently at high titers, and can achieve high transgene expression in a short time frame

inflammation can be an obstacle to the use of adenoviral vectors for gene therapy in general, we have confirmed that ex vivo AdIL-10 gene therapy is superior to in vivo therapy in terms of less vector-related inflammation and superior lung function after transplantation [101]. Further safety and immunologic benefits of ex vivo AdIL-10 gene therapy will be elucidated in a preclinical pig survival model toward its clinical application in the near future.

IL-10 also plays a key immunomodulatory role in the acquired immune response in transplantation (Fig. 10.4). T cells infiltrate the transplanted graft rapidly after reperfusion and contribute to various posttransplant injuries and graft dysfunction [20]. T-helper (Th) 1 cells secrete pro-inflammatory cytokines such as interferon (IFN)-γ and IL-12 and exacerbate the injury, whereas Th2 cells release anti-inflammatory cytokines such as IL-10, IL-4, and IL-13 and play a protective role in posttransplant injury. Hence, the administration of IL-10 during EVLP could induce a favorable shift toward Th2 cells during the early phase of reperfusion, which could eventually reduce the incidence or severity of ischemia-reperfusion injury, acute rejection, and chronic lung allograft dysfunction (CLAD) [4, 35, 36]. Furthermore, it has been reported that CD4+ T cells could induce IL-10 producing Type 1 T regulatory 1 (Tr1) cells in the presence of IL-10 and induce antigen-specific tolerance that is the ultimate goal of organ transplantation [58, 89]. These promising immunomodulatory effects of ex vivo IL-10 gene therapy need to be further elucidated in preclinical and clinical studies.

10.15 Mesenchymal Stem Cell (MSC)-Based Therapy

MSCs are multipotent adult cells, identified by criteria established by the International Society of Cellular Therapy, with the capacity to differentiate into mesenchymal lineages, including bone, cartilage, fat, and fibroblasts [21]. The constitutive low expression levels of major histocompatibility complex (MHC) Class I and II antigens and the lack of T cell co-stimulatory molecules, such as CD80 and CD86, empower allogeneic MSCs to be well tolerated by the host immune system, suggestive of their potential use for organ transplantation [83]. The major beneficial effects of MSCs derive from their capacity to migrate to the site of injury, interact with various host cells, and release a range of optimal paracrine soluble factors which could regulate alveolar fluid clearance (AFC), epithelial and endothelial permeability, and immune responses, suppress inflammation and cell apoptosis, and improve bacterial clearance [56, 69]. Therefore, there are several potential benefits of administration of MSCs during EVLP: (1) repair of donor-related injuries, such as inflammation, infection, and pulmonary edema before transplantation, and (2) immunomodulation of subsequent ischemia-reperfusion-induced lung injury, related to innate and adaptive immune responses after transplantation.

Matthay's group at the University of California has studied the potential therapeutic role of MSCs administered during EVLP, using human lungs declined for transplantation [29, 55, 57, 70]. Either intrabronchial or intravenous administration of MSCs during EVLP restored AFC, decreased pulmonary edema, suppressed inflammation, and reduced bacterial load through the secretion of keratinocyte growth factor (KGF) in an acute lung injury model induced by endotoxin or *Escherichia coli* bacteria [55, 57]. Furthermore, they have recently reported the similar therapeutic effects of 5×10^6 MSCs or microvesicles derived from MSCs added to the perfusate in declined donor lungs subjected to prolonged cold preservation [29, 70]. Although AFC and pulmonary edema were significantly improved by the MSC administration during 4–6 h of EVLP in their studies, there was no effect of MSCs on physiological lung function during EVLP. They also rightly suggested that the effects of MSCs should be evaluated using a clinical-grade EVLP system with appropriate perfusate and perfusion-ventilation strategies.

We have recently investigated the optimal route and dose of MSCs to treat pig lungs after 18 h of cold ischemic storage during 12 h of EVLP [76]. The administration of MSCs into the perfusate through the pulmonary artery was significantly associated with significantly superior retention of MSCs in the pulmonary parenchyma, compared to the intrabronchial administration. The administered dose of 150×10^6 MSCs (5×10^6 cells/kg) was found to be the optimal dose and was associated with a significant increase in vascular endothelial growth factor (VEGF) levels in lung tissues and a decrease in EVLP perfusate IL-8 levels [76]. Further transplant studies will be required to determine the anti-inflammatory and immunomodulatory effects of MSCs delivered during EVLP.

10.16 The Organ Repair Center Concept

The current advancement in normothermic perfusion technology will ultimately be incorporated into a national and international organ transplantation strategy [99]. One strategy is to develop a small number of specialized "organ assessment and regeneration centers," where there is a concentration of personnel with expertise in using normothermic ex vivo organ perfusion to deliver advanced repair therapies. In this hub-and-spoke model, injured or marginal donor organs would be transported to an "organ assessment and regeneration center" for organ rehabilitation. When the assessed and repaired organs are deemed acceptable for transplant, they could then be allocated to the most suitable recipients in other transplant centers. Currently prolonged organ preservation, using the conventional static cold storage combined with the normothermic ex vivo perfusion, could allow for this broad and individualized matching. In fact, in the first clinical case of its kind, we described the first remote EVLP assessment and treatment of marginal donor lungs. Donor lungs retrieved in the USA were assessed and treated during 4 h of EVLP in Toronto and then transported to a transplant center in Chicago, followed by successful transplantation. The total ischemic time was 15 h in this case [100]. Furthermore, we have recently demonstrated in a pig EVLP-transplant model that an extended cold preservation time of 10 h following 6 h of EVLP did not affect posttransplant function, which indicates that the combination of static cold storage and normothermic ex vivo perfusion techniques could safely extend the total preservation time for donor lungs using cold when required for protection and normothermia when required for assessment and repair [102].

10.17 Conclusion

EVLP has been successfully translated into clinical practice with expansion of donor lung pool leading to favorable posttransplant outcomes in a growing number of transplant centers worldwide. More than evaluating the donor lung function, EVLP provides an excellent platform to treat lung injury and facilitate lung repair. EVLP can provide the opportunity to accurately define the specific diagnosis using biomarkers and deliver appropriately targeted treatment to each injured donor lung. It is anticipated that EVLP technology will be further developed to preserve lungs for extended periods, perhaps days, in organ repair centers, which would provide a larger time window to allow for more complex targeted therapies and repairs toward organ regeneration, providing a revolutionary transformation in organ transplantation and tissue engineering in the near future.

References

1. Aigner, C., Slama, A., Hötzenecker, K., Scheed, A., Urbanek, B., Schmid, W., Nierscher, F. J., Lang, G., & Klepetko, W. (2012). Clinical ex vivo lung perfusion – Pushing the limits. *American Journal of Transplantation, 12*, 1839–1847. doi:10.1111/j.1600-6143.2012.04027.x.

2. Andreasson, A., Karamanou, D. M., Perry, J. D., Perry, A., Özalp, F., Butt, T., Morley, K. E., Walden, H. R., Clark, S. C., Prabhu, M., Corris, P. A., Gould, K., Fisher, A. J., & Dark, J. H. (2014). The effect of ex vivo lung perfusion on microbial load in human donor lungs. *The Journal of Heart and Lung Transplantation, 33*, 910–916. doi:10.1016/j.healun.2013.12.023.
3. Avlonitis, V. S., Krause, A., Luzzi, L., Powell, H., Phillips, J. A., Corris, P. A., Gould, F. K., & Dark, J. H. (2003). Bacterial colonization of the donor lower airways is a predictor of poor outcome in lung transplantation. *European Journal of Cardio-Thoracic Surgery, 24*, 601–607.
4. Boehler, A., Chamberlain, D., Xing, Z., Slutsky, A. S., Jordana, M., Gauldie, J., Liu, M., & Keshavjee, S. (1998). Adenovirus-mediated interleukin-10 gene transfer inhibits post-transplant fibrous airway obliteration in an animal model of bronchiolitis obliterans. *Human Gene Therapy, 9*, 541–551. doi:10.1089/hum.1998.9.4-541.
5. Boffini, M., Ricci, D., Bonato, R., Fanelli, V., Attisani, M., Ribezzo, M., Solidoro, P., Del Sorbo, L., Ranieri, V. M., & Rinaldi, M. (2014). Incidence and severity of primary graft dysfunction after lung transplantation using rejected grafts reconditioned with ex vivo lung perfusion. *European Journal of Cardio-Thoracic Surgery, 46*, 789–793. doi:10.1093/ejcts/ezu239.
6. Bonde, P. N., Patel, N. D., Borja, M. C., Allan, S. H., Barreiro, C. J., Williams, J. A., Thakur, N. A., Orens, J. B., & Conte, J. V. (2006). Impact of donor lung organisms on post-lung transplant pneumonia. *The Journal of Heart and Lung Transplantation, 25*, 99–105. doi:10.1016/j.healun.2005.06.026.
7. Chang, R. S., Wright, K., & Effros, R. M. (1981). Role of albumin in prevention of edema in perfused rabbit lungs. *Journal of Applied Physiology, 50*, 1065–1070.
8. Chastre, J., & Fagon, J.-Y. (2002). Ventilator-associated pneumonia. *American Journal of Respiratory and Critical Care Medicine, 165*, 867–903. doi:10.1164/ajrccm.165.7.2105078.
9. Comellas, A. P., & Briva, A. (2009). Role of endothelin-1 in acute lung injury. *Translational Research, 153*, 263–271. doi:10.1016/j.trsl.2009.02.007.
10. Cypel, M., Kaneda, H., Yeung, J. C., Anraku, M., Yasufuku, K., De Perrot, M., Pierre, A., Waddell, T. K., Liu, M., & Keshavjee, S. (2011a). Increased levels of interleukin-1β and tumor necrosis factor-α in donor lungs rejected for transplantation. *The Journal of Heart and Lung Transplantation, 30*, 452–459. doi:10.1016/j.healun.2010.11.012.
11. Cypel, M., & Keshavjee, S. (2011). Extracorporeal lung perfusion. *Current Opinion in Organ Transplantation, 16*, 469–475. doi:10.1097/MOT.0b013e32834ab15a.
12. Cypel, M., Levvey, B., Van Raemdonck, D., Erasmus, M., Dark, J., Love, R., Mason, D., Glanville, A. R., Chambers, D., Edwards, L. B., Stehlik, J., Hertz, M., Whitson, B. A., Yusen, R. D., Puri, V., Hopkins, P., Snell, G., & Keshavjee, S. (2015). International Society for Heart and Lung Transplantation donation after circulatory death registry report. *The Journal of Heart and Lung Transplantation, 34*, 1278–1282. doi:10.1016/j.healun.2015.08.015.
13. Cypel, M., Liu, M., Rubacha, M., Yeung, J. C., Hirayama, S., Anraku, M., Sato, M., Medin, J., Davidson, B. L., de Perrot, M., Waddell, T. K., Slutsky, A. S., & Keshavjee, S. (2009a). Functional repair of human donor lungs by IL-10 gene therapy. *Science Translational Medicine, 1*, 4ra9. doi:10.1126/scitranslmed.3000266.
14. Cypel, M., Rubacha, M., Yeung, J., Hirayama, S., Torbicki, K., Madonik, M., Fischer, S., Hwang, D., Pierre, A., Waddell, T. K., De Perrot, M., Liu, M., & Keshavjee, S. (2009b). Normothermic ex vivo perfusion prevents lung injury compared to extended cold preservation for transplantation. *American Journal of Transplantation, 9*, 2262–2269. doi:10.1111/j.1600-6143.2009.02775.x.
15. Cypel, M., Yeung, J. C., Hirayama, S., Rubacha, M., Fischer, S., Anraku, M., Sato, M., Harwood, S., Pierre, A., Waddell, T. K., de Perrot, M., Liu, M., & Keshavjee, S. (2008). Technique for prolonged normothermic ex vivo lung perfusion. *The Journal of Heart and Lung Transplantation, 27*, 1319–1325. doi:10.1016/j.healun.2008.09.003.
16. Cypel, M., Yeung, J. C., Liu, M., Anraku, M., Chen, F., Karolak, W., Sato, M., Laratta, J., Azad, S., Madonik, M., Chow, C.-W., Chaparro, C., Hutcheon, M., Singer, L. G., Slutsky, A. S., Yasufuku, K., de Perrot, M., Pierre, A. F., Waddell, T. K., & Keshavjee, S. (2011b).

Normothermic ex vivo lung perfusion in clinical lung transplantation. *The New England Journal of Medicine, 364*, 1431–1440. doi:10.1056/NEJMoa1014597.
17. Cypel, M., Yeung, J. C., Machuca, T., Chen, M., Singer, L. G., Yasufuku, K., de Perrot, M., Pierre, A., Waddell, T. K., & Keshavjee, S. (2012). Experience with the first 50 ex vivo lung perfusions in clinical transplantation. *The Journal of Thoracic and Cardiovascular Surgery, 144*, 1200–1206. doi:10.1016/j.jtcvs.2012.08.009.
18. de Perrot, M., Liu, M., Waddell, T. K., & Keshavjee, S. (2003a). Ischemia-reperfusion-induced lung injury. *American Journal of Respiratory and Critical Care Medicine, 167*, 490–511. doi:10.1164/rccm.200207-670SO.
19. De Perrot, M., Sekine, Y., Fischer, S., Waddell, T. K., McRae, K., Liu, M., Wigle, D. A., & Keshavjee, S. (2002). Interleukin-8 release during early reperfusion predicts graft function in human lung transplantation. *American Journal of Respiratory and Critical Care Medicine, 165*, 211–215. doi:10.1164/ajrccm.165.2.2011151.
20. de Perrot, M., Young, K., Imai, Y., Liu, M., Waddell, T. K., Fischer, S., Zhang, L., & Keshavjee, S. (2003b). Recipient T cells mediate reperfusion injury after lung transplantation in the rat. *Journal of Immunology, 171*, 4995–5002.
21. Dominici, M., Le Blanc, K., Mueller, I., Slaper-Cortenbach, I., Marini, F., Krause, D., Deans, R., Keating, A., Prockop, D., & Horwitz, E. (2006). Minimal criteria for defining multipotent mesenchymal stromal cells. The International Society for Cellular Therapy position statement. *Cytotherapy, 8*, 315–317. doi:10.1080/14653240600855905.
22. Dong, B., Stewart, P. W., & Egan, T. M. (2013). Postmortem and ex vivo carbon monoxide ventilation reduces injury in rat lungs transplanted from non-heart-beating donors. *The Journal of Thoracic and Cardiovascular Surgery, 146*, 429–436.e1. doi:10.1016/j.jtcvs.2012.11.005.
23. Dong, B. M., Abano, J. B., & Egan, T. M. (2009). Nitric oxide ventilation of rat lungs from non-heart-beating donors improves posttransplant function. *American Journal of Transplantation, 9*, 2707–2715. doi:10.1111/j.1600-6143.2009.02840.x.
24. Elliott, M. R., Chekeni, F. B., Trampont, P. C., Lazarowski, E. R., Kadl, A., Walk, S. F., Park, D., Woodson, R. I., Ostankovich, M., Sharma, P., Lysiak, J. J., Harden, T. K., Leitinger, N., & Ravichandran, K. S. (2009). Nucleotides released by apoptotic cells act as a find-me signal to promote phagocytic clearance. *Nature, 461*, 282–286. doi:10.1038/nature08296.
25. Eltzschig, H. K., & Eckle, T. (2011). Ischemia and reperfusion – From mechanism to translation. *Nature Medicine, 17*, 1391–1401. doi:10.1038/nm.2507.
26. Fischer, S., Liu, M., MacLean, A. A., de Perrot, M., Ho, M., Cardella, J. A., Zhang, X. M., Bai, X. H., Suga, M., Imai, Y., & Keshavjee, S. (2001). In vivo transtracheal adenovirus-mediated transfer of human interleukin-10 gene to donor lungs ameliorates ischemia-reperfusion injury and improves early posttransplant graft function in the rat. *Human Gene Therapy, 12*, 1513–1526. doi:10.1089/10430340152480249.
27. Fisk, R. L., Symes, J. F., Aldridge, L. L., & Couves, C. M. (1970). The pathophysiology and experimental therapy of acid pneumonitis in ex vivo lungs. *Chest, 57*, 364–370.
28. Frank, J. A., Briot, R., Lee, J. W., Ishizaka, A., Uchida, T., & Matthay, M. A. (2007). Physiological and biochemical markers of alveolar epithelial barrier dysfunction in perfused human lungs. *American Journal of Physiology. Lung Cellular and Molecular Physiology, 293*, L52–L59. doi:10.1152/ajplung.00256.2006.
29. Gennai, S., Monsel, A., Hao, Q., Park, J., Matthay, M. A., & Lee, J. W. (2015). Microvesicles derived from human mesenchymal stem cells restore alveolar fluid clearance in human lungs rejected for transplantation. *American Journal of Transplantation, 15*, 2404–2412. doi:10.1111/ajt.13271.
30. Giaid, A., Yanagisawa, M., Langleben, D., Michel, R. P., Levy, R., Shennib, H., Kimura, S., Masaki, T., Duguid, W. P., & Stewart, D. J. (1993). Expression of endothelin-1 in the lungs of patients with pulmonary hypertension. *The New England Journal of Medicine, 328*, 1732–1739. doi:10.1056/NEJM199306173282402.

31. Gkaliagkousi, E., Ritter, J., & Ferro, A. (2007). Platelet-derived nitric oxide signaling and regulation. *Circulation Research, 101*, 654–662. doi:10.1161/CIRCRESAHA.107.158410.
32. Gomez-de-Antonio, D., Campo-Cañaveral, J. L., Crowley, S., Valdivia, D., Cordoba, M., Moradiellos, J., Naranjo, J. M., Ussetti, P., & Varela, A. (2012). Clinical lung transplantation from uncontrolled non-heart-beating donors revisited. *The Journal of Heart and Lung Transplantation, 31*, 349–353. doi:10.1016/j.healun.2011.12.007.
33. Haam, S., Lee, S., Paik, H. C., Park, M. S., Song, J. H., Lim, B. J., & Nakao, A. (2015). The effects of hydrogen gas inhalation during ex vivo lung perfusion on donor lungs obtained after cardiac death. *European Journal of Cardio-Thoracic Surgery, 48*, 542–547. doi:10.1093/ejcts/ezv057.
34. Hashimoto, K., Besla, R., Zamel, R., Kim, H., Saito, T., Azad, S., Waddell, T. K., Cypel, M., Liu, M., & Keshavjee, S. (2015). The impact of cell death signals on short and long term outcome in human lung transplantation. *The Journal of Heart and Lung Transplantation, 34*, 134.
35. Hirayama, S., Sato, M., Liu, M., Loisel-Meyer, S., Yeung, J. C., Wagnetz, D., Cypel, M., Zehong, G., Medin, J. A., & Keshavjee, S. (2011). Local long-term expression of lentivirally delivered IL-10 in the lung attenuates obliteration of intrapulmonary allograft airways. *Human Gene Therapy, 22*, 1453–1460. doi:10.1089/hum.2010.225.
36. Hirayama, S., Sato, M., Loisel-Meyer, S., Matsuda, Y., Oishi, H., Guan, Z., Saito, T., Yeung, J., Cypel, M., Hwang, D. M., Medin, J. A., Liu, M., & Keshavjee, S. (2013). Lentivirus IL-10 gene therapy down-regulates IL-17 and attenuates mouse orthotopic lung allograft rejection. *American Journal of Transplantation, 13*, 1586–1593. doi:10.1111/ajt.12230.
37. Huang, C.-S., Kawamura, T., Toyoda, Y., & Nakao, A. (2010). Recent advances in hydrogen research as a therapeutic medical gas. *Free Radical Research, 44*, 971–982. doi:10.3109/10715762.2010.500328.
38. Inci, I., Ampollini, L., Arni, S., Jungraithmayr, W., Inci, D., Hillinger, S., Leskosek, B., Vogt, P., & Weder, W. (2008). Ex vivo reconditioning of marginal donor lungs injured by acid aspiration. *The Journal of Heart and Lung Transplantation, 27*, 1229–1236. doi:10.1016/j.healun.2008.07.027.
39. Inci, I., Yamada, Y., Hillinger, S., Jungraithmayr, W., Trinkwitz, M., & Weder, W. (2014). Successful lung transplantation after donor lung reconditioning with urokinase in ex vivo lung perfusion system. *The Annals of Thoracic Surgery, 98*, 1837–1838. doi:10.1016/j.athoracsur.2014.01.076.
40. Inci, I., Zhai, W., Arni, S., Inci, D., Hillinger, S., Lardinois, D., Vogt, P., & Weder, W. (2007). Fibrinolytic treatment improves the quality of lungs retrieved from non-heart-beating donors. *The Journal of Heart and Lung Transplantation, 26*, 1054–1060. doi:10.1016/j.healun.2007.07.033.
41. Ingemansson, R., Eyjolfsson, A., Mared, L., Pierre, L., Algotsson, L., Ekmehag, B., Gustafsson, R., Johnsson, P., Koul, B., Lindstedt, S., Lührs, C., Sjöberg, T., & Steen, S. (2009). Clinical transplantation of initially rejected donor lungs after reconditioning ex vivo. *The Annals of Thoracic Surgery, 87*, 255–260. doi:10.1016/j.athoracsur.2008.09.049.
42. Israni, A. K., Zaun, D. A., Rosendale, J. D., Snyder, J. J., & Kasiske, B. L. (2015). OPTN/SRTR 2013 Annual Data Report: Deceased organ donation. *American Journal of Transplantation, 15*(Suppl 2), 1–13. doi:10.1111/ajt.13202.
43. Itano, H., Zhang, W., Ritter, J. H., McCarthy, T. J., Mohanakumar, T., & Patterson, G. A. (2000). Adenovirus-mediated gene transfer of human interleukin 10 ameliorates reperfusion injury of rat lung isografts. *The Journal of Thoracic and Cardiovascular Surgery, 120*, 947–956. doi:10.1067/mtc.2000.109240.
44. Jirsch, D. W., Fisk, R. L., & Couves, C. M. (1970). Ex vivo evaluation of stored lungs. *The Annals of Thoracic Surgery, 10*, 163–168.
45. Kaneda, H., Waddell, T. K., De Perrot, M., Bai, X. H., Gutierrez, C., Arenovich, T., Chaparro, C., Liu, M., & Keshavjee, S. (2006). Pre-implantation multiple cytokine mRNA expression

analysis of donor lung grafts predicts survival after lung transplantation in humans. *American Journal of Transplantation, 6*, 544–551. doi:10.1111/j.1600-6143.2005.01204.x.

46. Kanekal, S., Plopper, C., Morin, D., & Buckpitt, A. (1990). Metabolic activation and bronchiolar Clara cell necrosis from naphthalene in the isolated perfused mouse lung. *The Journal of Pharmacology and Experimental Therapeutics, 252*, 428–437.
47. Kaplan, E., Diehl, J. T., Peterson, M. B., Somerville, K. H., Daly, B. D., Connolly, R. J., Cooper, A. G., Seiler, S. D., & Cleveland, R. J. (1990). Extended ex vivo preservation of the heart and lungs. Effects of acellular oxygen-carrying perfusates and indomethacin on the autoperfused working heart-lung preparation. *The Journal of Thoracic and Cardiovascular Surgery, 100*, 687–697. discussion 697–698.
48. Khalifé-Hocquemiller, T., Sage, E., Dorfmuller, P., Mussot, S., Le Houérou, D., Eddahibi, S., & Fadel, E. (2014). Exogenous surfactant attenuates lung injury from gastric-acid aspiration during ex vivo reconditioning in pigs. *Transplantation, 97*, 413–418. doi:10.1097/01.TP.0000441320.10787.c5.
49. Kondo, T., Chen, F., Ohsumi, A., Hijiya, K., Motoyama, H., Sowa, T., Ohata, K., Takahashi, M., Yamada, T., Sato, M., Aoyama, A., & Date, H. (2015). β2-Adrenoreceptor agonist inhalation during ex vivo lung perfusion attenuates lung injury. *The Annals of Thoracic Surgery, 100*, 480–486. doi:10.1016/j.athoracsur.2015.02.136.
50. Kraft, S. A., Fujishima, S., McGuire, G. P., Thompson, J. S., Raffin, T. A., & Pearl, R. G. (1995). Effect of blood and albumin on pulmonary hypertension and edema in perfused rabbit lungs. *Journal of Applied Physiology, 78*, 499–504.
51. Kramer, G., Erdal, H., Mertens, H. J. M. M., Nap, M., Mauermann, J., Steiner, G., Marberger, M., Bivén, K., Shoshan, M. C., & Linder, S. (2004). Differentiation between cell death modes using measurements of different soluble forms of extracellular cytokeratin 18. *Cancer Research, 64*, 1751–1756.
52. Kubes, P., Suzuki, M., & Granger, D. N. (1991). Nitric oxide: An endogenous modulator of leukocyte adhesion. *Proceedings of the National Academy of Sciences of the United States of America, 88*, 4651–4655.
53. Langleben, D., DeMarchie, M., Laporta, D., Spanier, A. H., Schlesinger, R. D., & Stewart, D. J. (1993). Endothelin-1 in acute lung injury and the adult respiratory distress syndrome. *The American Review of Respiratory Disease, 148*, 1646–1650. doi:10.1164/ajrccm/148.6_Pt_1.1646.
54. Laumonier, T., Walpen, A. J., Maurus, C. F., Mohacsi, P. J., Matozan, K. M., Korchagina, E. Y., Bovin, N. V., Vanhove, B., Seebach, J. D., & Rieben, R. (2003). Dextran sulfate acts as an endothelial cell protectant and inhibits human complement and natural killer cell-mediated cytotoxicity against porcine cells. *Transplantation, 76*, 838–843. doi:10.1097/01.TP.0000078898.28399.0A.
55. Lee, J. W., Fang, X., Gupta, N., Serikov, V., & Matthay, M. A. (2009). Allogeneic human mesenchymal stem cells for treatment of E. coli endotoxin-induced acute lung injury in the ex vivo perfused human lung. *Proceedings of the National Academy of Sciences of the United States of America, 106*, 16357–16362. doi:10.1073/pnas.0907996106.
56. Lee, J. W., Fang, X., Krasnodembskaya, A., Howard, J. P., & Matthay, M. A. (2011). Concise review: Mesenchymal stem cells for acute lung injury: Role of paracrine soluble factors. *Stem Cells, 29*, 913–919. doi:10.1002/stem.643.
57. Lee, J. W., Krasnodembskaya, A., McKenna, D. H., Song, Y., Abbott, J., & Matthay, M. A. (2013). Therapeutic effects of human mesenchymal stem cells in ex vivo human lungs injured with live bacteria. *American Journal of Respiratory and Critical Care Medicine, 187*, 751–760. doi:10.1164/rccm.201206-0990OC.
58. Levings, M. K., & Roncarolo, M. G. (2000). T-regulatory 1 cells: A novel subset of CD4 T cells with immunoregulatory properties. *The Journal of Allergy and Clinical Immunology, 106*, S109–S112.
59. Linacre, V., Cypel, M., Machuca, T. N., Nakajima, D., Hashimoto, K., Zamel, R., Chen, M., Iskender, I., Waddell, T. K., Liu, M., & Keshavjee, S. (2014). A positive left atrial pressure is important during ex vivo lung perfusion. *The Journal of Heart and Lung Transplantation, 33*, 27–28.

60. Lindbergh, C. A. (1935). An apparatus for the culture of whole organs. *The Journal of Experimental Medicine, 62*, 409–431. doi:10.1084/jem.62.3.409.
61. Lu, G. M., Zhao, Y., Zhang, L. J., & Schoepf, U. J. (2012). Dual-energy CT of the lung. *AJR. American Journal of Roentgenology, 199, S40–S53*.
62. Luc, J. G. Y., Bozso, S. J., Freed, D. H., & Nagendran, J. (2015). Successful repair of donation after circulatory death lungs with large pulmonary embolus using the lung organ care system for ex vivo thrombolysis and subsequent clinical transplantation. *Transplantation, 99*, e1–e2. doi:10.1097/TP.0000000000000485.
63. Machuca, T. N., Cypel, M., Yeung, J. C., Bonato, R., Zamel, R., Chen, M., Azad, S., Hsin, M. K., Saito, T., Guan, Z., Waddell, T. K., Liu, M., & Keshavjee, S. (2014). Protein expression profiling predicts graft performance in clinical ex vivo lung perfusion. *Annals of Surgery*. doi:10.1097/SLA.0000000000000974.
64. Machuca, T. N., Cypel, M., Zhao, Y., Grasemann, H., Tavasoli, F., Yeung, J. C., Bonato, R., Chen, M., Zamel, R., Chun, Y.-M., Guan, Z., de Perrot, M., Waddell, T. K., Liu, M., & Keshavjee, S. (2015). The role of the endothelin-1 pathway as a biomarker for donor lung assessment in clinical ex vivo lung perfusion. *The Journal of Heart and Lung Transplantation, 34*, 849–857. doi:10.1016/j.healun.2015.01.003.
65. Machuca, T. N., Hsin, M. K., Ott, H. C., Chen, M., Hwang, D. M., Cypel, M., Waddell, T. K., & Keshavjee, S. (2013). Injury-specific ex vivo treatment of the donor lung: Pulmonary thrombolysis followed by successful lung transplantation. *American Journal of Respiratory and Critical Care Medicine, 188*, 878–880. doi:10.1164/rccm.201302-0368LE.
66. Marik, P. E. (2001). Aspiration pneumonitis and aspiration pneumonia. *The New England Journal of Medicine, 344*, 665–671. doi:10.1056/NEJM200103013440908.
67. Martins, S., de Perrot, M., Imai, Y., Yamane, M., Quadri, S. M., Segall, L., Dutly, A., Sakiyama, S., Chaparro, A., Davidson, B. L., Waddell, T. K., Liu, M., & Keshavjee, S. (2004). Transbronchial administration of adenoviral-mediated interleukin-10 gene to the donor improves function in a pig lung transplant model. *Gene Therapy, 11*, 1786–1796. doi:10.1038/sj.gt.3302357.
68. Mathur, A., Baz, M., Staples, E. D., Bonnell, M., Speckman, J. M., Hess, P. J., Klodell, C. T., Knauf, D. G., Moldawer, L. L., & Beaver, T. M. (2006). Cytokine profile after lung transplantation: Correlation with allograft injury. *The Annals of Thoracic Surgery, 81*, 1844–1849.; discussion 1849–1850. doi:10.1016/j.athoracsur.2005.11.053.
69. Matthay, M. A., Thompson, B. T., Read, E. J., McKenna, D. H., Liu, K. D., Calfee, C. S., & Lee, J. W. (2010). Therapeutic potential of mesenchymal stem cells for severe acute lung injury. *Chest, 138*, 965–972. doi:10.1378/chest.10-0518.
70. McAuley, D. F., Curley, G. F., Hamid, U. I., Laffey, J. G., Abbott, J., McKenna, D. H., Fang, X., Matthay, M. A., & Lee, J. W. (2014). Clinical grade allogeneic human mesenchymal stem cells restore alveolar fluid clearance in human lungs rejected for transplantation. *American Journal of Physiology. Lung Cellular and Molecular Physiology, 306*, L809–L815. doi:10.1152/ajplung.00358.2013.
71. McAuley, D. F., Frank, J. A., Fang, X., & Matthay, M. A. (2004). Clinically relevant concentrations of beta2-adrenergic agonists stimulate maximal cyclic adenosine monophosphate-dependent airspace fluid clearance and decrease pulmonary edema in experimental acid-induced lung injury. *Critical Care Medicine, 32*, 1470–1476.
72. Meers, C., Van Raemdonck, D., Verleden, G. M., Coosemans, W., Decaluwe, H., De Leyn, P., Nafteux, P., & Lerut, T. (2010). The number of lung transplants can be safely doubled using extended criteria donors; a single-center review. *Transplant International, 23*, 628–635. doi:10.1111/j.1432-2277.2009.01033.x.
73. Meers, C. M., De Wever, W., Verbeken, E., Mertens, V., Wauters, S., De Vleeschauwer, S. I., Vos, R., Vanaudenaerde, B. M., Verleden, G. M., & Van Raemdonck, D. E. M. (2011). A porcine model of acute lung injury by instillation of gastric fluid. *The Journal of Surgical Research, 166*, e195–e204. doi:10.1016/j.jss.2010.10.015.
74. Mitaka, C., Hirata, Y., Nagura, T., Tsunoda, Y., & Amaha, K. (1993). Circulating endothelin-1 concentrations in acute respiratory failure. *Chest, 104*, 476–480.

75. Moore, K. W., de Waal, M. R., Coffman, R. L., & O'Garra, A. (2001). Interleukin-10 and the interleukin-10 receptor. *Annual Review of Immunology, 19*, 683–765. doi:10.1146/annurev.immunol.19.1.683.
76. Mordant, P., Nakajima, D., Kalaf, R., Iskender, I., Maahs, L., Behrens, P., Coutinho, R., Lyer, R., Davies, J. E., Cypel, M., Liu, M., Waddell, T. K., & Keshavjee, S. (2015). Ex vivo administration of human mesenchymal stem cells ameliorates injury in pig lungs through an increase in parenchymal VEGF and a decrease in circulating IL-8. *The Journal of Heart and Lung Transplantation, 34*, 267.
77. Moreno, I., Vicente, R., Ramos, F., Vicente, J. L., & Barberá, M. (2007). Determination of interleukin-6 in lung transplantation: Association with primary graft dysfunction. *Transplantation Proceedings, 39*, 2425–2426. doi:10.1016/j.transproceed.2007.07.056.
78. Motoyama, H., Chen, F., Hijiya, K., Kondo, T., Ohsumi, A., Yamada, T., Sato, M., Aoyama, A., Bando, T., & Date, H. (2014). Plasmin administration during ex vivo lung perfusion ameliorates lung ischemia-reperfusion injury. *The Journal of Heart and Lung Transplantation, 33*, 1093–1099. doi:10.1016/j.healun.2014.06.004.
79. Mutlu, G. M., & Factor, P. (2008). Alveolar epithelial beta2-adrenergic receptors. *American Journal of Respiratory Cell and Molecular Biology, 38*, 127–134. doi:10.1165/rcmb.2007-0198TR.
80. Nakajima, D., Cypel, M., Bonato, R., Machuca, T. N., Iskender, I., Hashimoto, K., Linacre, V., Chen, M., Coutinho, R., Azad, S., Martinu, T., Waddell, T. K., Hwang, D. M., Husain, S., Liu, M., & Keshavjee, S. (2016). Ex vivo perfusion treatment of infection in human donor lungs. *American Journal of Transplantation*. doi:10.1111/ajt.13562.
81. Nakajima, D., Ohsumi, A., Iskender, I., Kalaf, R., Chen, M., Coutinho, R., Kanou, T., Maahs, L., Behrens, P., Sakamoto, J., Hsin, M. K., Azad, S., Waddell, T. K., Cypel, M., Liu, M., & Keshavjee, S. (2015). Lung lavage and surfactant administration for the ex vivo pre-transplant treatment of donor lungs injured due to gastric acid aspiration. *The Journal of Heart and Lung Transplantation, 34*, 92.
82. Oto, T., Rabinov, M., Griffiths, A. P., Whitford, H., Levvey, B. J., Esmore, D. S., Williams, T. J., & Snell, G. I. (2005). Unexpected donor pulmonary embolism affects early outcomes after lung transplantation: A major mechanism of primary graft failure? *The Journal of Thoracic and Cardiovascular Surgery, 130*, 1446. doi:10.1016/j.jtcvs.2005.07.025.
83. Patel, S. A., Sherman, L., Munoz, J., & Rameshwar, P. (2008). Immunological properties of mesenchymal stem cells and clinical implications. *Archivum Immunologiae et Therapiae Experimentalis (Warsz), 56*, 1–8. doi:10.1007/s00005-008-0001-x.
84. Peták, F., Habre, W., Hantos, Z., Sly, P. D., & Morel, D. R. (2002). Effects of pulmonary vascular pressures and flow on airway and parenchymal mechanics in isolated rat lungs. *Journal of Applied Physiology, 92*, 169–178.
85. Porzecanski, I., & Bowton, D. L. (2006). Diagnosis and treatment of ventilator-associated pneumonia. *Chest, 130*, 597–604. doi:10.1378/chest.130.2.597.
86. Quadri, S. M., Segall, L., de Perrot, M., Han, B., Edwards, V., Jones, N., Waddell, T. K., Liu, M., & Keshavjee, S. (2005). Caspase inhibition improves ischemia-reperfusion injury after lung transplantation. *American Journal of Transplantation, 5*, 292–299. doi:10.1111/j.1600-6143.2004.00701.x.
87. Raghavendran, K., Nemzek, J., Napolitano, L. M., & Knight, P. R. (2011). Aspiration-induced lung injury. *Critical Care Medicine, 39*, 818–826. doi:10.1097/CCM.0b013e31820a856b.
88. Remy-Jardin, M., Faivre, J. B., Pontana, F., Molinari, F., Tacelli, N., & Remy, J. (2014). Thoracic applications of dual energy. *Seminars in Respiratory and Critical Care Medicine, 35*, 64–73.
89. Roncarolo, M. G., Bacchetta, R., Bordignon, C., Narula, S., & Levings, M. K. (2001). Type 1 T regulatory cells. *Immunological Reviews, 182*, 68–79.
90. Ruiz, I., Gavaldà, J., Monforte, V., Len, O., Román, A., Bravo, C., Ferrer, A., Tenorio, L., Román, F., Maestre, J., Molina, I., Morell, F., & Pahissa, A. (2006). Donor-to-host transmission of bacterial and fungal infections in lung transplantation. *American Journal of Transplantation, 6*, 178–182. doi:10.1111/j.1600-6143.2005.01145.x.

91. Sage, E., Mussot, S., Trebbia, G., Puyo, P., Stern, M., Dartevelle, P., Chapelier, A., & Fischler, M. (2014). Lung transplantation from initially rejected donors after ex vivo lung reconditioning: The French experience. *European Journal of Cardio-Thoracic Surgery, 46*, 794–799. doi:10.1093/ejcts/ezu245.
92. Salama, M., Andrukhova, O., Hoda, M. A., Taghavi, S., Jaksch, P., Heinze, G., Klepetko, W., & Aharinejad, S. (2010). Concomitant endothelin-1 overexpression in lung transplant donors and recipients predicts primary graft dysfunction. *American Journal of Transplantation, 10*, 628–636. doi:10.1111/j.1600-6143.2009.02957.x.
93. Seeger, W., Schneider, U., Kreusler, B., von Witzleben, E., Walmrath, D., Grimminger, F., & Neppert, J. (1990). Reproduction of transfusion-related acute lung injury in an ex vivo lung model. *Blood, 76*, 1438–1444.
94. Steen, S., Sjöberg, T., Pierre, L., Liao, Q., Eriksson, L., & Algotsson, L. (2001). Transplantation of lungs from a non-heart-beating donor. *Lancet, 357*, 825–829. doi:10.1016/S0140-6736(00)04195-7.
95. Tikkanen, J. M., Cypel, M., Machuca, T. N., Azad, S., Binnie, M., Chow, C.-W., Chaparro, C., Hutcheon, M., Yasufuku, K., de Perrot, M., Pierre, A. F., Waddell, T. K., Keshavjee, S., & Singer, L. G. (2015). Functional outcomes and quality of life after normothermic ex vivo lung perfusion lung transplantation. *The Journal of Heart and Lung Transplantation, 34*, 547–556. doi:10.1016/j.healun.2014.09.044.
96. Vénéreau, E., Ceriotti, C., & Bianchi, M. E. (2015). DAMPs from cell death to new life. *Frontiers in Immunology, 6*, 422. doi:10.3389/fimmu.2015.00422.
97. Wallinder, A., Ricksten, S.-E., Silverborn, M., Hansson, C., Riise, G. C., Liden, H., Jeppsson, A., & Dellgren, G. (2014). Early results in transplantation of initially rejected donor lungs after ex vivo lung perfusion: A case-control study. *European Journal of Cardio-Thoracic Surgery, 45*, 40–44.; discussion 44–45. doi:10.1093/ejcts/ezt250.
98. Warnecke, G., Moradiellos, J., Tudorache, I., Kühn, C., Avsar, M., Wiegmann, B., Sommer, W., Ius, F., Kunze, C., Gottlieb, J., Varela, A., & Haverich, A. (2012). Normothermic perfusion of donor lungs for preservation and assessment with the organ care system lung before bilateral transplantation: A pilot study of 12 patients. *Lancet (London, England), 380*, 1851–1858. doi:10.1016/S0140-6736(12)61344-0.
99. Whitson, B. A., & Black, S. M. (2014). Organ assessment and repair centers: The future of transplantation is near. *World Journal of Transplantation, 4*, 40–42. doi:10.5500/wjt.v4.i2.40.
100. Wigfield, C. H., Cypel, M., Yeung, J., Waddell, T., Alex, C., Johnson, C., Keshavjee, S., & Love, R. B. (2012). Successful emergent lung transplantation after remote ex vivo perfusion optimization and transportation of donor lungs. *American Journal of Transplantation, 12*, 2838–2844. doi:10.1111/j.1600-6143.2012.04175.x.
101. Yeung, J. C., Wagnetz, D., Cypel, M., Rubacha, M., Koike, T., Chun, Y.-M., Hu, J., Waddell, T. K., Hwang, D. M., Liu, M., & Keshavjee, S. (2012). Ex vivo adenoviral vector gene delivery results in decreased vector-associated inflammation pre- and post-lung transplantation in the pig. *Molecular Therapy, 20*, 1204–1211. doi:10.1038/mt.2012.57.
102. Yew Hsin, M. K., Iskander, I., Nakajima, D., Chen, M., Kim, H., Dos Santos, P. R., Sakamoto, J., Lee, J., Hashimoto, K., Harmantas, C., Hwang, D., Waddell, T., Liu, M., Keshavjee, S., & Cypel, M. (2015). Extension of donor lung preservation with hypothermic storage after normothermic ex vivo lung perfusion. *The Journal of Heart and Lung Transplantation*. doi:10.1016/j.healun.2015.05.017.
103. Zamel, R., Machuca, T. N., Yeung, J. C., Bonato, R., Bai, X.-H., Waddell, T. K., Liu, M., Cypel, M., & Keshavjee, S. (2013). Discovery of mRNA biomarkers predicting donor lung failure. *The Journal of Heart and Lung Transplantation, 32*, 45–46.
104. Zeerleder, S., Mauron, T., Lämmle, B., & Wuillemin, W. A. (2002). Effect of low-molecular weight dextran sulfate on coagulation and platelet function tests. *Thrombosis Research, 105*, 441–446.
105. Zych, B., Popov, A. F., Stavri, G., Bashford, A., Bahrami, T., Amrani, M., De Robertis, F., Carby, M., Marczin, N., Simon, A. R., & Redmond, K. C. (2012). Early outcomes of bilateral sequential single lung transplantation after ex-vivo lung evaluation and reconditioning. *The Journal of Heart and Lung Transplantation, 31*, 274–281. doi:10.1016/j.healun.2011.10.008.

Chapter 11
3D Bioprinting in Transplantation

Armando Salim Munoz-Abraham, Christopher Ibarra, Raghav Agarwal, John Geibel, and David C. Mulligan

Abstract The continued rise in patients suffering from organ failure has raised the need for additional sources for replacement organs to improve the quality of life for these patients. Organ transplantation is the standard of care for end-stage organ disease, and as such, there has been an ever-increasing need worldwide to find suitable grafts for those patients.

In the last decade, the field of biomedical engineering has made important technological advances in tissue bioengineering through the use of three-dimensional bioprinting. These innovative advances have contributed to developing biocompatible materials and supporting scaffolds that allow the production of functional tissues and printed organ models. 3D printing in medicine could eventually allow the application of printed tissues and organs to replace damaged or irreparable grafts from trauma or disease. Using these new and emerging additive-manufacturing technologies, it is hoped to be able to implant printed synthetics for end-stage organ disease (ESOD) and help with the shortage of viable organs for transplantation.

Multiple bioprinter configurations for tissue printing along with printing techniques have emerged to revolutionize the creation of 3D biostructures. Current advances of tissue bioengineering strive to allow for self-assembly of cells and tissues to become a reality, which would augment the possibility of generating new graft models. Around the world, scientists have developed vascular grafts, liver, kidney, and heart models that are in various stages of development and in some cases have been implanted in animal models. Many years of work are still to come in order for these basic models to be useful for human implantation.

A.S. Munoz-Abraham, MD, MBEE • J. Geibel, MD, DSc, MSc, AGAF
D.C. Mulligan, MD, FACS (✉)
Department of Surgery, Yale School of Medicine, 310 Cedar Street, New Haven, CT 06510, USA
e-mail: david.mulligan@yale.edu

C. Ibarra, MD
Section of Transplantation and Immunology, Department of Surgery, Yale School of Medicine, 333 Cedar Street, FMB 121, New Haven, CT 06510, USA

R. Agarwal
Undergraduate Student Volunteer – Class of 2018, Arizona State University, Tempe, AZ, USA

S.N. Nadig, J.A. Wertheim (eds.), *Technological Advances in Organ Transplantation*, DOI 10.1007/978-3-319-62142-5_11

Ultimately, the goal of developing bioprinted tissues and organs is to overcome the shortage of available grafts. Furthermore, these replacement tissues could be made of cells from the donor, thereby reducing the risk of rejection and the levels of immunosuppressive agents being used.

11.1 Introduction

For more than three decades, the field of organ transplantation has made great strides on improved organ graft outcomes, saving thousands of patients with end-stage organ disease (ESOD) [27]. Improved surgical techniques and new antirejection drugs have helped pave the way toward increased success in organ transplantation. Yet, even with these advancements, the waiting list for transplantation remains extremely long and outpaces the number of donations. Organs continue to be a scarce resource, and even with potential changes in organ allocation in the United States and in some areas in the world, the number of people in need far exceeds the availability of these limited assets.

However, recent innovations in technology and new developing fields in science can potentially bring a change to the unmet organ demand. The field of biomedical engineering is among the most important areas, already bringing new solutions to medicine and will potentially provide game-changing results to the transplantation world. Creation of functional tissues in the lab setting became a reality by the end of 1980s. With this basis, the concept of organogenesis was born with the idea of creating not only tissues but also whole organs with the use of cells and a support system to allow them to thrive. Since then, tissue graft assembly employing synthetic materials for the creation of synthetic scaffolds became a reality. This process called decellularization functions as the support system that designates specific roles to living cells [29].

Most recently, a new concept, known as 3D bioprinting (3DB), where the use of a 3D printer is capable of transferring cellular material onto an extracellular matrix in a tridimensional fashion, has become the most modern technology for tissue bioengineering. Multiple research centers around the world are adopting this technique for research with the idea of building a biostructure that resembles and works as a real organ [21]. This chapter will focus on the origins of 3DB and the basics of how bioprinting works and will describe the current efforts and early developments in 3DB.

11.2 Background

Charles Hull described a technique known as stereolithography, which dates back to the 1980s and is considered the early origins of 3D printing. Stereolithography is a form of printing, where a laser is used to solidify a polymer material being extruded from a needle to form a solid 3D structure. The instructions for the design of the 3D-printed structure come from computer software connected directly to the printer [17].

The field of medicine originally adopted 3D printing from the basic anatomical standpoint. One of the early purposes for the use of 3D printing was for presurgical evaluation based on imaging models that would allow an integral visualization of the organs [6]. The ability to reproduce an organ in a 3D digital format based on computed tomography (CT), magnetic resonance imaging (MRI), or ultrasound images led to the idea of printing synthetic organs (e.g., hepatic models) with the primary purpose of reproducing the internal anatomy of the organ to determine surgical approaches [25]. The digitized information of the CT or MRI is transferred to standard 3D printer computer software that allows printing the organ as desired [31]. In the field of transplantation, this allows an ideal preoperative evaluation of patients that are candidates for living donor liver transplantation, for example, the assessment of the liver anatomy, its biliary system, and blood supply before taking the donor to the operating room [1].

Parallel to printing synthetic organs for presurgical evaluation, 3D printing of synthetic organs or bone structures has been adopted by fields such as orthopedics; ears, nose, and throat; and facial and dental surgery, for the creation of anatomic molds, devices, and implants that can replace anatomical parts [14, 25, 32].

With the creation of 3D-printed synthetic organs, a step further was taken. 3D-printed synthetic scaffolds were created, seeded with live cells, and used for implantation in the human body with the ability to function as a tissue or organ. The first successful attempt at this approach was on patients with bladder cancer, with the use of a synthetic scaffold of human bladder by the Wake Forest group [2, 22]. Decellularization came along, a technique that involves the use of natural or biological scaffolds obtained from allogenic or xenogeneic organs or tissues. Along with the application of a detergent that removes the cellular elements, decellularization maintains the extracellular matrix and provides a support structure that can be seeded with new cells and create a new functional organ [29].

Finally, in the early 2000s, the group of Forgacs et al. from the University of Missouri-Columbia and Mironov from the University of South Carolina began working on the idea of 3D organ printing with a three-step principle: preprocessing or development of blueprints for organs, processing or actual organ printing, and post-processing or organ conditioning and accelerated organ maturation [12, 19]. The first patent application for a bioprinting platform was filed based on this initial work, and in 2007, Organovo, Inc. was established as the first 3D bioprinting company. Based on the research performed by Mironov and Forgacs' group, they soon devised the first commercially available 3D bioprinter. In 2010, with the support of an NIH grant, Organovo was able to print the first fully cellular blood vessel [20].

With this brief review on the recent history of 3DB, we continue to the next section where we describe the basic principles of how 3DB works.

11.3 Basic Principles

The bases for reconstruction of tissues and organs through 3DB consist of a set of techniques that transfer biologically active material onto a substrate. These techniques should include a high-resolution tridimensional printer able to inject or deposit cellular structures and biocompatible materials (e.g., agar gels) that can support building blocks of cells. The use of microstructures with simple geometry planning and orientation, along with computerized technology, helps create a macrostructure that is functional [19]. The basic concept of 3DB allows the building process to be able to create cellular patterns, which are the systematical organization of cell-to-cell interactions and produce mechanical and chemical signaling. These patterns confined in a tridimensional structure hope to achieve cellular functionality and become a viable tissue or whole organ.

Important concepts to understand the basics of 3DB are described in this following section.

11.3.1 Basic Components of 3DB

11.3.1.1 3D Bioprinter

Hardware

- Printer head: Metal plate with attached printheads that are remotely controlled by a series of motors that allow them to function along the x, y, and z axes. The printhead function is to inject or deposit the bioink (biomaterial composed of living cells, intended to create the 3D structure) or biogel (extracellular matrix, agar gel to provide support to the cells). The printheads (syringe shaped) contain either a glass capillary that works as a fine needle that will deposit or inject the biomaterial. Depending on the manufacturer, the printer heads can contain a reservoir where the bioink/biogel is contained. Other manufacturers rely on continuous reloading of glass capillaries or needles.
- Printing platform: Flat surface where the biogel container (or petri dish) is placed to allow the printer heads to deposit the bioink/biogel (Fig. 11.1).

Biologic Material

- Bioink: Biomaterial composed of living cells intended to be the main functioning cells of the 3D structure.
- Biogel: Supporting extracellular matrix that is present in the container where the 3D biostructure is created. This biogel is also injected or deposited by the printer head to support the cellular layers in order to support the tridimensional structure.

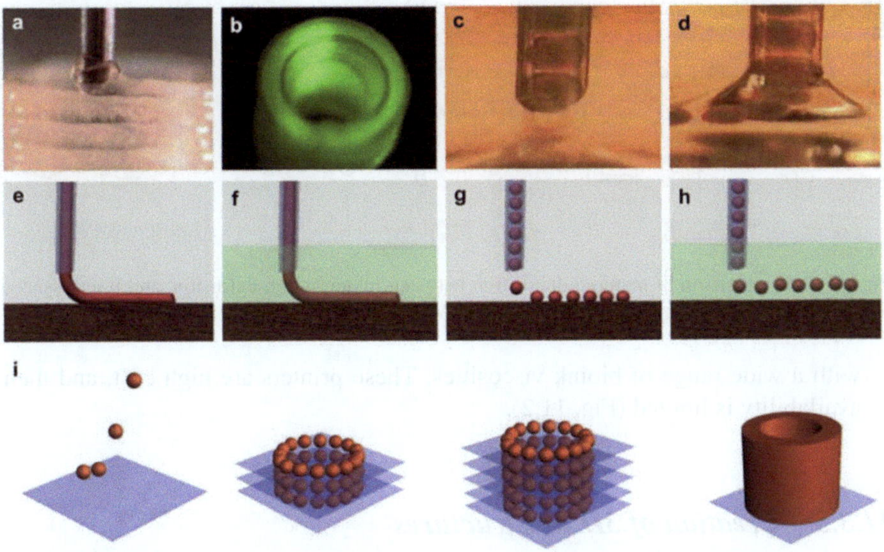

Fig. 11.1 Various methods of gel placement on hydrogel scaffolds (Permission to use image granted by the author under license number 4003381457926 [20])

11.3.2 Types of Bioprinters

Currently, the 3DB process can be achieved through three different modalities:

- *Micro-extrusion bioprinting*: Characterized by a temperature-controlled biomaterial dispensing system, based on a standard 3D printer, contains a fiber-optic light-illuminated deposit in area for photo-initiator activator and a piezoelectric humidifier. The system generates continuous biomaterial beads that are deposited in two dimensions; layers are placed along the *x*- and *z*-axes and then move higher in the *y*-axis. The final product is a tridimensional structure. The process is guided by computational software. The micro-extrusion printers have proven its value for creation of aortic valves and vascular structures [30].
- *Inkjet bioprinting:* Works by thermal or acoustic forces that promote the ejection of the bioink onto a scaffold or biogel base [9]. Thermal inkjet printers produce pulses of pressure via electricity to heat the printhead and thus stimulate droplets of bioink to fall from the nozzle into the biogel. These printers are high speed and low cost and they are widely available. Acoustic inkjet bioprinters have a piezoelectric crystal that creates an acoustic wave in the printhead that stimulates the deployment of cells into the biogel base. Printers such as these are very precise and uniform in terms of the bioink droplet deposition size. Overall, inkjet bioprinters are limited to low cell density deposition in order to prevent nozzle clogging. They have proven their value for printing functional skin and cartilage.
- *Laser-assisted bioprinting*: Consists of a pulsed laser beam that promotes the deployment of the bioink into the biogel plate. The laser bioprinter is compatible

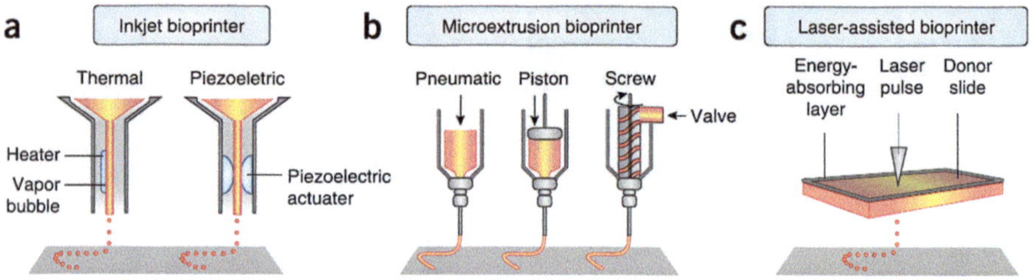

Fig. 11.2 Comparison of material deposition between inkjet, micro-extrusion, and laser-assisted bioprinters (Reprint permission by Murphy and Atala [22]. Copyright (2016))

with a wide range of bioink viscosities. These printers are high cost, and their availability is limited (Fig. 11.2).

11.3.3 Creation of 3D Biostructures

Although each type of printer might have slight differences in the steps of creating a tridimensional biostructure, the common basic steps of fabricating a 3D tissue or organ are described below [4]:

1. *Preprocessing*: Creation of a biological structure in a computer software, also considered the "blueprint."
2. *Processing*: Consists of preparing the materials that will be required to create the biostructure, as well as the steps to produce it.
   (a) *Layer of hydrogel or hydrogel container* – Series of steps to create an extracellular matrix structure that is contained in a petri dish or container that will serve as the foundation base or support where the desired printer tissue will be deposited.
   (b) *Bioink* – Cultured cells or tissue spheroids that are obtained from standardized methods for tissue/cell cultures. The bioink is loaded into the printer heads or printer head needles (glass capillaries). The bioink is eventually disposed on the hydrogel container.
   (c) *Hydrogel* – The same material used to support the cells that are in the container will also be used for dispensation in between the layers of bioink in order to support the multiple cellular layers.
   (d) *Bioink dispensation process* – The bioink or hydrogel is dispensed repeatedly depending on the amount of desired layers based on the blueprint design. As the layers are deposited, the bioink will fuse, eventually forming a 3D structure containing cells and extracellular matrix.
3. *Maturation*: The 3D biostructure is placed in an incubator and left to mature for a certain period of time (will depend on the materials and type of tissue created).

4. *Application*: Once the printed biostructure is mature, it can be used for pharmaceutical drug testing, in vitro models or animal models. Eventually the printed tissues/organs will potentially be used for human transplantation.

11.4 Current Research on 3DB

The field of tissue engineering and the recent application of 3D printing to medicine can potentially provide game-changing solutions to transplantation in the near future. Several examples of tissue and organ replacements are being described, designed, and applied around the world [12, 21]. As already mentioned in this chapter, multiple approaches are being investigated such as the use of natural or artificial scaffolds, decellularized organs to ultimately build 3DB. The key for success seems to be providing an ideal microenvironment allowing integration of tissue cells and development of a full organ. Although scaffolds provide an adequate structure for a 3D organ, they can be problematic in terms of immunological reactions and degradation of the extracellular matrix. Thus, the ideal approach is to rely on self-assembly and self-organization of the cells and tissues, which can be achieved with 3DB [12]. The field of whole-organ bioprinting is still in early stages, and many years of complex research will be required to obtain substantial results. Fortunately, multiple institutions around the world are focusing on different aspects of tissue engineering and 3DB process, which could lead to future applications. We continue to describe some of the relevant aspects that are being researched by different groups around the world including the aforementioned scaffolds, decellularization, and the advances achieved in some areas of real 3DB.

11.4.1 Steps Toward 3D Bioprinting

The advances on scaffolding and decellularization are described below as they have set the foundation for ideas to further develop the 3DB grounds.

One of the earliest stories of success in scaffolding development was achieved by one of the leading groups in 3D printing in medicine. The Wake Forest group [2] was able to engineer a three-dimensional bladder scaffold intended for patients with nonfunctional bladders. A biodegradable 3D scaffold was created and then seeded with autologous bladder urothelial and muscle cells grown in vitro, and once mature, the structure was anastomosed to the native bladder. Another example by Mannoor et al., with the use of cybernetics and tissue engineering, proved they were able to create a three-dimensional bionic ear, by printing a hydrogel matrix, seeding cells, and inter-wined conducting polymers with infused silver nanoparticles. This allowed in vitro culturing of cartilage tissue around an inductive coil antenna in the ear that was capable of detecting sound waves [16] (Fig. 11.3).

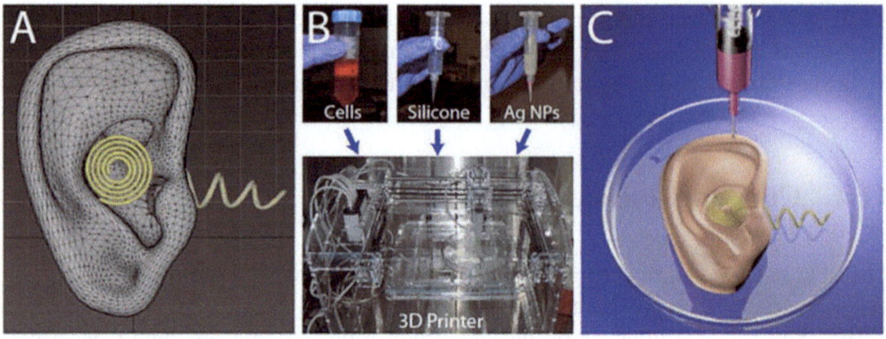

Fig. 11.3 (**a**) Computer software used to design the bionic ear, (**b**) material mixtures used in 3D printer, (**c**) printing of a bionic ear (Reprinted with permission from Mannoor et al. [16]. Copyright (2016) American Chemical Society)

Another school of thought focuses on the use of scaffolds as template for developing 3D tissues/organs based on a decellularization technique, which is based on the use of detergents that are perfused through the vasculature of the organ/tissue to remove cellular elements, but the extracellular matrix remains intact. The scaffold obtained after decellularization can then be seeded combined with cellular components with different techniques (including 3D bioprinters) in order to incorporate epithelial cells (liver, lung, heart) and endothelial (intra-organ vasculature). Song et al. developed an in vitro model and were able to incorporate endothelial and epithelial cells into a decellularized kidney scaffold with promising results [28]. Yagi et al. described a technique on a porcine liver model that can support functional hepatocytes maintaining the vascular and biliary network of the original organ [34].

Various approaches that preceded 3DB have been developed for vascular grafts. Niklason et al. developed a human-cultured smooth muscle cell model with a polyglycolic acid scaffold to generate a functional artery, demonstrating the importance of pulsatile stretching and improving strength of the vessel, but limitations were noted with the polymer remnants [8]. Dahl et al. reported a human vascular graft using cadaveric smooth muscle cells and the implantation onto baboon vessels after decellularization; these vessels were able to maintain blood flow for 6 months; after their animal model, a clinical trial in human was started in Poland in 2012 in patients with kidney failure as AV grafts [7].

Although animal tissues and organs are widely available for the use of scaffold techniques, it comes with the caveat that these scaffolds can potentially host infectious agents or trigger an immune response. Unrecovered organ donor organs could serve as a source of scaffolds as well, but ethical issues could potentially set a barrier. In terms of synthetic scaffolds, limitations due to polymer/synthetic available materials will remain. Overall, the use of scaffolds has been great in setting the scenario for creating 3D biological tissues and organs, but the use of these synthetic or natural scaffolds comes with their own limitation. Ideally the 3D biostructure should be allowed a natural interaction between the epithelial cells and the support cells, with a natural maturation process of the tissue in order to become fully functional.

11.4.2 True 3D Bioprinting Advances

The use of scaffoldings to develop 3D biostructures has set the scenario for real 3D bioprinting, since, for an organ to really achieve its functions, it requires an intricate network of tridimensional systems in order to thrive. Thus, the idea of creating tissues and organs from scratch to create a native functional biostructure is now feasible with the transformation of 3D printers to 3D bioprinters. Here we review the work of several institutions on bioprinting technologies and their strategies.

11.4.2.1 Vascular

One of the most important challenges of successful bioprinting for organ creation is the establishment of vascular networks that are essential to the organ, since without adequate organ supply, the desired tissue cannot survive. Kolesky et al. described a method for fabricating 3D biostructures with vasculature, multiple types of cells, and ECM, with the use of four different bioinks in vitro [15]. They designed a 2D vascular network with bifurcations found in natural biostructures that provide nutrient transportation and waste removal. Their model was originally proposed for drug screening, wound healing, and angiogenesis, but it also sets the scenario for manufacturing 3D tissues and organs.

In the field of transplantation, one of the determinant factors for a successful graft implantation is to have optimal vascular structures attached to the donor organ in order to perform an adequate anastomosis and to prevent thrombosis, stenosis, or leaks in the implanted organ. Often the vasculature of these organs is not usable due to their length, diameter, and integrity or due to inherent disease of the vessel. With the development of bioengineered constructs and the special interest in developing vessels for many vascular diseases, synthetic vascular grafts were created for repairing or replacing vessels. However, synthetic grafts often come with complications, thrombosis, and infection in the majority of cases. Tissue-specific engineered vascular grafts offer an attractive alternative.

In terms of 3DB vascular structures, some centers have developed their own models. Norotte et al. [23] described a layer-by-layer printing technique with the use of multicellular spheroids containing smooth cells and fibroblasts along with agarose rods, resulting in single- and double-layer small diameter vascular structures. This is one of the early examples of fully biological vascular tubular grafts and provides high expectations for the advancement in the vascular areas that will benefit the cardiovascular and transplantation field. Potential applications of 3D-bioprinted vascular grafts in the field of transplantation are meso-rex bypass from the superior mesenteric vein to left portal vein for the treatment of portal vein thrombosis. These vascular grafts could also prove useful in living donor liver transplantation where many times there is a need to extend vasculature for the drainage of segments V and VIII of the liver to the middle hepatic vein or vena cava to prevent outflow issues in right lobe grafts (Fig. 11.4).

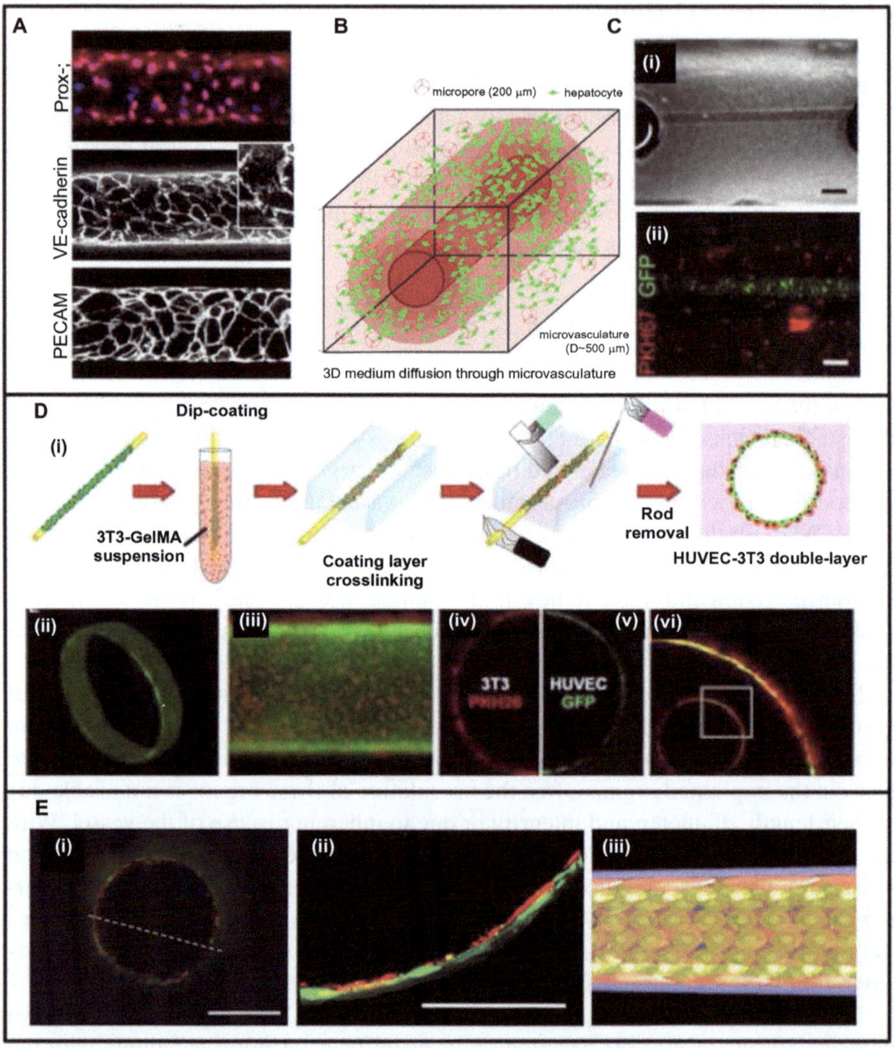

Fig. 11.4 (**A**) computer software used to design the bionic ear, (**B**) material mixtures used in 3D printer, C)printing of a bionic ear. (Reprinted with permission from Mannoor et al. [16]. Copyright (2016) American Chemical Society)

11.4.2.2 Liver

The importance of the liver in terms of its role in drug metabolism has fueled multiple interests around the world to develop liver biostructures to test the biopharmacology of many drugs. Research to develop 3D liver structures will eventually benefit the field of transplantation. To date, only reports of early data in in vitro models are available. Robbins et al. [26] with the use of the NovoGen MMX

Bioprinter ™ printed a metabolically functional 3D liver biostructure that showed cell-to-cell interaction, protein production, and enzymatic activity [24]. Their structure also contained stellate cells as well as endothelial cells. Chang et al. [5] created an in vitro device that had a 3D liver biostructure housed in a chamber that resembled the natural microenvironment of the hepatocyte, including a perfusion system that allowed assessing the metabolic function of the biostructure as well as the interaction with drugs. This model was developed with NASA support to assess drug pharmacokinetic profiles in planetary environments. As mentioned above, the success of these 3D liver biostructures relies on providing a vascular network. Miller et al. [18] were able to print 3D filament networks made of carbohydrate glass in a cylindrical shape that were lined with endothelial cells and perfused with blood. This model was tested in a rat hepatocyte model, maintaining the metabolic function of the cells.

11.4.2.3 Intestine

Although GI tract models have been developed with diverse bioengineering techniques, the 3D bioprinting world has not focused on the development of GI tract models. The complexity of creating an intestine requires the creation of vasculature, neural, and lymphoid tissue along with epithelial tissue with absorptive and secretory functions. At our institution, efforts are being focused on the creation of a muscular graft onto which all the functions previously mentioned are added [33].

11.4.2.4 Kidney

Drug nephrotoxicity is estimated to cause 25% of acute renal failure. However, this is just an estimated number, and the real percentage is hard to determine. As in liver, the creation of 3D-bioprinted kidney models is being fueled by the need to understand better the interaction between the kidney and multiple drugs. This again will benefit the field of kidney transplantation with the subsequent development of fabricated human kidney tissue models.

King et al. [13] created an in vitro model of multicellular, 3D-bioprinted proximal tubules. In their model, the interface between tubular epithelium and renal interstitial cells was observed. Also an extensive endothelial network was described. Homan et al. [11] created a 3D-bioprinted human renal proximal tubule model in vitro within an extracellular matrix and housed in perfusable tissue chips that allowed the model to survive for more than 2 months. Their model exhibits enhanced epithelial morphology and functional properties. Cyclosporine (known nephrotoxin but also used for immunosuppression in transplantation) demonstrated disruption of the epithelial barrier in a dose-dependent manner, proving the utility of the in vitro model. Even with these advances, the development of a kidney model is far from being accomplished (Fig. 11.5).

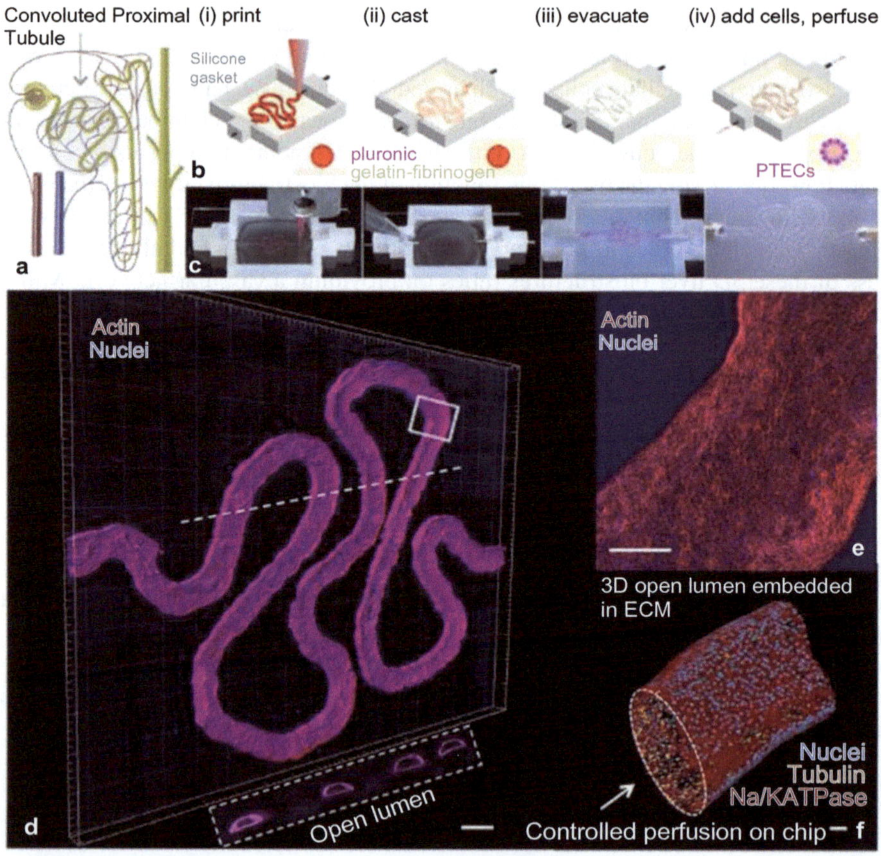

Fig. 11.5 3D image of a printed convoluted proximal tubule (Reprinted with permission from Homan et al. [11]. Copyright (2016) Scientific Reports, Nature Publishing Group)

11.4.2.5 Heart

The complexity of the heart tissue poses a barrier that only few researchers are willing to confront. Zhang and Zhang [35] created a hybrid strategy based on 3D bioprinting and scaffolding with an Organovo NovoGen MMX Bioprinter (Organovo Holdings) [24]. First with the use of bioink containing endothelial cells, they injected microfibrous hydrogel scaffolds. The endothelial cells migrated toward the periphery to form a layer of endothelium. This endothelial layer was then seeded with cardiomyocytes in order to generate aligned myocardium capable of spontaneous and synchronous contractions. Even though cell migration was achieved, the overall composite demonstrated a lack of structure and functionality. This is one of the earliest demonstrations of 3D-bioprinted heart tissue [36] (Fig. 11.6).

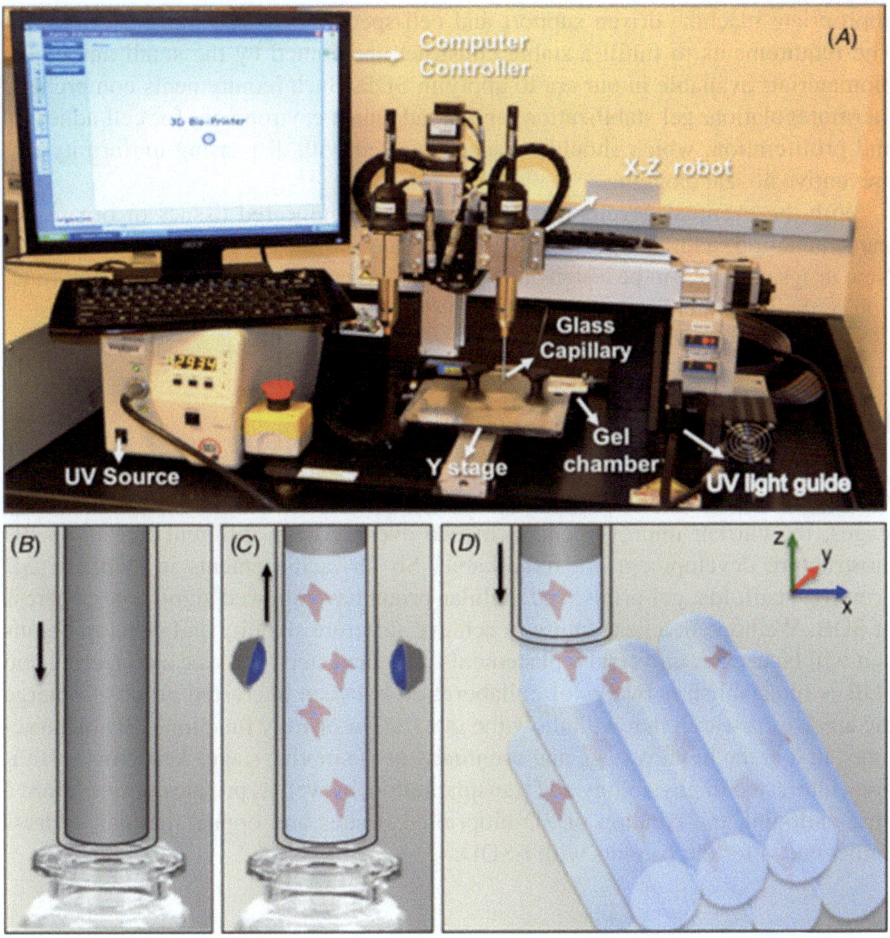

Fig. 11.6 (**a**) Photograph of an Organovo Bioprinter setup, (**b**) metallic cylinder immersed in a vial containing cells and a hydrogel precursor. (**c**) Upward aspiration of the cell-laden hydrogel. (**c**) Precise printing of cell-laden hydrogel fibers [3]

11.5 Barriers to Overcome

Every organ with potential 3DB creation presents with its own complex barriers. Replicating the cell-to-cell interaction requires establishment of a vascular network and neurohumoral response. Efforts to achieve organogenesis, as it occurs in the human body, must be attempted. More "out of the box" approaches might need to be performed along with much needed interinstitutional collaborations.

Currently, bioprinters are able to deliver biomaterials, micromolecules, and viable cells to generate experimental constructs, but lack sufficient organizational integrity and stability for surgical graft replacement. Experimental biomaterial mixtures are still being tested to find the adequate median for cell carriers and to provide appropriate machine-driven support and cell-specific cues and eliminate toxicity. The requirements to fulfill a stable construct are limited by the small amount of biomaterials available in our era to apply in 3DB. Such requirements comprehend thermoregulation, gel stabilization, and an adequate environment for cell adhesion and proliferation, which should be accomplished with dispensing uniformity and preventive nozzle clogging.

With the eventual development of mature bioengineered tissues or organs for transplantation, addressing the implantation stage will be a complex task as well. New drugs will have to be tested and developed to promote the adaptation and function of the implanted tissue or organ to fully achieve what is desired.

11.6 Conclusion

Although most of the aforementioned advances in the 3DB field are in the early stages, the current improvements achieved every year in different aspects of 3D biostructure development are remarkable. So far, achievements in synthetic 3D printing, scaffolds, gel prints, and cellular prints have allowed significant progress of 3DB. We hope in a near future to achieve structure stability and develop organs that will be used as universal replacements. The next step for tissue engineering and 3DB is to establish networks of collaboration between interested groups to merge the areas of research that will allow the integration of fully functional 3D biostructures, first at the in vitro stage and eventually at the in vivo stage. Academic institutions interested in physiology and transplantation, as well as private entities, share a final endpoint, the creation of 3D-bioprinted tissues and organs that can address unmet needs for the patients with ESOD.

References

1. Alkhouri, N., & Zein, N. N. (2016). Three-dimensional printing and pediatric liver disease. *Current Opinion in Pediatrics, 28*, 626–630.
2. Atala, A., Bauer, S. B., Soker, S., Yoo, J. J., & Retik, A. B. (2006). Tissue-engineered autologous bladders for patients needing cystoplasty. *Lancet, 367*, 1241–1246.
3. Bertassoni, L. E., Cardoso, J. C., Manoharan, V., Cristino, A. L., Bhise, N. S., Araujo, W. A., Zorlutuna, P., Vrana, N. E., Ghaemmaghami, A. M., Dokmeci, M. R., & Khademhosseini, A. (2014). Direct-write bioprinting of cell-laden methacrylated gelatin hydrogels. *Biofabrication, 6*, 024105.
4. Catros, S., Fricain, J. C., Guillotin, B., Pippenger, B., Bareille, R., Remy, M., Lebraud, E., Desbat, B., Amedee, J., & Guillemot, F. (2011). Laser-assisted bioprinting for creating on-demand patterns of human osteoprogenitor cells and nano-hydroxyapatite. *Biofabrication, 3*, 025001.
5. Chang, C. C., Boland, E. D., Williams, S. K., & Hoying, J. B. (2011). Direct-write bioprinting three-dimensional biohybrid systems for future regenerative therapies. *Journal of Biomedical Materials Research. Part B, Applied Biomaterials, 98*, 160–170.

6. Chia, H. N., & Wu, B. M. (2015). Recent advances in 3D printing of biomaterials. *Journal of Biological Engineering, 9*, 4.
7. Gui, L., Boyle, M. J., Kamin, Y. M., Huang, A. H., Starcher, B. C., Miller, C. A., Vishnevetsky, M. J., & Niklason, L. E. (2014). Construction of tissue-engineered small-diameter vascular grafts in fibrin scaffolds in 30 days. *Tissue Engineering. Part A, 20*, 1499–1507.
8. Gui, L., & Niklason, L. E. (2014). Vascular tissue engineering: Building Perfusable vasculature for implantation. *Current Opinion in Chemical Engineering, 3*, 68–74.
9. Guillotin, B., Souquet, A., CATROS, S., Duocastella, M., Pippenger, B., Bellance, S., Bareille, R., Remy, M., Bordenave, L., Amedee, J., & Guillemot, F. (2010). Laser assisted bioprinting of engineered tissue with high cell density and microscale organization. *Biomaterials, 31*, 7250–7256.
10. Hasan, A., PAUL, A., Vrana, N. E., Zhao, X., Memic, A., Hwang, Y. S., Dokmeci, M. R., & Khademhosseini, A. (2014). Microfluidic techniques for development of 3D vascularized tissue. *Biomaterials, 35*, 7308–7325.
11. Homan, K. A., Kolesky, D. B., Skylar-Scott, M. A., Herrmann, J., Obuobi, H., Moisan, A., & Lewis, J. A. (2016). Bioprinting of 3D convoluted renal proximal tubules on Perfusable chips. *Scientific Reports, 6*, 34845.
12. Jakab, K., Norotte, C., Marga, F., Murphy, K., Vunjak-Novakovic, G., & Forgacs, G. (2010). Tissue engineering by self-assembly and bio-printing of living cells. *Biofabrication, 2*, 022001.
13. King, S., Creasey, O., Presnell, S., & Nguyen, D. (2015). Design and characterization of a multicellular, three-dimensional (3D) tissue model of the human kidney proximal tubule. The FASEB Journal, 29(1 Supplement), LB426.
14. Klammert, U., Vorndran, E., Reuther, T., Muller, F. A., Zorn, K., & Gbureck, U. (2010). Low temperature fabrication of magnesium phosphate cement scaffolds by 3D powder printing. *Journal of Materials Science. Materials in Medicine, 21*, 2947–2953.
15. Kolesky, D. B., Truby, R. L., Gladman, A. S., Busbee, T. A., Homan, K. A., & Lewis, J. A. (2014). 3D bioprinting of vascularized, heterogeneous cell-laden tissue constructs. *Advanced Materials, 26*, 3124–3130.
16. Mannoor, M. S., Jiang, Z. W., James, T., Kong, Y. L., Malatesta, K. A., Soboyejo, W. O., Verma, N., Gracias, D. H., & Mcalpine, M. C. (2013). 3D printed bionic ears. *Nano Letters, 13*, 2634–2639.
17. McCallen, J. D., Schaefer, A., Lee, P., Hing, L., & Lai, S. K. (2017). Stereolithography-Based 3D Printed "Pillar Plates" that Minimizes Fluid Transfers During Enzyme Linked Immunosorbent Assays. Annals of biomedical engineering, 45(4), 982–989.
18. Miller, J. S., Stevens, K. R., Yang, M. T., Baker, B. M., Nguyen, D. H., Cohen, D. M., Toro, E., Chen, A. A., Galie, P. A., Yu, X., Chaturvedi, R., Bhatia, S. N., & Chen, C. S. (2012). Rapid casting of patterned vascular networks for perfusable engineered three-dimensional tissues. *Nature Materials, 11*, 768–774.
19. Mironov, V. (2003). Printing technology to produce living tissue. *Expert Opinion on Biological Therapy, 3*, 701–704.
20. Mironov, V., Visconti, R. P., Kasyanov, V., Forgacs, G., Drake, C. J., & Markwald, R. R. (2009). Organ printing: Tissue spheroids as building blocks. *Biomaterials, 30*, 2164–2174.
21. Munoz-Abraham, A. S., Rodriguez-Davalos, M. I., Bertacco, A., Wengerter, B., Geibel, J. P., & Mulligan, D. C. (2016). 3D printing of organs for transplantation: Where are we and where are we heading? *Current Transplantation Reports, 3*, 93–99.
22. Murphy, S. V., & Atala, A. (2014). 3D bioprinting of tissues and organs. *Nature Biotechnology, 32*, 773–785.
23. Norotte, C., Marga, F. S., Niklason, L. E., & Forgacs, G. (2009). Scaffold-free vascular tissue engineering using bioprinting. *Biomaterials, 30*, 5910–5917.
24. Organovo. 2015. *The bioprinting process.*
25. Rengier, F., Mehndiratta, A., Von Tengg-Kobligk, H., Zechmann, C. M., Unterhinninghofen, R., Kauczor, H. U., & Giesel, F. L. (2010). 3D printing based on imaging data: Review of medical applications. *International Journal of Computer Assisted Radiology and Surgery, 5*, 335–341.

26. Robbins, J. B., Gorgen, V., Min, P., Shepherd, B. R., & Presnell, S. C. (2013). A novel in vitro three-dimensional bioprinted liver tissue system for drug development. *FASEB Journal, 27.*
27. Segev, D. L., Muzaale, A. D., Caffo, B. S., Mehta, S. H., Singer, A. L., Taranto, S. E., Mcbride, M. A., & Montgomery, R. A. (2010). Perioperative mortality and long-term survival following live kidney donation. *JAMA, 303*, 959–966.
28. Song, J. J., Guyette, J. P., Gilpin, S. E., Gonzalez, G., Vacanti, J. P., & Ott, H. C. (2013). Regeneration and experimental orthotopic transplantation of a bioengineered kidney. *Nature Medicine, 19*, 646–651.
29. Soto-Gutierrez, A., Wertheim, J. A., Ott, H. C., & Gilbert, T. W. (2012). Perspectives on whole-organ assembly: Moving toward transplantation on demand. *The Journal of Clinical Investigation, 122*, 3817–3823.
30. Stanton, M. M., Samitier, J., & Sanchez, S. (2015). Bioprinting of 3D hydrogels. *Lab on a Chip, 15*, 3111–3115.
31. Sun, W., Starly, B., Darling, A., & Gomez, C. (2004). Computer-aided tissue engineering: Application to biomimetic modelling and design of tissue scaffolds. *Biotechnology and Applied Biochemistry, 39*, 49–58.
32. Utrecht, U. 3D-Printed Skull Impanted In Patient. 2014 [cited 2015 2015]; Available from: http://www.umcutrecht.nl/en/Research/News/3D-printed-skull-implanted-in-patient.
33. Wengerter, B. C., Emre, G., Park, J. Y., & Geibel, J. (2016). Three-dimensional printing in the intestine. *Clinical Gastroenterology and Hepatology, 14*, 1081–1085.
34. Yagi, H., Fukumitsu, K., Fukuda, K., Kitago, M., Shinoda, M., Obara, H., Itano, O., Kawachi, S., Tanabe, M., Coudriet, G. M., Piganelli, J. D., Gilbert, T. W., Soto-Gutierrez, A., & Kitagawa, Y. (2013). Human-scale whole-organ bioengineering for liver transplantation: A regenerative medicine approach. *Cell Transplantation, 22*, 231–242.
35. Zhang, X. Y., & Zhang, Y. D. (2015). Tissue engineering applications of three-dimensional Bioprinting. *Cell Biochemistry and Biophysics, 72*, 777–782.
36. Zhang, Y. S., Arneri, A., Bersini, S., Shin, S. R., Zhu, K., Goli-Malekabadi, Z., Aleman, J., Colosi, C., Busignani, F., Dell'erba, V., Bishop, C., Shupe, T., Demarchi, D., Moretti, M., Rasponi, M., Dokmeci, M. R., Atala, A., & Khademhosseini, A. (2016). Bioprinting 3D microfibrous scaffolds for engineering endothelialized myocardium and heart-on-a-chip. *Biomaterials, 110*, 45–59.

Chapter 12
Xenotransplantation

J.A. Shah, B. Ekser, and P.A. Vagefi

Abstract Shortages in the number of available donor organs continue to force the transplant community to seek alternative options in an effort to meet the high demand. Cross species, or xenotransplantation, using swine as potential donors, has long been hypothesized as a potential attractive strategy for solving the organ shortage crisis due to the supply of available donors, as well as anatomical and physiological similarities between swine and humans. Early studies with wild-type swine donors were limited due to shortened survival as a result of acute humoral xenograft rejection due to circulating preformed antibodies. The eventual development of α-1,3-galactosyltransferase knock-out swine donors in the early 2000s has been critical in advancing preclinical xenotransplantation research, and more recently through significant improvements in genetic engineering technology such as CRISPR/Cas9, the development of multitransgenic swine donors has allowed xenotransplantation to progress closer to becoming a clinical reality. Here, we provide a brief overview of early clinical xenotransplantation experience, followed by major technological advances and current barriers to solid organ (kidney, liver, heart, and lung) and islet cell xenotransplantation.

12.1 Introduction

The continued need for an increased supply of donor organs remains a critical area of research in the field of organ transplantation. Despite the utilization of "extended criteria" and living donors, in 2015 alone approximately 34 patients each day were removed from the waiting list in the United States due to either dying while waiting for an organ or being too sick to be transplanted at the time an organ became available [1]. Furthermore, despite efforts to increase the number

J.A. Shah, DO • P.A. Vagefi, MD, FACS (✉)
Center for Transplantation Sciences, Massachusetts General Hospital/Harvard Medical School, 55 Fruit Street, White 521c, Boston, MA 02114, USA
e-mail: pvagefi@mgh.harvard.edu

B. Ekser, MD, PhD
Transplant Division, Department of Surgery, Indiana University School of Medicine, Indianapolis, IN, USA

S.N. Nadig, J.A. Wertheim (eds.), *Technological Advances in Organ Transplantation*, DOI 10.1007/978-3-319-62142-5_12

of transplants performed annually, totaling over 650,000 since 1988 [1], the number of patients added to the wait-list continues to outpace those that ultimately achieve transplantation. When we look at transplantation statistics from around the world, this discrepancy between supply and demand only continues to worsen. Indeed, the disparity between those transplanted versus those waiting for organs continues to be a prominent issue, one which faces many ethical, social, and humanitarian concerns.

One potential solution to this crisis is cross-species transplantation, or xenotransplantation, utilizing organs from pigs (Fig. 12.1). Pigs remain the current best option for clinical xenotransplantation given their anatomical and physiological similarities to humans [2–4], favorable breeding characteristics [5], and their ability to undergo genetic modification, thus overcoming immunological hurdles due to their phylogenetic distance. Prior to the use of pigs, much of the early work pioneered in the field of xenotransplantation was performed using a more concordant species, nonhuman primates (NHPs), in an effort to overcome many of the immunological barriers inhibiting long-term survival. However, concerns over deadly infectious complications led to a moratorium in 1999 [6] banning further clinical NHP xenotransplantation research. Extensive preclinical work using pigs for xenotransplantation has been performed since the enactment of these guidelines, with slow, but steady progress being made in an effort to one day make clinical xenotransplantation a reality. This chapter will focus on providing the history and an update on progress made in solid organ and islet cell xenotransplantation, without discussing tissue (i.e. cornea) or other cell type (i.e. hepatocyte or neuronal cell) xenotransplantation.

12.2 Brief History of Clinical Xenotransplantation

The first experiences with clinical xenotransplantation date back to the early 1960s, with the first kidney xenotransplantation being performed by Reemtsma et al. from Tulane University in 1963 using a rhesus monkey [7], with subsequent experience utilizing a chimpanzee, allowing for the recipient to return to work and remain healthy for 9 months before succumbing to pneumonia. In 1964, the world's first heart xenotransplantation was attempted by Hardy et al. at the University of Mississippi Medical Center, again using a chimpanzee as the source for the donor organ, with survival lasting a couple of hours [8]. Subsequent efforts using a baboon heart in a neonate by Bailey et al. from Loma Linda University Medical Center in 1984 allowed for 20-day survival with graft loss due to worsening humoral rejection. With improvements in allotransplantation outcomes due to significant advancements in immunosuppression, mainly the widespread use of tacrolimus, in 1993 Starzl et al. from the University of Pittsburgh successfully transplanted a baboon liver into a patient with cirrhosis due to chronic hepatitis B virus infection. The patient survived for 70 days, ultimately succumbing to diffuse subarachnoid hemorrhage due to invasive aspergillosis [9].

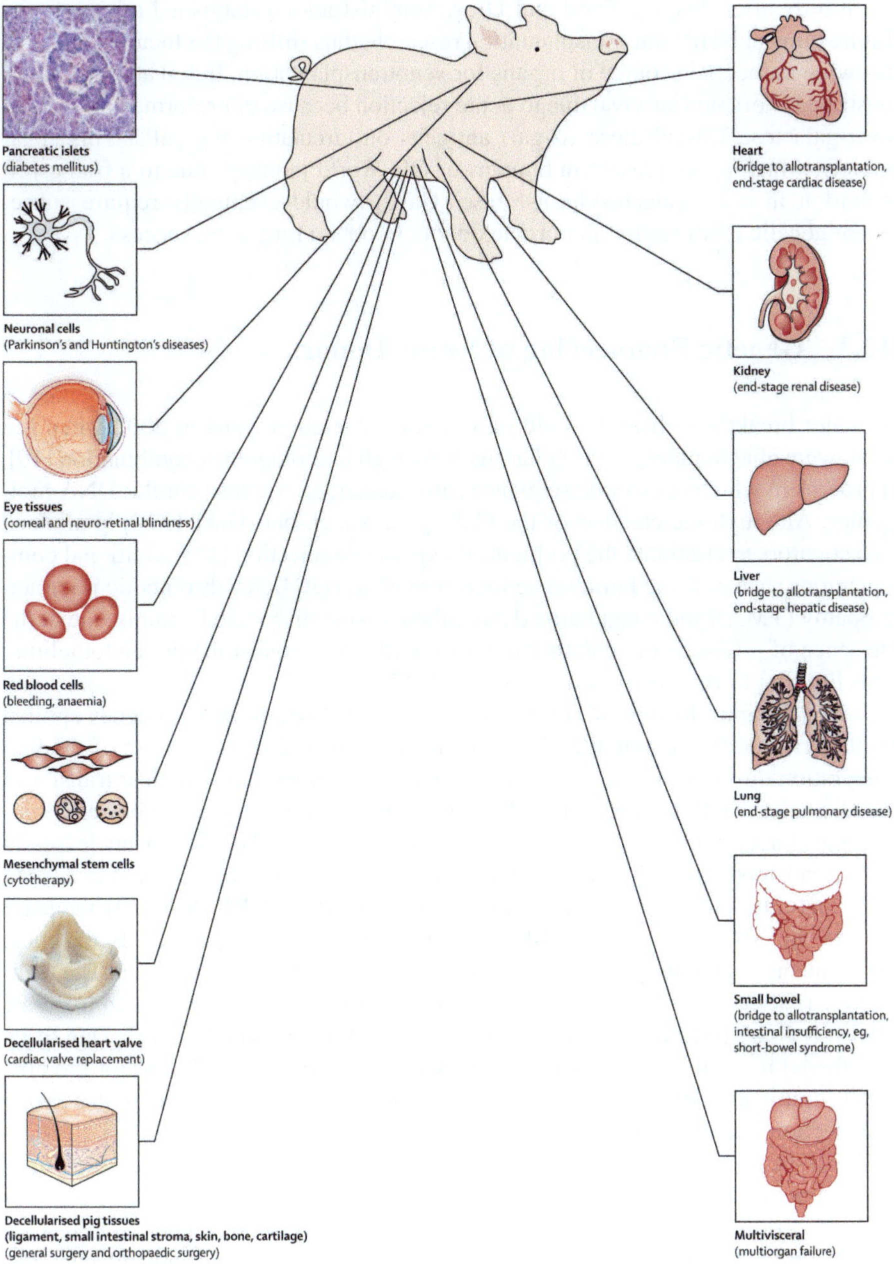

Fig. 12.1 Pigs as potential donors for xenotransplantation. (Adopted from: Ekser et al. [3]. Permission to use image granted by the author under license number 4235430612018)

Shortly after this, the Food and Drug Administration established a ban on any future clinical NHP xenotransplantation research, thus shifting the focus of research to swine as the main source of organs for xenotransplantation. Initial attempts demonstrated shortened survival due to acute rejection because of preformed antibodies to α-galactose-1,3-galactose (α-gal) antigens on circulating pig cells. This sugar moiety, which is not present in humans or Old World monkeys due to a frameshift mutation in α-1,3-galactosyltransferase (GalT), would eventually require subsequent genetic modification in porcine donors to obtain long-term success.

12.3 Genetic Engineering of Swine Donors

A major breakthrough in the field of xenotransplantation came in 2002 when Lai et al. were able to delete the α-gal antigen through homologous recombination [10], a process in which nucleotide sequences are exchanged between similar DNA molecules. Although the creation of the GalT gene knock-out (GalT-KO) pig allowed investigators to eliminate the problem of hyperacute rejection [11], additional complications such as acute humoral xenograft rejection (AHXR), thrombotic microangiopathy (TMA), and coagulation dysregulation were discovered, mainly due to the presence of preformed circulating non-Gal antibodies against the pig endothelium, thus limiting further long-term survival [12, 13] .

Since the introduction of homologous recombination, other approaches applied to further modify the genome of donor pigs have included the process of random integration through embryo microinjection in 1997, somatic cell nuclear transfer in 2000, nuclease editing through double-strand breaks and nonhomologous end joining zinc-finger nucleases in 2010, transcription activator-like effector nucleases in 2011, and more recently, the development of the clustered randomly interspaced short palindromic repeats and the associated protein 9 (CRISPR/Cas9) nuclease method in 2013 [14]. The CRISPR/Cas 9 method currently represents the most efficient and widespread technology used for gene editing in donor swine. The technology relies on a bacterial (*Streptococcus pyogenes*) defense mechanism to create CRISPR RNA (crRNA) and subsequent double-strand breaks [15, 16]. Due to its untethered method of DNA cleavage, the CRISPR system has allowed for the efficient combination of multiple knockouts through targeting several loci with a single transfection, creating multiple genetic knockouts on a single cell line [17].

With the development of these aforementioned methods, multitransgenic donor pigs have been created to overcome complications related to human complement regulation proteins (hCRP; CD46 – membrane cofactor protein, CD55, CD59 – protectin), anticoagulation, and anti-inflammatory gene expression (human tissue factor pathway inhibitor – hTFPI, human thrombomodulin – hTBM, human endothelial protein C receptor – hEPCR, human A20 – hA20, and human signal regulatory protein α (SIRPα) – CD47) [18] (Table 12.1). In addition, the deletion of potentially deleterious porcine retroviruses (PERVs) shown by Ramsoondar et al. [19] and Dieckhoff et al. [20] may eliminate potential infectious complications that have

Table 12.1 Porcine genetic modifications for xenotransplantation

Complement regulation by human complement regulatory gene expression
CD46 (membrane cofactor protein)
CD55 (decay-accelerating factor)
CD59 (protectin or membrane inhibitor of reactive lysis)
Gal or non-Gal antigen "masking" or deletion
Human H-transferase gene expression (expression of blood type "O" antigen)
Endo-β-galactosidase C (reduction of Gal antigen expression)
α1,3-Galactosyltransferase gene-knockout (GalT-KO)
Cytidine monophosphate-*N*-acetylneuraminic acid hydroxylase (CMAH)
B1,4-*N*-acetylgalactosaminyltransferase (β4GalNT2) gene-knockout (β4GalNT2-KO)
Suppression of cellular immune response by gene expression of downregulation
CIITA-DN (MHC class II transactivator knockdown, resulting in swine leukocyte antigen class II knockdown)
Class I MHC-knockout (MHC-IKO)
HLA-E/human β2-microglobulin (inhibits human natural killer cell cytotoxicity)
Human FAS ligand (CD95L)
Human *N*-acetylglucoaminyltransferase III (GnT-III) gene
Porcine CTLA4-Ig (cytotoxic T-lymphocyte antigen 4 or CD152)
Human TRAIL (tumor necrosis factor-α-related apoptosis-inducing ligand)
Anticoagulation and anti-inflammatory gene expression or deletion
von Willebrand factor (vWF)-deficient (natural mutant)
Human tissue factor pathway inhibitor (TFPI)
Human thrombomodulin
Human endothelial protein C receptor (EPCR)
Human ectonucleoside triphosphate diphosphohydrolase-1 (CD39)
Anticoagulation, anti-inflammatory, and antiapoptotic gene expression
Human tumor necrosis factor-α-induced protein 3 (A20)
Human heme oxygenase-1 (HO-1)
Human CD47 (species-specific interaction with SIRP-α inhibits phagocytosis)
Porcine asialoglycoprotein receptor 1 gene-knockout (ASGR1-KO; decreases platelet phagocytosis)
Human signal regulatory protein-α (SIRP-α; decreases platelet phagocytosis by "self" recognition)
Prevention of porcine endogenous retrovirus (PERV) activation
PERV siRNA

Adopted from: Cooper et al. [88]

been raised as a potential barrier for successful clinical pig xenotransplantation. More recently, Yang et al. [21] showed the genome-wide inactivation of PERV virus deleting 62 copies of genes from porcine kidney epithelial cells using CRISPR/cas9 technology. In the following sections, we will discuss the progress that has been made with each specific solid organ in the field of xenotransplantation using organs from the genetically modified pig donors that have been previously mentioned.

12.4 Kidney Xenotransplantation

Much of the successes with preclinical models of pig-to-NHP xenotransplantation have relied on the advances made in genetic engineering of the donor pigs with survival increasing as genetically modified donors became available. In 1989, long before the development of GalT-KO donors, early experience with genetically unmodified (wild-type) pig kidneys was poor with survival limited to only 23 days [22]. However, with the introduction of GalT-KO/human complement decay-accelerating factor (hDAF; CD55) donors in 2004, survival improved to 90 days [23]. Around the same time, Yamada et al. from Boston, utilizing a tolerance approach involving the cotransplantation of donor-specific thymic tissue from GalT-KO donors at the time of pig-to-baboon renal xenotransplantation, achieved survival of 83 days following GalT-KO pig kidney xenotransplantation [24]. Further genetic modification eventually led to the development of multitransgenic GalT-KO/CD55/CD46/CD39/hTBM/EPCR pig donors, with survival in a baboon reaching 136 days. In this report, induction immunosuppression consisted of thymoglobulin (ATG), anti-CD20 monoclonal antibody (mAb), cobra venom factor (CVF) for complement inactivation, and maintenance therapy with anti-CD40 mAb, rapamycin, corticosteroids as well as anti-TNF-α and anti-IL-6R mAbs. Despite the improved survival following the production of GalT-KO pigs, longer survival would be necessary in order for kidney xenotransplantation to become a destination therapy. The utilization of newer clinically applicable immunosuppressive agents, such as costimulation blockade of the CD28/B7 or CD40/CD154 pathways [25] as well as use of an anti-CD154 mAb [24], CTLA4-Ig, and anti-CD40 [26], may eventually help in achieving prolonged survival to a point where human trials are warranted.

It deserves to be mentioned that with the development of the both GalT-KO donors and newer methods to dampen the recipient immune response, both AHXR and hyperacute rejection appear to be avoidable, thus leading to the discovery of additional barriers, namely the development of TMA [27] within the xenografts and systemic consumptive coagulopathy [28] within the recipients. Research efforts focused on these barriers may allow for improved outcomes. The likely etiology of these coagulation disturbances are related to the incompatibilities between the donor and recipient endothelium, with the recipient coagulation system becoming activated upon graft reperfusion. Once activated, recipient platelets and tissue factor pathways begin to increase local and systemic inflammatory responses, releasing pro-inflammatory cytokines and leading to activation of the coagulation cascade and production of thrombin, the latter leading to TMA within the graft. The production of multitransgenic pigs (hCRPs, TBM, TFPI, EPCR, and CD39) have demonstrated the ability to regulate the procoagulant activity following xenotransplantation [29], and further improvements have been observed following the administration of antiplatelet and anticoagulant administration [30].

With progress in overcoming both the immune and coagulation disturbances following pig-to-NHP xenotransplantation, physiological incompatibilities also continue to be problematic; however, continued work has demonstrated that many of these barriers can be overcome. Although the initial incompatibilities between pigs and NHP's with regards to uric acid metabolism, the renin–angiotensin–aldosterone system, and erythropoietin production [4] seem to be surmountable, it is the development of postoperative proteinuria and hypoalbuminemia that appears to be the greater limiting factor following xenotransplantation [24]. However, with the use of genetically modified donors, this phenomenon also appears to be controllable as the use of GalT-KO/CD55/CD46/CD39/hTBM/EPCR pig donors appeared to prevent protein losing nephropathy following xenotransplantation [31, 32]. Furthermore, the use of rituximab has also been shown to delay and control the onset of proteinuria, possibly by preventing damage to podocytes due to the loss of sphingomyelin phosphodiesterase acid-like 3b activity within the Bowman's capsule [33].

Lastly, if kidney xenotransplantation were to become a widely accepted practice, investigation into the growth of the xenograft following transplantation is warranted. Indeed early research by Soin et al. has demonstrated that hDAF pig kidney xenografts grow according to the rate of the donor pig for 2 weeks following xenotransplantation, but then growth slows down considerably, thus preventing complications related to constriction and compartment syndrome [34]. Overall, these results thus far provide optimism that the immunological and physiological incompatibilities are surmountable and that clinical kidney xenotransplantation will one day become a reality.

Most recently, the selection of recipient baboons with low antibody titer and the use of anti-CD154 mAb included immunosuppressive regimen extended the survival of GalT-KO.hCD55 pig-to-NHP life-supporting kidney xenotransplantation up to 310 days [35] . In this experiment, there was no documented proteinuria, as was seen previously in shorter survival experiments.

12.5 Liver Xenotransplantation

Early experiences following pig-to-NHP liver xenotransplantation (LXT) date back to the early 1960s, with work by Calne et al. demonstrating survival of only a few hours following reperfusion of the xenoliver due to catastrophic hemorrhage [36]. Following the development of genetically engineered donors, Ramirez et al. were able to improve survival to 8 days with the use of pig livers expressing human CD55, with death occurring due to sepsis and coagulopathy [37]. Despite this improvement, thrombocytopenia following xenograft reperfusion remained a critical barrier to success. With the introduction of GalT-KO pigs, the Pittsburgh group demonstrated in a large study of 10 orthotopic liver xenotransplantations (2 GalT-KO and 8 GalT-KO/CD46) adequate liver function, with the exception of cholestasis occurring due to interspecies incompatibilities, in all of their recipients of a xenoliver [38] with survival ranging from 4 to 7 days. Of importance, the work done by the Pittsburgh group demonstrated sufficient production of important hepatic-produced pig proteins (albumin, plasminogen, fibrinogen, and haptoglobin) in baboon blood [39] along with adequate production of coagulation factors with clinically applicable immunosuppressive regimen; however, recipients still succumbed to consumptive coagulopathy and internal bleeding due to profound thrombocytopenia.

Following this work, the Massachusetts General Hospital group was able to improve survival to 9 days with the use of GalT-KO donors and immunosuppression consisting of induction with ATG, CVF, and anti-CD154 mAb and maintenance with tacrolimus. The administration of aminocaproic acid as well as the transfusion of platelets allowed for partial control of coagulopathy and limited thrombocytopenia [40] with evidence of adequate liver function but decreased coagulation factor production following xenotransplantation. Given these findings, the Boston group went on to develop a novel approach of exogenous human coagulation factor administration following pig-to-baboon LXT. In a series of six orthotopic pig-to-baboon LXTs performed using GalT-KO donors and immunosuppression consisting of thymoglobulin and CVF induction and maintenance therapy with corticosteroids and tacrolimus, a continuous, low-dose administration of either human factor VIIa or human prothrombin concentrate complex appeared to control coagulopathy, maintain circulating platelets and prevent TMA without the need for the administration of antifibrinolytic agents or platelet transfusions [41], and survival up to 7 days was achieved. Furthermore, the effects of exogenous human coagulation factor administration appeared to delay and decrease the overall blood transfusion requirement following LXT with experimental baboons requiring an average of 208 mL of blood compared with historical controls that required 626 mL over a similar time period. Additionally, for the first time, xenograft histology in recipients of continuous coagulation factors appeared to demonstrate the absence of TMA when compared with recipients of bolus administration, which demonstrated widespread TMA and graft necrosis [41] (Fig. 12.2).

Given these initial encouraging results, the Boston group then added belatacept for costimulation blockade to their immunosuppressive regimen given the encouraging results reported following heart xenotransplantation [26] and were able to

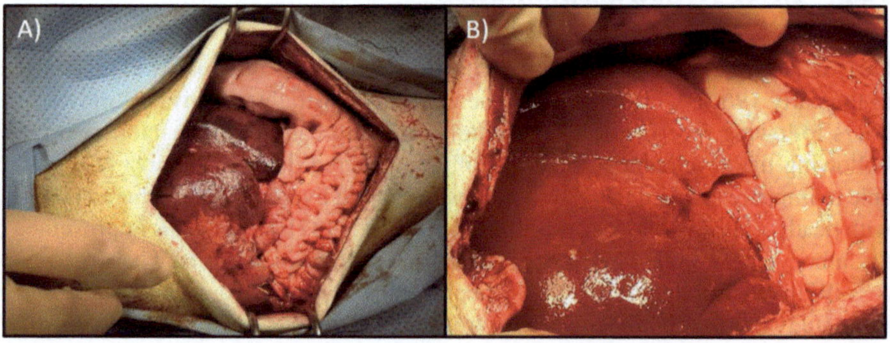

Fig. 12.2 Comparison of intraoperative porcine livers following pig-to-baboon liver xenotransplantation and varying coagulation factor protocols. (**a**) Bolus administration of exogenous human coagulation factors (postoperative day 1), demonstrating diffuse necrosis and widespread thrombosis. (**b**) Continuous, low-dose administration of exogenous human coagulation factors (postoperative day 4), demonstrating a pink, healthy appearing liver without evidence of necrosis or thrombosis

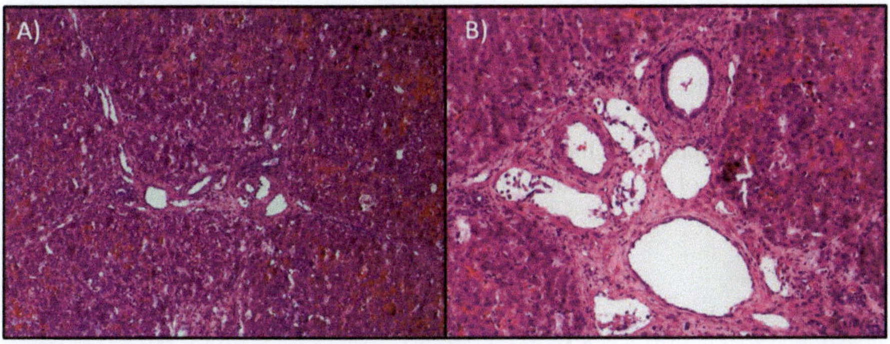

Fig. 12.3 Postoperative day 25 histology following pig-to-baboon liver xenotransplantation, (**a**) Hematoxylin & Eosin staining (100×), and (**b**) Hematoxylin & Eosin staining (200×), demonstrating intact hepatic architecture, no evidence of rejection, mild (30%) focal hepatic necrosis, and absence of thrombotic microangiopathy

achieve 25-day survival, along with the spontaneous recovery of circulating platelets on postoperative day (POD) 12, reaching a maximum of 614,000 on POD 21 [42]. Of note, the recipient baboon never required re-exploration due to bleeding, which was commonly seen in previous studies of LXT, and required euthanasia on POD 25 due to worsening cholestasis and bleeding due to plantar ulcers. As previously demonstrated, the need for blood transfusions was again limited over the 25-day survival, and liver function tests remained normal or near-normal, with the exception of a mild increase in the transaminases on POD 7 due to a mild case transaminitis due to presumed mild rejection as evidenced by improving LFT's following a 3-day pulse of corticosteroids. POD 25 histology again demonstrated the absence of TMA and no evidence of rejection (Fig. 12.3). The cholestasis observed was similar to that experienced by the Pittsburgh group. [39], indicating a difference in bile viscosity as a cause, rather than a mechanical obstruction. Indeed work by Kobayashi et al. has demonstrated that pig and human bile have similar characteristics and may not be an issue following pig-to-human LXT [43]. More recently, the replacement of belatacept with anti-CD40 mAb for costimulation blockade has allowed for even further prolongation of survival to 29 days, representing a new world record for hepatic xenotransplantation, with recipient euthanasia again required due to worsening plantar ulceration and posttransplant histology demonstrating an absence of inflammation, necrosis, or TMA (Shah et al., *American Journal of Transplantation* – In Press).

Although the etiology of the thrombocytopenia occurring following reperfusion of the xenoliver remains yet to be elicited, multiple groups have made progress in identifying several key responsible factors. It is thought that the thrombocytopenia and consumptive coagulopathy occurring as early as 2 h following xenograft reperfusion is due to activation of recipient platelets [44] and not due to immunological phenomenon. Additionally, differences in human and porcine platelet oligosaccharides causing activation of the liver sinusoidal endothelial cells can lead to platelet phagocytosis [45], as well as platelet phagocytosis due to activation of porcine aortic endothelial cells [46].

Evidence suggests that the immunological hurdles preventing long-term success following pig LXT are indeed surmountable [47], and with the introduction of multitransgenic GalT-KO donors, especially those expressing TBM, TFPI, CD55, CD46, and CD39, may allow for successful hepatic xenotransplantation utilizing conventional immunosuppression. Indeed, consistent achievement of 4-week survival in a preclinical model of pig-to-baboon LXT could justify human trials with pig livers serving as a bridge to allotransplantation for patients with acute hepatic failure.

12.6 Heart Xenotransplantation

Work in the field of heart xenotransplantation, unlike kidneys and livers, has been performed in three categories: a heterotopic abdominal approach, a heterotopic intrathoracic approach, and an orthotopic (intrathoracic) approach, with survival varying depending on the site of transplantation. Progress in the field of immunopharmacology as well as experience with genetically engineered donors has allowed for >2-year survival in a heterotopic abdominal fashion, 50 days in a heterotopic intrathoracic fashion, and 57 days in an orthotopic fashion [48]. Barriers to prolonged survival rest mainly on hyperacute rejection due to complement-mediated vascular injury and thrombocytopenia, both of which have been extensively researched in an effort to bring cardiac xenotransplantation to a clinical reality.

Heterotopic abdominal cardiac xenotransplantation remains the primary model of research for pig-to-NHP cardiac xenotransplantation given its nonlife supporting nature and the ability to follow the graft even in the face of organ failure, with the organ anastomosed to the recipient inferior vena cava and the xenograft aorta to the recipient abdominal aorta. Early studies were limited due to anti-Gal antibodies causing damage to the recipient endothelium and initiating a robust immune response [49]; however, with the creation of CD55-expressing donors, survival improved to 27 days in 2004 [50] and further improved to 179 days, with a median survival of 78 days in 2005 following the creation of GalT-KO donors [51, 52] without any evidence of hyperacute rejection. Further improvements following the creation of GalT-KO donors expressing hCRPs came with immunosuppression regimens consisting of anti-CD154 and anti-CD40 with survival up to 236 days (median 71) with GalT-KO/CD46 donors and an anti-CD154-based regimen [53], 149 days (median 84 days) with GalT-KO/CD46 donors and an anti-CD40-based regimen [54], and up to 945 days (median 298 days) with GalT-KO/CD46/hTBM donors and an anti-CD40-based regimen [55], the latter of which surpassed survival well beyond 2 years. More recently, Iwase et al. have provided their initial experience with GalT-KO/CD46/hTBM donors, comparing three different costimulation blockade regimens, the results of which demonstrate that an induction immunosuppression regimen consisting of thymoglobulin and methylprednisolone and a maintenance regimen consisting of anti-CD40 mAb, belatacept, rapamycin/tacrolimus, methylprednisolone, and low molecular weight heparin is able to demonstrate up to 130-day survival without evidence of invoking a T-cell response, preventing the development of thrombocytopenia and delaying the development of TMA and consumptive coagulopathy [56].

Despite this progress, the aforementioned model is not life-supporting; thus, efforts at achieving success following life-supporting cardiac xenotransplantation are underway, both in an intrathoracic heterotopic ("piggy-back") model, where the donor organ is anastomosed to the left and right atria, the ascending aorta, and the pulmonary artery, and an orthotopic model utilizing a conventional technique. The intrathoracic approach was first described in 1977 by Barnard et al. [57] with the advantage of providing temporary support or as a bridge to transplantation, and research in a large animal model demonstrated survival of up to 50 days using a GalT-KO/CD46 donor with a mild rejection crisis amenable to treatment with immunosuppression and immunepheresis [58]. Further studies utilizing multitransgenic donors and protocols previously shown in a heterotopic abdominal model to prolong survival are currently underway [59].

With preclinical heterotopic cardiac xenotransplantation studies now achieving median survival greater than the recommended 3 months prior to the development of clinical cardiac xenotransplantation protocols, as suggested by the 2000 International Society for Heart and Lung Transplantation advisory committee [60], the focus is shifting toward replicating these results in an orthotopic, life-supporting model. Although studies with this model are limited given high perioperative mortality due to ischemia/reperfusion injury and cardiogenic shock, survival ranged between 1 day (GalT-KO/CD55 donors) and 57 days (Gal+/CD46/TPC α-gal polyethylene glycol polymer treatment) [61]. With progress made in controlling the recipient immune response following cardiac xenotransplantation [62], attention should now focus on developing methods to enhance xenograft resistance to ischemia/reperfusion injury, for example through the use of donors expressing CD39 [63] as well as trials utilizing immunosuppression regimens based on anti-CD40 mAb and multitransgenic GalT-KO donors in an orthotopic model, as continued success in a preclinical pig-to-NHP models could one day pave the way for future clinical applications.

12.7 Lung Xenotransplantation

Human lung allotransplantation remains limited due to the endothelium's susceptibility to cytokine release causing ischemia/reperfusion injury during the procurement and the transplantation period, as well as alveolar barotrauma due to mechanical ventilation [64, 65]. The applicability of lung xenotransplantation would be especially advantageous due to the ability to use organs from donors that are otherwise healthy and devoid of complications related to sepsis, hypotension, prolonged mechanical ventilation, etc. Initial experience with wild-type pig lung xenotransplantation, similar to the kidney, liver, and the heart, was met with disappointing results, mainly due to circulating antibodies against pig Gal antigens, the effects of which resulted in activation of the complement cascade and rapid rise in the pulmonary vascular resistance due to widespread thrombosis [66]. Initial results with GalT-KO donors was poor as well, with survival limited to just 225 min following pig-to-baboon lung xenotransplantation [67] due to tracheal edema caused by capillary leakage and increased pulmonary vascular resistance. As a result of these initial findings, most of the significant work done in the field of lung xenotransplantation has been done in an ex vivo model by the group from the University of Maryland using genetically modified porcine lungs perfused with human blood, thus closely imitating a clinically applicable scenario [68].

The University of Maryland ex vivo model relies on the left and right lung independently ventilated and perfused with human blood, thus allowing the delivery of experimental drug therapies and the ability to measure and study each lung against the other [69, 70], the results of which identified four key barriers to success: (a) the innate immune response, (b) coagulation dysregulation, (c) the adaptive immune response, and (d) the inflammatory response [68]. Extensive work in overcoming these four key barriers has demonstrated that the activity of the complement system appears to be blunted through the use of GalT-KO donors expressing hCRP (CD46 and CD55) [71]. Additionally, the administration of liposomal clodronate pretransplant appears to be effective in depleting porcine pulmonary macrophages and prolonging graft survival by decreasing thrombosis [72], and costimulation blockade through the administration of CTLA4-Ig may inhibit the T-cell response [73]. Furthermore, coagulation dysregulation and platelet activation/sequestration may be overcome through the depletion of pig VWF via the administration of desmopressin [74], use of transgenic pigs expressing hTFPI [75], human TBM, and hEPCR [73].

In an orthotopic, life-supporting model, a single lung is transplanted into the recipient NHP with the placement of either a snare around the pulmonary artery allowing intermittent occlusion or a flow probe into the recipient aorta and donor pulmonary artery providing researchers the ability to measure blood flow and evaluate graft function [68]. Using this model, consistent achievement of graft survival beyond 24 hours has been achieved, with maximum survival lasting 8 days following pig-to-NHP lung xenotransplantation using a multitransgenic GalT-KO/CD46/CD55/EPCR/TFPI donor and the administration of desmopressin. Further work in this life-supporting model is underway in an effort to further prolong survival and hopefully achieve consistently achieve 3-month survival, the latter of which has been recommended by the International Society for Heart and Lung Transplantation before considering human trials [60].

12.8 Islet Xenotransplantation

With the successful reports of allogeneic islet transplants coming from the Edmonton group in 2000, demonstrating successful elimination of exogenous insulin requirements for 1 year [76], there has been growing interest in the use of xenogeneic islets for the treatment of diabetes. Initial xenotransplantation results by Bottino et al.

have determined that adult pigs seem to demonstrate a beneficial advantage as donors given the increased yield of islets isolated and the rapid function (within hours) when compared with fetal and neonatal pigs, which deliver a lower yield and take a longer period to function following transplantation [77, 78]. These results, in conjunction with the development of transgenic donor animals, have accelerated the possibility of clinical pig islet xenotransplantation.

The earliest reports of preclinical pig-to-NHP islet xenotransplantation date back to 2006 with the use of wild-type pigs and demonstrating survival (i.e. insulin independence) for >6 months using an immunosuppression regimen consisting of anti-IL-2R, anti-CD154, and rapamycin [79, 80]. With the development of transgenic animals, survival improved to 396 days (CD46 donor) using thymoglobulin, anti-CD154, and mycophenolate mofetil [81] and 249 days (GalT-KO donor) using anti-CD154, anti-LFA-1, CTLA4-Ig, and mycophenolate mofetil (MMF) [82]. Although encouraging, these protocols cannot be applied to a clinical setting given the concern over thrombotic complications related to the use of anti-CD154. Furthermore, as seen in solid organ xenotransplantation where recipient thrombocytopenia and consumptive coagulopathy predominate, cellular islet xenotransplantation is uniquely complicated by an instant blood-mediated inflammatory reaction (IBMIR) when using wild-type donors, resulting in a considerable loss of islets immediately following transplantation [83]. With the advent of newer multitransgenic GalT-KO/CD55/CD59 donors, Hawthorne et al. have been able to demonstrate minimal IBMIR; however, with an inability to demonstrate prolonged survival, further studies are warranted [84]. It deserves to be mentioned that the development of GalT-KO donors has allowed for the elimination of immediate problems related to hyperacute rejection and that chronic rejection still remains a concern. Given this, Thompson et al. have demonstrated that through the use of costimulation blockade, the T-cell response can be effectively diminished and survival can be prolonged [85].

Of final note, novel strategies, using wild-type pig donors, such as the intraperitoneal (rather than an intraportal) administration of islets encapsulated in alginate and polyethylene glycol membranes have allowed for islet survival and insulin production for up to 804 days in the absence of immunosuppression [86], thus providing alternative avenues for further research in xenoislet transplantation. These strategies, in conjunction with continued efforts at establishing consistent prolonged graft survival, along with clinically acceptable immunosuppressive protocols will hope to soon bring make islet xenotransplantation a clinical reality.

12.9 Summary

Since the early experiences with wild-type donors, advances in xenotransplantation have made significant progress due to advances in gene editing and the development of multitransgenic pig donors, such as triple knock-out pigs [87]. In vitro studies using triple knock-out pigs (i.e. GalT-KO.CMAH-KO.B4GalNT2-KO) showed very promising results in response to minimizing antibody response even with highly sensitized human serums (Estrada et al.) and in vivo studies are currently ongoing. The use of these multitransgenic GalT-KO donors, along with improvements in available immunosuppressive agents, has allowed for progress in overcoming the otherwise robust immunological barriers preventing long-term survival following solid organ xenotransplantation. The addition of hCRPs as well as genes regulating coagulation will hopefully allow further success in controlling additional limiting factors such as thrombocytopenia and coagulopathy. It is imperative that for future success in the field of xenotransplantation research that increased funding opportunities become available to improve outcomes at large animal research facilities, as well as help facilitate collaborative efforts between xenotransplantation research institutions. It is hopeful that with the demonstration of consistently improved survival in a pig-to-NHP model, human clinical trials with xenotransplantation may be possible in the not so distant future, initially as a bridge to allotransplantation for those with acute organ failure, and hopefully eventually as destination therapy.

References

1. OPTN Stats OPTN. (2016). Organ Procurement and Transplantation Network. http://optn.transplant.hrsa.gov. Last accessed on 28 Mar 2016.
2. Cooper, D. K. C., Gollackner, B., & Sachs, D. H. (2002). Will the pig solve the transplantation backlog? *Annual Review of Medicine, 53*(53), 133–147.
3. Ekser, B., Ezzelarab, M., Hara, H., Van Der Windt, D. J., Wijkstrom, M., Bottino, R., et al. (2012). Clinical xenotransplantation: The next medical revolution? *Lancet, 379*(9816), 672–683.
4. Ibrahim, Z., Busch, J., Awwad, M., Wagner, R., Wells, K., & Cooper, D. K. C. (2006). Selected physiologic compatibilities and incompatibilities between human and porcine organ systems. *Xenotransplantation, 13*(6), 488–499.
5. Sachs, D. H. (1994). The pig as a xenograft donor. *Pathologie Biologie (Paris), 42*(3), 217–219.
6. Editorial. (1999). US guidelines on xenotransplantation. *Nature Medicine, 5*(5), 465.
7. Reemtsma, K., Mccracken, B. H., Schlegel, J. U., & Pearl, M. (1964). Heterotransplantation of the kidney: Two clinical experiences. *Science, 143*(3607), 700–702.
8. Hardy, J. D., Kurrus, F. D., Chavez, C. M., Neely, W. A., Eraslan, S., Turner, M. D., et al. (1964). Heart transplantation in man. Developmental studies and report of a case. *Journal of the American Medical Association, 188*(13), 1132–1140.
9. Starzl, T. E., Fung, J., Tzakis, A., Todo, S., Demetris, A. J., Marino, I. R., et al. (1993). Baboon-to-human liver transplantation. *Lancet, 341*(8837), 65–71.
10. Lai, L., Kolber-Simonds, D., Park, K. W., Cheong, H. T., Greenstein, J. L., Im, G. S., et al. (2002). Production of alpha-1,3-galactosyltransferase knockout pigs by nuclear transfer cloning. *Science, 295*(5557), 1089–1092.
11. Ekser, B., Rigotti, P., Gridelli, B., & Cooper, D. K. C. (2009). Xenotransplantation of solid organs in the pig-to-primate model. *Transplant Immunology, 21*(2), 87–92.
12. Byrne, G. W., McGregor, C. G. A., & Breimer, M. E. (2015). Recent investigations into pig antigen and anti-pig antibody expression. *International Journal of Surgery, 23*, 223–228.
13. Azimzadeh, A. M., Byrne, G. W., Ezzelarab, M., Welty, E., Braileanu, G., Cheng, X., et al. (2008). Development of a consensus protocol to quantify primate anti-non-gal xenoreactive antibodies using pig aortic endothelial cells. *Xenotransplantation, 21*(6), 555–566.

14. Butler, J. R., Ladowski, J. M., Martens, G. R., Tector, M., & Tector, A. J. (2015). Recent advances in genome editing and creation of genetically modified pigs. *International Journal of Surgery, 23*, 217–222.
15. Mali, P., Esvelt, K. M., & Church, G. M. (2013). Cas9 as a versatile tool for engineering biology. *Nature Methods, 10*(10), 957–963.
16. Feng, W., Dai, Y., Mou, L., Cooper, D. K. C., Shi, D., & Cai, Z. (2015). The potential of the combination of CRISPR/Cas9 and pluripotent stem cells to provide human organs from chimaeric pigs. *International Journal of Molecular Sciences, 16*(3), 6545–6556.
17. Li, P., Estrada, J. L., Burlak, C., Montgomery, J., Butler, J. R., Santos, R. M., et al. (2015). Efficient generation of genetically distinct pigs in a single pregnancy using multiplexed single-guide RNA and carbohydrate selection. *Xenotransplantation, 22*(1), 20–31.
18. Cooper, D. K. C., Ekser, B., & Tector, A. J. (2015). Immunobiological barriers to xenotransplantation. *International Journal of Surgery, 23*, 211–216.
19. Ramsoondar, J., Vaught, T., Ball, S., Mendicino, M., Monahan, J., Jobst, P., et al. (2009). Production of transgenic pigs that express porcine endogenous retrovirus small interfering RNAs. *Xenotransplantation, 16*(3), 164–180.
20. Dieckhoff, B., Petersen, B., Kues, W. A., Kurth, R., Niemann, H., & Denner, J. (2008). Knockdown of porcine endogenous retrovirus (PERV) expression by PERV-specific shRNA in transgenic pigs. *Xenotransplantation, 15*(1), 36–45.
21. Yang, L., Güell, M., Niu, D., George, H., Lesha, E., Grishin, D., et al. (2015 Nov 27). Genome-wide inactivation of porcine endogenous retroviruses (PERVs). *Science, 350*(6264), 1101–1104.
22. Lambrigts, D., Sachs, D. H., & Cooper, D. K. (1998 Sep 15). Discordant organ xenotransplantation in primates: World experience and current status. *Transplantation, 66*(5), 547–561.
23. Baldana, N., Rigotti, P., Calabrese, F., Cadrobbi, R., Dedja, A., Iacopetti, I., et al. (2004). Ureteral stenosis in HDAF pig-to-primate renal xenotransplantation: A phenomenon related to immunological events? *American Journal of Transplantation, 4*(4), 475–481.
24. Yamada, K., Yazawa, K., Shimizu, A., Iwanaga, T., Hisashi, Y., Nuhn, M., et al. (2005). Marked prolongation of porcine renal xenograft survival in baboons through the use of alpha1,3-galactosyltransferase gene-knockout donors and the Cotransplantation of vascularized Thymic tissue. *Nature Medicine, 11*(1), 32–34.
25. Buhler, L., Awwad, M., Basker, M., Gojo, S., Watts, A., Treter, S., et al. (2000). High-dose porcine hematopoeitic cell transplantation combined with CD40 ligand blockade in baboons prevents an induced anti-pig humoral response. *Transplantation, 69*(11), 2296–2304.
26. Iwase, H., Ekser, B., Satyananda, V., Bhama, J., Hara, H., Ezzelarab, M., et al. (2015). Pig-to-baboon heterotopic heart transplantation – exploratory preliminary experience with pigs transgenic for human thrombomodulin and comparison of three costimulation blockade-based regimens. *Xenotransplantation, 22*, 211–220.
27. Shimizu, A., Yamada, K., Yamamoto, S., Lavelle, J. M., Barth, R. N., Robson, S. C., et al. (2005). Thrombotic microangiopathic glomerulopathy in human decay accelerating factor-transgenic swine-to-baboon kidney xenografts. *Journals of the American Society of Nephrology, 16*(9), 2732–2745.
28. Ierino, F. L., Kozlowski, T., Siegel, J. B., Shimizu, A., Colvin, R. B., Banerjee, P. T., et al. (1998). Disseminated intravascular coagulation in association with the delayed rejection of pig-to-baboon renal xenografts. *Transplantation, 66*(11), 1439–1450.
29. Miwa, Y., Yamamoto, K., Onishi, A., Iwamoto, M., Yazaki, S., Haneda, M., et al. (2010). Potential value of human thrombomodulin and DAF expression for coagulation control in pig-to-human xenotransplantation. *Xenotransplantation, 17*(1), 26–37.
30. Iwase, H., Ezzelarab, M. B., Ekser, B., & Cooper, D. K. C. (2014). The role of platelets in coagulation dysfunction in xenotransplantation, and therapeutic options. *Xenotransplantation, 21*(3), 201–220.
31. Higginbotham, L., Mathews, D., Breeden, C. A., Song, M., Farris, A. B., Larsen, C. P., et al. (2015). Pre-transplant antibody screening and anti-CD154 costimulation blockade promote

long-term xenograft survival in a pig-to-primate kidney transplant model. *Xenotransplantation, 22*, 221–230.
32. Iwase, H., Liu, H., Wijkstrom, M., Zhou, H., Singh, J., Hara, H., et al. (2015). Pig kidney graft survival in a baboon for 136 days: Longest life-supporting organ graft survival to date. *Xenotransplantation, 22*(4), 302–309.
33. Tasaki, M., Shimizu, A., Hanekamp, I., Torabi, R., Villani, V., & Yamada, K. (2014). Rituximab treatment prevents the early development of proteinuria following pig-to-baboon xeno-kidney transplantation. *Journals of the American Society of Nephrology, 25*(4), 737–744.
34. Soin, B., Ostlie, D., Cozzi, E., Smith, K. G., Bradley, J. R., Vial, C., et al. (2000). Growth of porcine kidneys in their native and xenograft environment. *Xenotransplantation, 7*(2), 96–100.
35. Higginbotham, L., Mathews, D., Stephenson, A., Breeden, C., Larsen, C., & For, M. (2015). Long-term survival of pig-to-primate renal xenotransplant using costimulation-blockade immunosuppression. *Xenotransplantation, 22*(Supplement:S45).
36. Calne, R. Y., White, H. J., Herbertson, B. M., Millard, P. R., Davis, D. R., Salaman, J. R., et al. (1968). Pig-to-baboon liver xenografts. *Lancet, 1*(7553), 1176–1178.
37. Ramirez, P., Chavez, R., Majado, M., Munitiz, V., Muñoz, A., Hernandez, Q., et al. (2000). Life-supporting human complement regulator decay accelerating factor transgenic pig liver xenograft maintains the metabolic function and coagulation in the nonhuman primate for up to 8 days. *Transplantation, 70*(7), 989–998.
38. Ekser, B., Long, C., Echeverri, G. J., Hara, H., Ezzelarab, M., Lin, C. C., et al. (2010). Impact of thrombocytopenia on survival of baboons with genetically modified pig liver transplants. *American Journal of Transplantation, 10*, 273–285.
39. Ekser, B., Echeverri, G. J., Hassett, A. C., Yazer, M. H., Long, C., Meyer, M., et al. (2010). Hepatic function after genetically engineered pig liver transplantation in baboons. *Transplantation, 90*, 483–493.
40. Kim, K., Schuetz, C., Elias, N., Veillette, G. R., Wamala, I., Varma, M., et al. (2012). Up to 9-day survival and control of thrombocytopenia following alpha1,3-galactosyl transferase knockout swine liver xenotransplantation in baboons. *Xenotransplantation, 19*, 256–264.
41. Navarro-Alvarez, N., Shah, J. A., Zhu, A., Ligocka, J., Yeh, H., Elias, N., et al. (2016). The effects of exogenous administration of human coagulation factors following pig-to-baboon liver xenotransplantation. *American Journal of Transplantation, 16*(6), 1715–1725.
42. Shah, J. A., Navarro-Alvarez, N., DeFazio, M., Rosales, I., Elias, N., Yeh, H., et al. (2016). A bridge to somewhere: 25-day survival following pig-to-baboon liver xenotransplantation. *Annals of Surgery, 263*(6), 1069–1071.
43. Kobayashi, T., Taniguchi, S., Ye, Y., Niekrasz, M., Nour, B., & Cooper, D. K. (1998 Apr). Comparison of bile chemistry between humans, baboons, and pigs: Implications for clinical and experimental liver xenotransplantation. *Laboratory Animal Science, 48*(2), 197–200.
44. Ekser, B., Lin, C. C., Long, C., Echeverri, G. J., Hara, H., Ezzelarab, M., et al. (2012 Aug). Potential factors influencing the development of thrombocytopenia and consumptive coagulopathy after genetically modified pig liver xenotransplantation. *Transplant International, 25*(8), 882–896.
45. Paris, L. L., Chihara, R. K., Sidner, R. A., Joseph Tector, A., & Burlak, C. (2012). Differences in human and porcine platelet oligosaccharides may influence phagocytosis by liver sinusoidal cells in vitro. *Xenotransplantation, 19*(1), 31–39.
46. Peng, Q., Yeh, H., Wei, L., Enjyoj, K., Machaidze, Z., Csizmad, E., et al. (2012). Mechanisms of xenogeneic baboon platelet aggregation and phagocytosis by porcine liver sinusoidal endothelial cells. *PloS One, 7*(10), 1–7.
47. Ekser, B., Markmann, J. F., & Tector, A. J. (2015). Current status of pig liver xenotransplantation. *International Journal of Surgery, 23*, 1–7.
48. Mohiuddin, M. M., Reichart, B., Byrne, G. W., & McGregor, C. G. A. (2015). Current status of pig heart xenotransplantation. *International Journal of Surgery, 23*, 234–239.
49. Buhler, L., Friedman, T., Iacomini, J., & Cooper, D. K. (1999). Xenotransplantation – state of the art – update 1999. *Frontiers in Bioscience, 4*(D4), 16–32.

50. Kuwaki, K., Knosalla, C., Dor, F. J. M. F., Gollackner, B., Tseng, Y. L., Houser, S., et al. (2004). Suppression of natural and elicited antibodies in pig-to-baboon heart transplantation using a human anti-human CD154 mAb-based regimen. *American Journal of Transplantation, 4*(3), 363–372.
51. Kuwaki, K., Tseng, Y. L., Dor, F. J., Shimizu, A., Houser, S. L., Lancos, C. J., et al. (2005). Heart transplantation in baboons using alpha1,3-galactosyltransferase gene- knockout pigs as donors: Initial experience. *Nature Medicine, 11*(1), 29–31.
52. Tseng, Y.-L., Kuwaki, K., Dor, F. J. M. F., Shimizu, A., Houser, S., Hisashi, Y., et al. (2005). alpha1,3-galactosyltransferase gene-knockout pig heart transplantation in baboons with survival approaching 6 months. *Transplantation, 80*(10), 1493–1500.
53. Mohiuddin, M. M., Corcoran, P. C., Singh, A. K., Azimzadeh, A., Hoyt, R. F., Thomas, M. L., et al. (2012). B-cell depletion extends the survival of GTKO.hCD46Tg pig heart xenografts in baboons for up to 8 months. *American Journal of Transplantation, 12*(3), 763–771.
54. Mohiuddin, M. M., Singh, A. K., Corcoran, P. C., Hoyt, R. F., Thomas, M. L., Lewis, B. G. T., et al. (2014). Role of anti-CD40 antibody-mediated costimulation blockade on non-gal antibody production and heterotopic cardiac xenograft survival in a GTKO.hCD46Tg pig-to-baboon model. *Xenotransplantation, 21*(1), 35–45.
55. Mohiuddin, M. M., Singh, A. K., Corcoran, P. C., Thomas Iii, M. L., Clark, T., Lewis, B. G., et al. (2016). Chimeric 2C10R4 anti-CD40 antibody therapy is critical for long-term survival of GTKO.hCD46.hTBM pig-to-primate cardiac xenograft. *Nature Communications, 7*, 11138.
56. Iwase, H., Ekser, B., Satyananda, V., Bhama, J., Hara, H., Ezzelarab, M., et al. (2015). Pig-to-baboon heterotopic heart transplantation – exploratory preliminary experience with pigs transgenic for human thrombomodulin and comparison of three costimulation blockade-based regimens. *Xenotransplantation, 23*, 211–220.
57. Barnard, C. N., Losman, J. G., Curcio, C. A., Sanchez, H. E., Wolpowitz, A., & Barnard, M. S. (1977). The advantage of heterotopic cardiac transplantation over orthotopic cardiac transplantation in the management of severe acute rejection. *The Journal of Thoracic and Cardiovascular Surgery, 74*(6), 918–924.
58. Bauer, A., Postrach, J., Thormann, M., Blanck, S., Faber, C., Wintersperger, B., et al. (2010). First experience with heterotopic thoracic pig-to-baboon cardiac xenotransplantation. *Xenotransplantation, 17*(3), 243–249.
59. Mohiuddin, M. M., Singh, A. K., Corcoran, P. C., Hoyt, R. F., Thomas, M. L., Ayares, D., et al. (2014 Sep). Genetically engineered pigs and target-specific immunomodulation provide significant graft survival and hope for clinical cardiac xenotransplantation. *The Journal of Thoracic and Cardiovascular Surgery, 148*(3), 1106–1114.
60. Cooper, D. K., Keogh, A. M., Brink, J., Corris, P. A., Klepetko, W., Pierson, R. N., et al. (2000). Report of the Xenotransplantation Advisory Committee of the International Society for Heart and Lung Transplantation: The present status of xenotransplantation and its potential role in the treatment of end-stage cardiac and pulmonary diseases. *The Journal of Heart and Lung Transplantation, 19*(12), 1125–1165.
61. Byrne, G. W., Du, Z., Sun, Z., Asmann, Y. W., & McGregor, C. G. A. (2011). Changes in cardiac gene expression after pig-to-primate orthotopic xenotransplantation. *Xenotransplantation, 18*(1), 14–27.
62. McGregor, C. G. A., Ricci, D., Miyagi, N., Stalboerger, P. G., Du, Z., Oehler, E. A., et al. (2012). Human CD55 expression blocks hyperacute rejection and restricts complement activation in gal knockout cardiac xenografts. *Transplantation, 93*(7), 686–692.
63. Wheeler, D. G., Joseph, M. E., Mahamud, S. D., Aurand, W. L., Mohler, P. J., Pompili, V. J., et al. (2012). Transgenic swine: Expression of human CD39 protects against myocardial injury. *Journal of Molecular and Cellular Cardiology, 52*(5), 958–961.
64. den Hengst, W. A., Gielis, J. F., Lin, J. Y., Van Schil, P. E., De Windt, L. J., & Moens, A. L. (2010). Lung ischemia-reperfusion injury: A molecular and clinical view on a complex pathophysiological process. *American Journal of Physiology, Heart and Circulatory Physiology, 299*(5), H1283–H1299.

65. Ranieri, V., Suter, P., Tortorella, C., De Tullio, R., Dayer, J., Brienza, A., et al. (1999). Effect of mechanical ventilation on inflammatory mediators in patients with acute respiratory distress syndrome: A randomized controlled trial. *Journal of the American Medical Association, 282*(1), 54–61.
66. Pierson, R. N. (2009). Antibody-mediated xenograft injury: Mechanisms and protective strategies. *Transplant Immunology, 21*(2), 65–69.
67. Nguyen, B. N. H., Azimzadeh, A. M., Zhang, T., Wu, G., Shuurman, H. J., Sachs, D. H., et al. (2007). Life-supporting function of genetically modified swine lungs in baboons. *The Journal of Thoracic and Cardiovascular Surgery, 133*(5), 1354–1363.
68. Kubicki, N., Laird, C., Burdorf, L., Pierson, R. N., & Azimzadeh, A. M. (2015). Current status of pig lung xenotransplantation. *International Journal of Surgery, 23*, 247–254.
69. Burdorf, L., Azimzadeh, A. M., & Pierson, R. N. (2012). Xenogeneic lung transplantation models. *Methods in Molecular Biology, 885*(4), 169–189.
70. Sanchez, P. G., Bittle, G. J., Burdorf, L., Pierson, R. N., & Griffith, B. P. (2012). State of art: Clinical ex vivo lung perfusion: Rationale, current status, and future directions. *The Journal of Heart and Lung Transplantation, 31*(4), 339–348.
71. Harris, D. G., Quinn, K. J., French, B. M., Schwartz, E., Kang, E., Dahi, S., et al. (2015). Meta-analysis of the independent and cumulative effects of multiple genetic modifications on pig lung xenograft performance during ex vivo perfusion with human blood. *Xenotransplantation, 22*(2), 102–111.
72. Collins, B. J., Blum, M. G., Parker, R. E., Chang, A. C., Blair, K. S., Zorn, G. L., et al. (2001). Thromboxane mediates pulmonary hypertension and lung inflammation during hyperacute lung rejection. *Journal of Applied Physiology, 90*(6), 2257–2268.
73. Cooper, D. K. C., Ekser, B., Burlak, C., Ezzelarab, M., Hara, H., Paris, L., et al. (2013). Clinical lung xenotransplantation – what donor genetic modifications may be necessary? *Xenotransplantation, 19*(3), 144–158.
74. Kim, Y. T., Lee, H. J., Lee, S. W., Kim, J. Y., Wi, H. C., Park, S. J., et al. (2008). Pre-treatment of porcine pulmonary xenograft with desmopressin: A novel strategy to attenuate platelet activation and systemic intravascular coagulation in an ex-vivo model of swine-to-human pulmonary xenotransplantation. *Xenotransplantation, 15*(1), 27–35.
75. Chen, D., Riesbeck, K., McVey, J. H., Kemball-Cook, G., Tuddenham, E. G., Lechler, R. I., et al. (1999). Regulated inhibition of coagulation by porcine endothelial cells expressing P-selectin-tagged hirudin and tissue factor pathway inhibitor fusion proteins. *Transplantation, 68*(6), 832–839.
76. Shapiro, A. M., Lakey, J. R., Ryan, E. A., Korbutt, G. S., Toth, E., Warnock, G. L., et al. (2000 Jul 27). Islet transplantation in seven patients with type 1 diabetes mellitus using a glucocorticoid-free immunosuppressive regimen. *The New England Journal of Medicine, 343*(4), 230–238.
77. Bottino, R., Balamurugan, A. N., Smetanka, C., Bertera, S., He, J., Rood, P. P. M., et al. (2007). Isolation outcome and functional characteristics of young and adult pig pancreatic islets for transplantation studies. *Xenotransplantation, 14*(1), 74–82.
78. Eventov-Friedman, S., Tchorsh, D., Katchman, H., Shezen, E., Aronovich, A., Hecht, G., et al. (2006). Embryonic pig pancreatic tissue transplantation for the treatment of diabetes. *PLoS Medicine, 3*(7), 1165–1177.
79. Hering, B. J., Wijkstrom, M., Graham, M. L., Hårdstedt, M., Aasheim, T. C., Jie, T., et al. (2006). Prolonged diabetes reversal after intraportal xenotransplantation of wild-type porcine islets in immunosuppressed nonhuman primates. *Nature Medicine, 12*(3), 301–303.
80. Cardona, K., Korbutt, G. S., Milas, Z., Lyon, J., Cano, J., Jiang, W., et al. (2006). Long-term survival of neonatal porcine islets in nonhuman primates by targeting costimulation pathways. *Nature Medicine, 12*(3), 304–306.
81. Van Der Windt, D. J., Bottino, R., Casu, A., Campanile, N., Smetanka, C., He, J., et al. (2009). Long-term controlled normoglycemia in diabetic non-human primates after transplantation with hCD46 transgenic porcine islets. *American Journal of Transplantation, 9*(12), 2716–2726.

82. Thompson, P., Badell, I. R., Lowe, M., Cano, J., Song, M., Leopardi, F., et al. (2011 Dec). Islet xenotransplantation using gal-deficient neonatal donors improves engraftment and function. *American Journal of Transplantation, 11*(12), 2593–2602.
83. Park, C. G., Bottino, R., & Hawthorne, W. J. (2015). Current status of islet xenotransplantation. *International Journal of Surgery, 23*, 261–266.
84. Hawthorne, W. J., Salvaris, E. J., Phillips, P., Hawkes, J., Liuwantara, D., Burns, H., et al. (2014). Control of IBMIR in neonatal porcine islet xenotransplantation in baboons. *American Journal of Transplantation, 14*(6), 1300–1309.
85. Thompson, P., Cardona, K., Russell, M., Badell, I. R., Shaffer, V., Korbutt, G., et al. (2011). CD40-specific costimulation blockade enhances neonatal porcine islet survival in nonhuman primates. *American Journal of Transplantation, 11*(5), 947–957.
86. Dufrane, D., Goebbels, R.-M., & Gianello, P. (2010). Alginate macroencapsulation of pig islets allows correction of streptozotocin-induced diabetes in primates up to 6 months without immunosuppression. *Transplantation, 90*(10), 1054–1062.
87. Estrada, J. L., Martens, G., Li, P., Adams, A., Newell, K. A., Ford, M. L., et al. (2015). Evaluation of human and non-human primate antibody binding to pig cells lacking GGTA1/CMAH/β4GalNT2 genes. *Xenotransplantation, 22*(3), 194–202.
88. Cooper, D. K., Ekser, B., Ramsoondar, J., et al. (2016). The role of genetically engineered pigs in xenotransplantation. *The Journal of Pathology, 238*(2), 288–299.

Chapter 13
Big Data and Kidney Transplantation: Basic Concepts and Initial Experiences

David J. Taber, Amit K. Mathur, and Titte R. Srinivas

Abstract We live in a data-rich world that is ever expanding, and the field of medicine has become particularly enriched with data from the electronic health record (EHR) and from sensors such as EKG monitors, glucometers, and pacemakers. Big Data is a term that is now frequently encountered in both the lay press and the technical literature and is best defined by the extreme volume, variety, or velocity of data. Large relational databases alone do not equate to Big Data (Table 13.2 and see discussion that follows). The magnitude of the data explosion that we live in consciously or unconsciously is underscored, which is outlined throughout this chapter. As a specific example this ever-growing field can have, we will use our recent inquiry into predicting kidney transplant outcomes using a big data approach and discuss the applicability of big data techniques in clinical transplantation.

Abbreviations

AUC-ROC	Area Under the Curve-Receiver Operating Characteristic Curve
BK	BK Virus
BMI	Body Mass Index
BP	Blood Pressure
CI	Confidence Interval
CMV	Cytomegalovirus
DGF	Delayed Graft Function

This chapter is an in-depth review of content published in the American Journal of Transplantation. Srinivas et al. Big Data, Predictive Analytics, and Quality Improvement in Kidney Transplantation: A Proof of Concept. AJT. 17:3;March 2017. 671–681.

D.J. Taber, PharmD, MS, BCPS
Medical University of South Carolina, Division of Transplant Surgery, Charleston, SC, USA

A.K. Mathur, MD
Mayo Clinic, Division of Transplant Surgery, Scottsdale, AZ, USA

T.R. Srinivas, MD, FAST (✉)
Transplant Nephrology, Intermountain Medical Center, Transplant Services,
5169 S Cottonwood St. Suite 320 Eccles Outpt. Bldg, Murray, UT 84107, UK
e-mail: titte.srinivas@imail.org

S.N. Nadig, J.A. Wertheim (eds.), *Technological Advances in Organ Transplantation*, DOI 10.1007/978-3-319-62142-5_13

eGFR	Estimated Glomerular Filtration Rate
EHR	Electronic Health Record
GL	Graft Loss
HGB	Hemoglobin
ICD-9	International Classification of Diseases
KDRI	Kidney Donor Risk Index
Max	Maximum
MI	Myocardial Infarction
NLP	Natural Language Processing
OR	Odds Ratio
PCR	Polymerase Chain Reaction
SBP	Systolic Blood Pressure
SRTR	Scientific Registry of Transplant Recipients
Tx Database	Transplant Database
UNOS	United Network for Organ Sharing

13.1 Introduction

The field of transplantation is particularly rich in data such as data resident in large national databases (Tables 13.1 and 13.2). However, even though these databases capture donor and recipient data at the time of transplant, graft loss, and death, they lack patient-level data that reflect longitudinal clinical evolution [1]. National transplant registry data do not contain real-time dynamic data, thus, not fulfilling all of the requirements of Big Data (Table 13.2). With the massive amounts of patient-level data embedded within the EHRs, analyses could potentially incorporate dynamically evolving clinically relevant patient-level data that could be used to target care and prevent graft loss and death. We tested this hypothesis through incorporation of manually abstracted dynamic patient-level data into predictive models aimed at 30-day readmissions after kidney transplantation and were able to show a significant increase in predictive efficacy over what is possible with national data [2, 3]. However, such an approach is seldom feasible outside the research setting as several data elements that reside in EHRs are unstructured, such as clinician notes, pathology reports, and radiology reports (The term "unstructured data" is applied to

Table 13.1 Scope of Big Data in health care [5]

Walmart's Data warehouse hosts upwards of 2.5 petabytes of information, a volume estimated as 160 times larger than the U.S. Library of Congress.
US Health care data, in 2011, were estimated at 150 exabytes. Five exabytes (1018 gigabytes) of data would contain all the words ever spoken by human beings on earth. It has been estimated that big data in US health care will soon reach zetabyte (1021) scale and even yottabytes (1024 Gigabytes) very shortly
Kaiser Permanente, the California-based health network with more than 9 million covered lives manages between 26.5 and 44 petabytes of patient data just from electronic health records (EHR), including images and annotations. The scope of their data approximates the holdings of 4440 Libraries of Congress

Table 13.2 Large structured databases that inform on transplantation [25]

Database	Population addressed	Strengths	Limitations	Structured data	Unstructured data
UNOS	Transplant candidates, transplant recipients, live and deceased donors	Represents entire US transplant population; longitudinal follow-up	Poor graft loss ascertainment (see SRTR); comorbidity data may be limited	Yes	No
USRDS	ESRD patients; 5% sample of Medicare claims and prescription data	Data on ESRD patients regardless of whether they had access to transplantation or were transplanted; claims data allow linking of outcome to claims (approximation of cost)	Claims data restricted to a 5% sample of Medicare beneficiaries	Yes	No
SRTR	Live and deceased donors, transplant candidates, transplant recipients	Represents entire US transplant population; some longitudinal follow-up; graft loss ascertained through multiple sources with consequent increase in accuracy	Comorbidities may be missing; patient-level data are sparse; real time clinical evolution cannot be tracked	Yes	No

data that cannot be readily applied in analyses without prior abstraction). Patient-level data in EHRs are also characterized by their volume, velocity, and variety, all of which fit the definition of Big Data [4]. The field of Natural language processing (NLP) encompasses computational methods that bridge the interaction between computers and human (natural) language. NLP techniques can be used to extract information from text fields and export them to data sets as structured variables without the need for manual abstraction. In addition to this fundamental ability of computers to be able to read human language, computing systems that would be expected to have relevance to medicine need to be able to ingest not only words but also make sense of the various corpora of information that constitute what we collectively and colloquially refer to as the knowledge base in medicine. For computing systems to be able to do this, a process termed "deep learning," several core requirements need to be met. One such example is afforded by a proprietary solution that is marketed by International Business Machines Corporation (IBM), termed Watson [25]. Watson is endowed with massive parallelism, wherein a large number of computing processes work in parallel to optimize analytical speed and performance.

Using massive parallelism enables the system to scrutinize vast sources of information, evaluate different interpretations of analyses, and test alternate hypotheses at extremely fast speeds. Algorithms are also deployed in such a manner as to provide accurate solutions that are based on the corpus of knowledge previously ingested. Shallow learning provides answers to limited questions. In our case, we used a small application to interrogate the EHR to identify acute rejection episodes based on pathology reports and answer the additional question as to grade and severity of rejection. We were also able to generate lesion scores based on text documentation of rejection grades. This ability to infer lesion scores and grades in isolation from NLP searches of the EHR constitutes a very narrow example of shallow learning. If, however, the system is able to also provide answers with varying degrees of confidence on how best to treat the rejection that is discovered by the NLP search, the expected side effects and the latest scientific paper describing a novel treatment or prognostic biomarker for the particular type of rejection at hand, the ability of the system to establish connections across multiple domains of knowledge is the fundamental underpinning of deep knowledge. In a deep learning system, answers to queries are then provided with multiple levels of confidence with their always being multiple answers. The hierarchy of these answers and their associated level of confidence can be optimized to a particular field of knowledge such as disease state, a tumor family, a genomic database, investment banking, or consumer choice [25].

With the recent availability of high-throughput Big Data approaches to collect and curate structured and unstructured data and the ability to couple such data with high-throughput statistical analytical solutions, we felt that the promise of incorporating large datasets in analyses in near real-time fashion and thereby bringing predictive models with improved accuracy to the bedside to target preventive or therapeutic interventions in the clinic was feasible in the clinical environment [4]. Furthermore, within kidney transplantation, structured laboratory data including creatinines, glomerular filtration rates (GFRs), and hemoglobin results are obtained frequently in a longitudinal manner, as are unstructured data components contained in text form (e.g., biopsy reports, dictated vital signs, clinician notes). Such data also fit the characteristics of Big Data, viz., volume, variety, and velocity.

We began with the premise that this abundance of patient-level data in the kidney transplant clinic afforded the ideal stage to prove the concept that incorporating longitudinally evolving structured and unstructured patient-level data into analyses using Big Data approaches could improve our ability to predict graft loss and mortality among kidney transplant recipients compared with what is possible with national data.

We directed our attention toward prediction of 1- and 3-year graft loss and mortality among kidney transplant recipients using dynamic models that utilized baseline clinical data and patient-level information acquired in the context of routine clinical care, and compared these models with those derived using only structured static variables. The population and methods pertinent to our study are described in detail elsewhere [24].

In brief, we examined adult solitary kidney transplant recipients ≥18 years of age transplanted at our institution, between 2007 and 2015. We excluded patients who (1) had a graft loss (GL) or death in first 7 days' post-transplant or (2) did not have a GL or death recorded nor any other data during the first year or the first three years' post-transplant for the 1 and 3 year models, respectively.

Data Sources Detailed structured data were directly acquired from electronic medical records and United Network for Organ Sharing (UNOS) database elements. Our Center's Transplant Database (Tx Database) (Velos, Inc., Fremont) was leveraged to extract key social determinants of health. Data sources and flow are depicted in figure. Hadoop clusters were deployed in the context of our EDW architecture to allow access of data with minimal latency across multiple applications including "deeper dive" analyses to validate results flowing from automated statistical algorithms. Natural language processing (NLP) was applied to unstructured text fields using proprietary NLP solutions in the proprietary IBM Watson Content Analytics (IBM Corporation, Armonk) to extract Banff scores and vital signs from records that predated automated electronic capture [5]. Using NLP algorithms, we extracted Banff lesion scores from the text of pathology reports. As an example, lesion scores transcribed as g0, t0, i2, t2, v0 in free text pathology reports were extracted and transferred to analytical databases (as an example, $g = 0$, $t = 0$, $i = 2$ and $t = 2$ would constitute Pathological grade, Type-IIa rejection). Furthermore, we applied semantic analyses wherein, lesions transcribed in error as "tII" were deemed semantically equivalent to t2 and were automatically applied in analyses without manual input.

Primary Outcome Measures Account of any graft loss (GL) in UNOS data within 1-year or 3-year post-transplant was, defined as a return to chronic dialysis, re-transplantation, or death. We used a 90-day exposure period for 1-year GL and mortality models, and a 1-year exposure period was used to derive 3-year GL and mortality models.

Covariates UNOS data elements were utilized for key demographic and transplant-related variables in accordance with published methodology used by the SRTR (see supplemental file). The Transplant database (Tx database) was to extract key social determinants of health (see supplemental file). EHR data were utilized to supplement obesity data and vital signs, as well as to provide comorbidities, cardiovascular events, laboratory data, transplant length of stay, and post-transplant acute care utilization data, both inpatient (INP) and emergency department (ED).

Means, standard deviations, maximums, and regressed slopes were used to portray dynamic variables, in effect, capturing effects of change, direction of change, and magnitude of change throughout the exposure period. We applied this approach to estimated GFRs, pulse rates, blood pressures, and hemoglobin levels to incorporate measures reflective of the biological function of the transplant kidney (see below).

Using enhanced ICD-9-CM codes, we derived comorbidity from a modified Elixhauser coding algorithm and selected Charlson comorbidities [6].

In conceiving our model build, we used severable variables as surrogates to capture the many risk domains for graft loss and mortality with a view to utility at the bedside [7]. For instance, KDRI and evolution of blood pressures were used as surrogates for kidney quality. Hemoglobin slopes, eGFR evolution, delayed graft function served as physiological surrogates for graft function. Post-transplant cardiovascular events such as arrhythmias and myocardial infarction captured cardiovascular risk. Immunological risk was depicted by rejection rates (Banff scores), CMV infection, BK infection, and tacrolimus trough concentrations. Social determinants of health were incorporated through demographics, caregiver status, education level, and income. Unplanned acute care utilization served as a measure of access to care, unresolved acuity, and post-transplant clinical course.

Three predictive risk models were developed, using baseline and follow-up data (up to 90-day post-transplant exposure period for the 1-year model (1 year GL), up to 365-day post-transplant exposure period for the 3-year models (3-year GL & 3-year mortality), from both structured and unstructured data formats. The Firth multivariable logistic regression method which accounts for low event densities was employed [8–10]. Statistical significance was determined at the two-sided 5% level. A combination of statistical and clinical information was used for variable selection. Clinical adjudication was used if discrepancies of variables were revealed between methods. To assess and adjust for potential model overfitting, we used the Harrell Optimism Correction [11]. Once the final model was selected, bootstrapping methodology (1000 iterations) was used for model internal validation, and AUC was used to determine and compare model accuracy. We used IBM Watson Content Analytics Suite IBM SPSS Modeler (Version 17) and IBM SPSS Statistics – Essentials for R (IBM Corporation, Armonk, NY) for the statistical analysis.

13.2 Results

13.2.1 Donor Quality, Blood Pressures, and Pulse Rates Impact Graft Loss

A one-unit increase in donor's KDRI score increased the patient's likelihood of a 3-year GL by over 3 times (3.06 OR; (1.49, 6.28) 95% CI). Greater variability in pulse pressures was associated with significantly higher GL odds at 1 and 3 years (1.14 OR each; (1.04, 1.25), and (1.06, 1.22) 95% CI, respectively) while controlling for systolic blood pressure. A one-unit increase in mean pulse rate over the first year post-transplant was associated with a 3% increase in odds of 3-year GL (1.03 OR; (1.00, 1.07) 95% CI) and a 4% increase in the odds of 3-year mortality (1.04 OR; (1.00, 1.07) 95% CI).

13.2.2 Post-transplant Clinical Evolution of Graft Function Predicts Graft Loss

Patients with delayed graft function (DGF) had substantially higher risk of death within 3 years of transplant (2.48 OR; (1.20, 5.00) 95% CI). Patients with improving GFR during the first 90 days after transplant had a significantly lower risk of 1-year graft loss. Higher peak values of eGFR during the early post-transplant period, the exposure periods, were associated with lower risk of 1- and 3-year graft loss rates (0.97 and 0.99 OR; (0.95, 0.99) and (0.98, 1.00) 95% CI, respectively).

Concurrently, an increasing GFR slope during the first year after transplant was associated with 13% lower odds of 3-year mortality (0.87 OR; (0.73, 0.98) 95% CI). Further, an increased hemoglobin slope starting at day 7 post-transplant also decreased the risk of graft loss, both in the short and long term (0.78 and 0.72 OR; (0.58, 0.93) and (0.56, 0.87) 95% CI, respectively). Finally, 1-year GL increased the odds of 3-year mortality by nearly threefold (2.77 OR; (1.13, 6.54) 95% CI).

13.2.3 Demographic and Waitlist Characteristics

Recipient age was significantly associated with lower odds of developing graft loss at 3 years; conversely, age slightly increased the risk of 3-year mortality (0.98 and 1.03 OR; (0.96, 1.00) and (1.00, 1.05) 95% CI, respectively). African Americans had a significantly lower risk of death at 3 years' post-transplant (0.46 OR; (0.26, 0.81) 95% CI). Female sex had more than 40% lower risk of graft loss at 3 (0.57 OR; (0.32, 0.99) 95% CI). Time on the waiting list was actually associated with lower likelihood of graft loss at 1 year, whereas recipient blood type-B patients had more than 3 times the odds of graft loss at 1 year (3.41 OR; (1.55, 7.35) 95% CI).

13.2.4 Associations of Immunological Risk Predict Graft Loss

Acute rejection at 1 year, as captured through acute Banff lesion scores, was associated with a higher 3 year GL (1.37 OR; (1.22, 1.54) 95% CI). Those with at least one positive cytomegalovirus PCR (CMV-PCR) (copy number > 500) by 90 days' post-transplant exhibited 75% lower relative odds of 1 year GL (0.20 OR; (0.02, 0.88) 95% CI) than those with no positive CMV-PCR. In contrast, we noted a trend for those with a positive BK virus PCR in the first year to have almost double the odds of 3-year mortality (1.93 OR; (0.90, 3.88) 95% CI).

13.2.5 Social Determinants of Health Impact Graft Loss

Caregiver status and insurance type were important correlates of graft loss. Those with private insurance associated with lower risk of 3-year mortality (0.46 OR; (0.21, 0.91) 95% CI) and patients that finished high school were significantly less likely to develop graft loss at 1-year past-transplant (0.47 OR; (0.23, 0.97) 95% CI). Caregiver support was also substantially associated with 3-year graft loss (0.40 OR; (0.23, 0.69) 95% CI).

13.2.6 Cardiovascular Risk and Post-transplant Cardiovascular Events Predict Graft Loss

As a general theme, we noted that both cardiovascular morbidity at transplant and that at post-transplant cardiovascular event density were important correlates of graft loss and mortality. Pretransplant cardiovascular comorbidity was associated with post-transplant cardiovascular events. Those that developed an acute myocardial infarctions (MI) within the first year had more than 11 times the odds of graft loss (11.14 OR; (2.15, 54.30) 95% CI). Concurrently, those that experienced a cardiac or vascular event early post-transplant had almost 2.5 times the odds of graft loss within 1 year of transplant. (2.48 OR; (1.06, 5.66) 95% CI), almost 3 times more likely for a 3 year GL (2.98 OR; (1.74, 5.10) 95% CI), and greater than 2 times more likely to experience 3-year mortality (2.23 OR; (1.21, 4.08) 95% CI).

13.2.7 Acute Care Utilization Increases Graft Loss Risk

Each additional day of the initial transplant hospitalization period was associated with higher odds of 3-year mortality (1.08 OR; (1.01, 1.26) 95% CI); prehospitalization after discharge within 90 days was also associated with mortality (1.42 OR; (1.03, 1.93) 95% CI).

13.2.8 Layering of Data Sources Augments Predictive Accuracy

Figure 13.1 demonstrates the added value of data sources and variable construction on the predictive accuracy of the iterative predictive models. For the 1-year GL model, if only UNOS data were used in the predictive model, the AUC-ROC was 0.716 (0.641, 0.790 95% CI; Model 1). With the addition of caregiver data from the transplant database, the predictive performance improved, with an AUC of 0.741 (0.669, 0.814 95% CI; Model 2). EHR comorbidity data further improved the accuracy of the models predictive capability, with an AUC of 0.769 (0.692, 0.845 95% CI; Model 3). Finally, with addition of trajectory and NLP variables to the model, the AUC significantly improved to 0.873 (0.807, 0.939, 95% CI; Model 4). Similar improvements in iterative model accuracy with layering of data sources were demonstrated for the outcomes of 3-year GL and 3-year mortality [24].

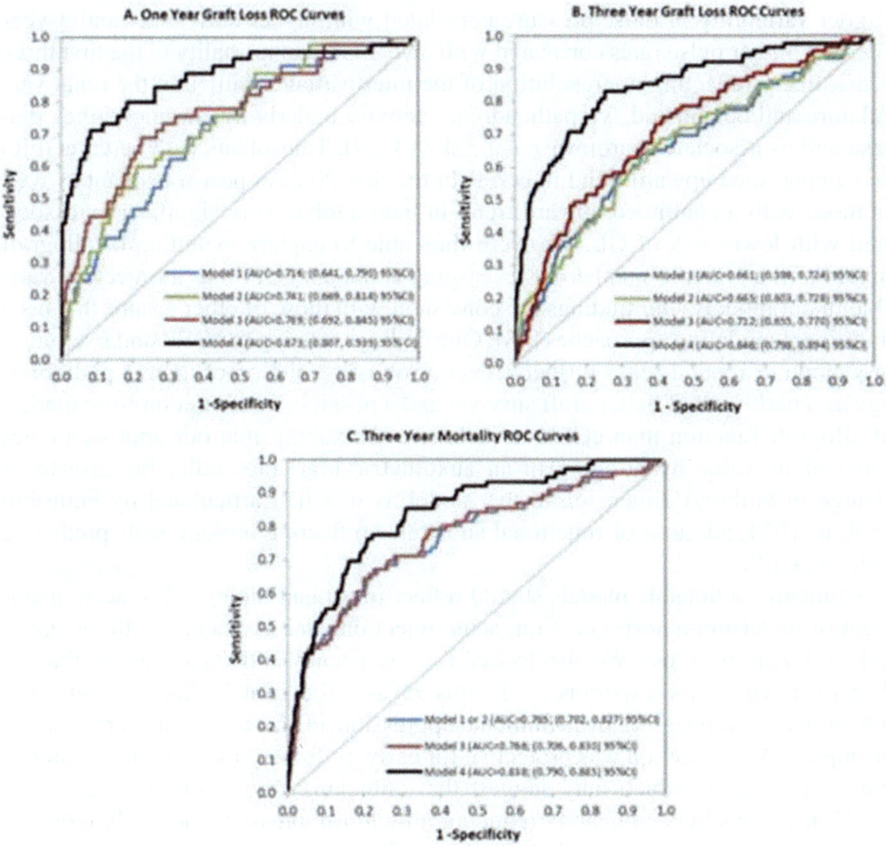

Fig. 13.1 Layering of data sources in the form of social determinants of health, comorbid conditions, NLP variables, and trajectory variables leads to significant improvements in model performance over that achievable with national data alone. (Reprinted with permission of the American Journal of transplantation. Copyright © 2018 American Journal of transplantation)

13.3 Discussion

At this time, our work has enable us to furnish proof of concept of an approach to modeling transplant outcomes using data that are available in the EHRs that substantially improve the accuracy of predictive models for GL and mortality over the current predictive ability of the national models is modest, with the most recently published c-statistic standing at 0.68 [12].

Our approach was able to capture the dominant mechanistic underpinnings of the biology of transplantation and how they relate to the environment of care that the transplanted organ and the patient interact with, in turn, discernible as clinical outcomes. The significance of higher KDRI as a strong correlate of graft loss reflects kidney quality; higher kidney quality associates with lower risk of GL [13].

Higher variability in pulse pressure correlated with higher odds of 1- and 3-year GL, and higher pulse rates correlated with 3 year GL and mortality in the first three years, likely reflecting nonresolution of the microvascular milieu in the renal vasculature and heightened sympathoadrenal activity underlying chronic kidney disease and its associated cardiovascular risk [14–17]. Transplants with better eGFRs and a continued upward trend in eGFR in the first 90 days post-transplant as well as those with a continued upward trend in hemoglobin were significantly associated with lower risk of GL. We were thus able to capture evolution of allograft function in a granular manner by leveraging abundant eGFRs as a trajectory variable in our models. Our findings are consistent with those of other groups that used traditional modeling approaches [18]. Our findings relating to eGFR and resolution of anemia of renal disease further reflect a robust resolution of uremic pathobiology as a harbinger of better graft survival and a possibly a better composite marker of allograft function than eGFR in isolation. We submit that our approach owes some of its value by bringing in an auxometric (variables reflecting growth or change in biology) dimension to the model as was first articulated by Feinstein et al. in 1974; addition of functional substrates reflecting biology adds predictive efficiency [7].

Clinically actionable models should reflect transplant biology. The accompaniment of under-immunosuppression, acute rejection, was associated with increased risk of 3-year graft loss. We also looked for consistency in the associations that we discerned with prior experiences. In this regard, the relationship between viral infections, a correlate of over-immunosuppression in these models, serves as an example. CMV infection was protective for early graft loss, a seemingly counterintuitive association which may suggest that clinicians are reacting to these early CMV infections by significantly reducing immunosuppression and likely reducing the risk of early graft loss [19]. As such, we feel that the interpretation of models such as ours should be rooted in biological plausibility.

Despite the primacy of rejection prophylaxis in the medical management of the transplant recipient and the relevance of histological lesions to outcome, large national databases are remarkably limited in their capture of rejection, its treatment, and post-rejection evolution of renal function [1, 20]. We were able to surmount this deficiency by incorporating data on histological lesion scores from pathological reports on allograft biopsies using NLP techniques.

Cardiovascular disease is a major driver of post-transplant mortality and costs. Our approach was able to consistently identify this association and its contribution to the force of mortality in a direction consistent with prior experiences. In addition, our results underscore the importance of robust social support and a greater degree of educational attainment as being protective for graft loss, which is again consistent with our prior observations and those of others [21, 22].

Taken together, our work suggests that several actionable windows of opportunity exist in mutable domains such as cardiovascular disease, immunological risk, or social determinants of health. The event dense and data-rich clinical environment of transplantation is served by a manpower structure that is already well positioned with care coordination resources unlike many other clinical scenarios. Such an approach can thus be customized to the unique demographic, clinical, and socioeconomic characteristics of patients at individual transplant centers so as to accurately and prospectively identify groups of patients at risk for graft loss or death by allowing post-transplant clinical evolution to inform the workflow.

The overall concept of such a workflow as applicable to the transplant clinic is presented in Fig. 13.2. We are currently in the process of building EHR-based workflows in which data structures and analytical solutions power an automated daily capture of model variables, calculate and update individual patient risk (probability of event), and push these model-derived risk scores (predictive analytics) in a clinician facing interface that is daily refreshed and drives clinical practice (Prescriptive analytics). This workflow is aimed at triggering actions by appropriately skilled teams (Primary Care vs. Transplant Focused) as opposed to making all tasks parts of the transplant team's workflow. Such an approach has been used successfully to reduce readmissions among heart failure patients at our institution [23].

There are several limitations to our work. The first is that it is a retrospective proof of concept. As such, validation across periods other than that studied and prospective experiences would be needed for external validation. Importantly, the impact of such an approach on actual patient outcomes will need careful study. Even more importantly, such an approach will need to be replicated by other centers with differing populations. Furthermore, linkage to prescription and payer databases could provide further insights into the impact of behavioral, economic, and transactional data on transplant outcomes. We are also looking at ways to link clinical situations to optimal treatment approaches that are evidence based and vetted by peer review.

We hope that several centers will be prompted to conduct inquiries along the lines of what we have reported as our approach is feasible in any center that has an EHR and some type of analytical capability along the lines described previously (Table 13.3, Fig. 13.2). In brief, the steps would include understanding the biology of the problem, formulating a research or quality improvement question around the problem, understanding the data needed to answer the underlying questions, an understanding in depth of the data structure within the institution, and design, testing, and deployment of analytical solutions based on this type of understanding. Such an approach will necessarily require transplant teams to work in extremely close collaboration with IT teams in institutions, and such collaborations will require project management, infrastructure, and dedicated personnel as part of robust institutional support to ensure success. We hope that the attainment of a working vocabulary as summarized in Table 13.3 will help transplant professionals explore this burgeoning field and conduct critical conversations with their informatics colleagues at their respective institutions to start fruitful forays into big data analytics (Table 13.4).

We further submit that such an approach is easily translated to other situations in medicine in which structured care is applied to well-managed populations such as oncology, heart failure, diabetes, and critical care units.

Prospectus:

In the most famous example popularized in the lay press, Watson handily beats human players in the game of Jeopardy. Structured questions such as the ones used in Jeopardy constitute the basis for the inductive and deductive logic that powers

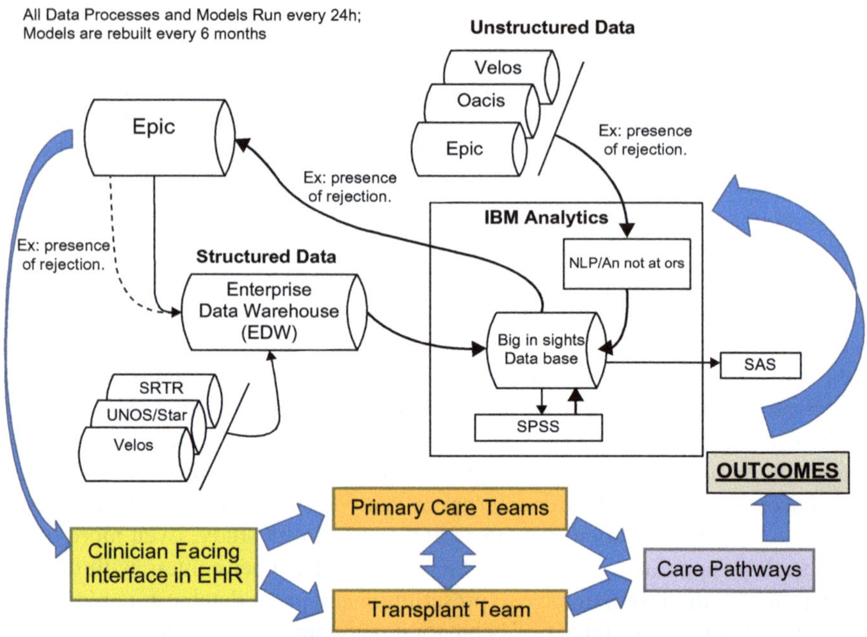

Fig. 13.2 Automated capture and curation of structured and unstructured data power analytical workflows. Quotidian push of model output to clinician facing interfaces triggers workflows appropriate to the individual patient's clinical needs. (Reprinted with permission of the American Journal of transplantation. Copyright © 2018 American Journal of transplantation)

clinical medicine. Although we are not yet at the stage where we have the ability to use deep learning systems in the clinic, we submit that transplantation is ripe for this approach as it is a field where the date that a disease starts is known as also the date that it ends and the disease state exists in an environment of care where the processes of care are well defined. Given the complexities involved, the existing systems of data and a practicing community that embraces, lives, and breathes innovation, the prospect of big data based innovative solutions powering the field to ever greater achievements and in turn, helping medicine at large elevate the human condition is not a farfetched expectation.

Table 13.3 Selected definitions of terms used in the context of big data [25]

Term	Definition	Comment
Algorithm	A step-by step description of a specific process, procedure or method	Can be embedded by programming or embedded within the system to learn from patterns that are observed as data are ingested and analyzed
Advanced analytics	Algorithms for complex analyses of either structured or unstructured data. Protocols may include sophisticated statistical models, machine learning, neural networks, text learning, and other advanced data mining techniques	Does not include database queries, reporting and visual tabulations of data
Big data	This is a *relative term* that describes data that are difficult to process with conventional technology due to extreme values in one of three attributes: *Volume* (How much data needs to be processed?), *Variety* (The complexity and heterogeneity of the data to be processed; e.g. combinations of unstructured text, structured data, images, sensor data as may be accrued in the context of clinical care and available in electronic form) and *Velocity* (the speed at which it is produced and/or arrives for processing; e.g.; laboratory data, organ offers, clinic notes in a transplant clinic)	As data management techniques, both hardware and software improve the threshold for what would constitute Big Data rises. A terabyte of slow moving data was once considered Big Data but is now easily managed in a palm-sized portable storage device. In the future, yottabyte magnitude data sets may be handled on desktops but for now are considered big data given complexities in processing
Data mining	The process of exploring and analyzing large amounts of data to find pattern	In the purest instance, it implies some form of automated deployment of algorithms to analyze data and present results
Enterprise Data Warehouse (EDW)	A large data store containing the organization's historical data. The primary purpose is to serve as a data system of record.	Most hospitals feed the EDW on a daily basis. In addition to patient-level data, data pertaining to financial transactions between the organization and patient and organization and payer reside here. This "backup system of record" can be a treasure trove for data analysis and data mining
ETL (Extract, Transform, Load)	A set of tools and procedures for locating and accessing data from a data store (data extraction), changing the structure or format of the data so that it can be used by the business or patient care application (data transformation) and sending the data to the business or analytical application (data load)	Understanding the structure of data within an institution will make clear what ETL procedures are needed to power Big Data projects

(continued)

Table 13.3 (continued)

Term	Definition	Comment
Infrastructure	Hardware and software elements that are necessary for the operation of a service, a dedicated application or an enterprise. The term encompasses basic computer hardware, networks, operating systems, and other software that applications run on top of.	An understanding of infrastructure in an institution is critical to determine feasibility of Big Data projects and will require clinical experts to collaborate with experts in IT to determine solutions. It is best that infrastructure that is newly erected be scalable in terms of ability to handle future data storage that an analysis needs.
Hadoop	A software framework that parallelizes data processing across computing nodes on large clusters of commodity hardware to speed up computations and optimize delays (latency) in processes. The two major components include a massively scalable distributed file system that can support petabytes of data and a computing engine that computes results in batch.	Hadoop clusters can feed deep dives into data to follow up on observed patterns.
Machine learning	A discipline grounded in computer science, statistics, and psychology that employs algorithms that learn or improve their performance based on exposure to patterns in data rather than by explicit a priori programming	(a) This is especially important in medicine as data will change with respect to both explanatory variables and outcomes based on changing population characteristics, era changes in treatment effects, and changes in processes of care (b) Clinicians need to validate model output in context to train the algorithms
Metadata	The definitions, mappings, and other characteristics used to describe how to find, access, and use an institution's data and software components	Understanding metadata and cataloging and mapping their anatomy and functions meticulously are critical to the design, execution, and sustained success of analytics initiatives in an institution.
Neural networks	Neural network algorithms are designed to emulate the human brain using variably weighted input nodes, mutilayered operations, and output nodes. Iterative processes are used to analyze data.	Interaction with clinicians needed to train models as output needs to be considered in context and weights re-adjusted till optimal results are adjudicated.

Table 13.3 (continued)

Term	Definition	Comment
Natural Language Processing (NLP)	NLP describes the capability of computer systems to process text written or recorded in a language written for human communication. NLP can be used to identify the semantics of words, phrases, and other linguistic units in documents or other unstructured sources within a knowledge base (corpus).	Context is critical to NLP as are statistical patterns and linkages. Example: When extracting Banff lesion scores from pathology reports in text form lesion scores transcribed as g0, t0, i2, t2, v0 need to be extracted and transferred to analytical databases (wherein, $g = 0$, $t = 0$, $i = 2$ and $t = 2$ would constitute Pathological grade, Type IIa rejection). Furthermore, values transcribed in error as "tII" can be deemed semantically equivalent to t2 and so extracted and analyzed.
Predictive analytics	A statistical or data mining solution consisting of algorithms and techniques that can be used on both structured and unstructured data to help determine future outcomes.	Applications include prediction and optimization. Just as in the clinical context, users should be aware of the constraints of internal and external validities of analytical results for optimal utility.
Structured data	Data that have a defined length and format.	Examples of structured data include numbers, dates, groupings of numbers, and words (strings) such as addresses. In the medical context, discrete numerical biochemical laboratory result values and manually inputted demographic data, ICD-10 codes, and CPT codes would be considered structured data
Unstructured data	Information that does not follow a specified data format.	Startifying
Supervised learning	An approach that teaches a system to detect or match patterns in data based on examples that it encounters during training with sample data	Can be valuable in stratifying patients by outcomes or defining characteristics.
Unsupervised learning	Machine learning approaches that use inferential statistical modeling algorithms to discover rather than detect patterns or similarities in data.	An unsupervised system should be able to identify a new pattern rather than merely match a set of patterns that were presented in training.

(continued)

Table 13.3 (continued)

Term	Definition	Comment
Watson	A proprietary cognitive system developed by IBM that combines several capabilities in NLP, machine learning, and analytics. A key function of is the ability to ingest large corpora of knowledge such as textbooks and journals and provide answers to questions with confidence levels attached to the answers with attendant reasoning.	Several derivatives exist, and a critical need that should be met regardless of the vendor used is to make sure that data acquisition, curation, analysis, and deployment of analytical output in the clinic can be automated and made available real time, bedside for clinical effectiveness.

Table 13.4 Types of data that contribute to the variety and velocity that describes big data [5]

1. *Web and social media data*: Clickstream and interaction data from social media such as Facebook, Twitter, Linkedin, and blogs. It can also include health plan websites, smartphone apps, etc.
2. *Machine-to-machine data:* Readings from sensors, meters, and other devices.
3. *Big transaction data:* Health care claims and other billing records increasingly available in semistructured and unstructured formats.
4. *Biometric data:* Fingerprints, genetics, handwriting, retinal scans, and similar types of data. This would also include X-rays and other medical images, blood pressure, pulse and pulse-oximetry readings, and other similar types of data.
5. *Human-generated data:* Unstructured and semistructured data such as electronic health records (EHRs), physicians' notes, email, and paper documents.

References

1. Kaplan, B., Schold, J., & Meier-Kriesche, H. U. (2003). Overview of large database analysis in renal transplantation. *American Journal of Transplantation, 3*, 1052–1056.
2. Taber, D. J., Palanisamy, A. P., Srinivas, T. R., et al. (2015). Inclusion of dynamic clinical data improves the predictive performance of a 30-day readmission risk model in kidney transplantation. *Transplantation, 99*, 324–330.
3. McAdams-Demarco, M. A., Grams, M. E., King, E., Desai, N. M., & Segev, D. L. (2014). Sequelae of early hospital readmission after kidney transplantation. *American Journal of Transplantation, 14*, 397–403.
4. IBM White Paper. (2016). 5 Steps to becoming a data-driven healthcare organization. In (pp. 1–7). Somers, NY: IBM Corporation.https://assets.sourcemedia.com/00/f5/b8f107cf478296eaa937a413581c/imw14682usen.PDF
5. Racusen, L. C., Solez, K., Colvin, R. B., et al. (1999). The Banff 97 working classification of renal allograft pathology. *Kidney International, 55*, 713–723.
6. Quan, H., Sundararajan, V., Halfon, P., et al. (2005). Coding algorithms for defining comorbidities in ICD-9-CM and ICD-10 administrative data. *Medical Care, 43*, 1130–1139.
7. Charlson, M. E., & Feinstein, A. R. (1974). The auxometric dimension. A new method for using rate of growth in prognostic staging of breast cancer. *JAMA, 228*, 180–185.
8. Firth, D. (1993). Bias reduction of maximum likelihood estimates. *Biometrika, 80*, 27–38.

9. Heinze, G. A. (2002). A solution to the problem of separation in logistic regression. *Statistics in Medicine, 21*, 2409–2419.
10. Heinze, G. A. (2006). A comparative investigation of methods for logistic regression with separated or nearly separated data. *Statistics in Medicine, 25*, 4216–4226.
11. Harrell, F. E., Jr., Lee, K. L., & Mark, D. B. (1996). Multivariable prognostic models: Issues in developing models, evaluating assumptions and adequacy, and measuring and reducing errors. *Statistics in Medicine, 15*, 361–387.
12. SRTR Risk Adjustment Model Documentationr Waiting List and Post-Transplant Outcomes. (2016). http://www.srtr.org/csr/current/modtabs.aspx. Accessed 16 June 2016.
13. Rao, P. S., Schaubel, D. E., Guidinger, M. K., et al. (2009). A comprehensive risk quantification score for deceased donor kidneys: The kidney donor risk index. *Transplantation, 88*, 231–236.
14. Amann, K., Wanner, C., & Ritz, E. (2006). Cross-talk between the kidney and the cardiovascular system. *Journals of the American Society of Nephrology, 17*, 2112–2119.
15. Chang, T. I., Tabada, G. H., Yang, J., Tan, T. C., & Go, A. S. (2016). Visit-to-visit variability of blood pressure and death, end-stage renal disease, and cardiovascular events in patients with chronic kidney disease. *Journal of Hypertension, 34*, 244–252.
16. Johnson, R. J., Rodriguez-Iturbe, B., Kang, D. H., Feig, D. I., & Herrera-Acosta, J. (2005). A unifying pathway for essential hypertension. *American Journal of Hypertension, 18*, 431–440.
17. Meier-Kriesche, H. U., Schold, J. D., Srinivas, T. R., Reed, A., & Kaplan, B. (2004). Kidney transplantation halts cardiovascular disease progression in patients with end-stage renal disease. *American Journal of Transplantation, 4*, 1662–1668.
18. Wan, S. S., Cantarovich, M., Mucsi, I., Baran, D., Paraskevas, S., & Tchervenkov, J. (2016). Early renal function recovery and long-term graft survival in kidney transplantation. *Transplant International, 29*, 619–626.
19. Elfadawy, N., Flechner, S. M., Liu, X., et al. (2013). CMV Viremia is associated with a decreased incidence of BKV reactivation after kidney and kidney-pancreas transplantation. *Transplantation, 96*, 1097–1103.
20. Gonzales, M. M., Bentall, A., Kremers, W. K., Stegall, M. D., & Borrows, R. (2016). Predicting individual renal allograft outcomes using risk models with 1-year surveillance biopsy and alloantibody data. *Journals of the American Society of Nephrology, 27*(10), 3165–3174.
21. Goldfarb-Rumyantzev, A. S., Rout, P., Sandhu, G. S., Khattak, M., Tang, H., & Barenbaum, A. (2010). Association between social adaptability index and survival of patients with chronic kidney disease. *Nephrology, Dialysis, Transplantation, 25*, 3672–3681.
22. Taber, D. J., Hamedi, M., Rodrigue, J. R., et al. (2016). Quantifying the race stratified impact of socioeconomics on graft outcomes in kidney transplant recipients. *Transplantation, 100*(7), 1550–1557.
23. Evans, R. S., Benuzillo, J., Horne, B. D., et al. (2016). Automated identification and predictive tools to help identify high-risk heart failure patients: Pilot evaluation. *Journal of the American Medical Informatics Association, 23*(5), 872–878.
24. Srinivas, T. R., Taber, D. J., Su, Z., et al. (2017). Big data, predictive analytics and quality improvement in kidney transplantation: A proof of concept. *American Journal of Transplantation, 17*, 671–681.
25. Hurwitz JS, Kaufman M, Bowles A in Cognitive Computing and Big Data Analytics. Wiley (Indianapolis) 2015.

MIX
Papier aus verantwortungsvollen Quellen
Paper from responsible sources
FSC® C105338

If you have any concerns about our products,
you can contact us on
ProductSafety@springernature.com

In case Publisher is established outside the EU,
the EU authorized representative is:
Springer Nature Customer Service Center GmbH
Europaplatz 3, 69115 Heidelberg, Germany

Printed by Libri Plureos GmbH
in Hamburg, Germany